Springer Series in Synergetics   Editor: Hermann Haken

Synergetics, an interdisciplinary field of research, is concerned with the cooperation of individual parts of a system that produces macroscopic spatial, temporal or functional structures. It deals with deterministic as well as stochastic processes.

Complex Systems – Operational Approaches

in Neurobiology, Physics, and Computers

Proceedings of the International Symposium
on Synergetics at Schloß Elmau, Bavaria,
May 6–11, 1985

Editor: H. Haken

With 208 Figures

Springer-Verlag Berlin Heidelberg New York Tokyo

Professor Dr. Dr. h.c. Hermann Haken

Institut für Theoretische Physik, Universität Stuttgart, Pfaffenwaldring 57/IV,
D-7000 Stuttgart 80, Fed. Rep. of Germany

ISBN-13: 978-3-642-70797-1   e-ISBN-13: 978-3-642-70795-7
DOI: 10.1007/978-3-642-70795-7

Library of Congress Cataloging in Publication Data. International Symposium on Synergetics (1985 : Schloss Elmau, Bavaria) Complex systems, operational approaches in neurobiology, physics, and computers. (Springer series in synergetics ; v. 31) Includes index. 1. System theory–Congresses. 2. Neurobiology –Congresses. 3. Physics–Congresses. 4. Computers–Congresses. I. Haken, H. II. Title. III. Series. Q295.I586 1985 003 85-22248

© Springer-Verlag Berlin Heidelberg 1985
Softcover reprint of the hardcover 1st edition 1985

2153/3130-543210

In Memoriam

WOLFGANG PRECHT

1938–1985

Preface

A great deal of the success of science has rested on its specific methods. One of which has been to start with the study of simple phenomena such as that of falling bodies, or to decompose systems into parts with well-defined properties simpler than those of the total system. In our time there is a growing awareness that in many cases of great practical or scientific interest, such as economics or the human brain, we have to deal with truly complex systems which cannot be decomposed into their parts without losing crucial properties of the total system. In addition, complex systems have many facets and can be looked at from many points of view.

Whenever a complicated problem arises, some scientists or other people are ready to invent lots of beautiful words, or to quote Goethe "denn immer wo Begriffe fehlen, dort stellt ein Wort zur rechten Zeit sich ein" ("whenever concepts are lacking, a word appears at the right time"). Quite often such a procedure gives not only the layman but also scientists working in fields different from that of the inventor of these new words the impression that this problem has been solved, and I am occasionally shocked to see how influential this kind of "linguistics" has become.

When I conceived the plan of this volume I was careful to emphasize *operational approaches* in the sense of general systems theory. Such approaches are, for instance, those in which quantitative and reproducible relations between events are established or where predictions are made which can be checked experimentally. Of course, operational approaches are much harder and less pretentious than "linguistics" but they are the only method which can promote science.

This volume consists of the invited papers presented at the International Symposium held at Schloss Elmau, Bavaria, May 6-11, 1985. Unfortunately Professor Yu. Klimontovich was unable to attend this meeting due to his illness, but I am glad that he submited a manuscript. The contributions to these proceedings deal with the theory of evolution, with brain activities in humans and animals, with the coordination of motion in biological systems and robots, with computers and computing, with typical phenomena of order and chaos in physical systems and with some basic theoretical methods. I hope that in this way this volume represents a typical cross section of the vast field of complex systems. All the participants found this symposium extremely useful in learning about each other's methods and approaches, and I do hope the reader of this volume will profit for his own work in a similar way.

Stuttgart, July 1985   *H. Haken*

Contents

Part IV　　Computers and Computing

Part V　　Theoretical Concepts

Part VI　　Physical Systems; Order and Chaos

Operational Approaches to Complex Systems.
An Introduction

H. Haken

Institut für Theoretische Physik, Universität Stuttgart, Pfaffenwaldring 57/IV,
D-7000 Stuttgart 80, Fed. Rep. of Germany

1) Complex systems

Let us first discuss why systems may be very complex. First of all, quite a number of systems contain very many elements. Examples are provided by the following table:

Table 1

brain	~	10^{11} - 10^{12}	neurons
world	~	10^{10}	people
laser	~	10^{18}	atoms
fluid	~	10^{23}	molecules/cm^3

Furthermore there are very many connections between the individual elements. If each element is connected with all others, the total number is proportional to n^2 where n is the number of elements. In practice the connections may be not so many but still numerous. When we take 1000 dendrites per neuron in the brain, the number of connections is still 10^{14} to 10^{15}. In order to describe the activities of such networks we have to process an enormous amount of information. Since the lifetime of each scientist and even of groups of them is limited, we have a problem in front of us which seems to be unsolvable in principle. Quite evidently we have to devise methods how to <u>compress</u> the enormous amount of <u>information</u> contained in such a system to an amount which can be handled by the <u>human mind</u>. In a way nature has been quite helpful in teaching us how such reductions can be achieved. Indeed, in practically all cases we can distinguish between a microscopic level with its numerous elements and a macroscopic level where we are confronted with the properties of the total system.

Depending on the problem, one of the two levels may be directly accessible to observation. For instance, in a crystal composed of atoms, the physical properties of the crystal are directly accessible to observation. On the other hand, when we consider human individuals as part of a society, these individuals are more or less directly accessible to observation, while the society is not. On the other hand, through scientific methods we are enabled to make more and more observations at the corresponding other levels, for instance on atoms in crystals by X-rays or by methods of demoscopy in society.

2) Examples

Let us list some systems which appear to us as being complex:

biology:	DNA
	cells
	organisms
	brain
economics:	companies
	national economy
	world economy

linguistics:	languages
sociology:	society
physics:	crystals
	plasmas
	fluids
	lasers
chemistry:	chemical reactions
computer science:	parallel computers

3) Operational approaches

Instead of trying to make more or less philosophical remarks on complex systems let us rather tackle the problem by briefly reviewing operational approaches. By ¯operational¯ we mean that these approaches are prescriptions how to deal with complex systems or how to make measurements on complex systems and to model their behavior.

a) the black box approach
This approach is well known to electronics,but actually medicine provides us with an example of that approach also, where drugs are administered to patients and then the patients¯ reactions are studied. In this approach the complex system is considered only with respect to its reactions to certain inputs. The reactions are also called outputs, or response, or behavior. Indeed, behaviorism is a theoretical concept based on the black box approach. Generally speaking, the output 0 is a function of the input

$$0 = f(J). \qquad (3.1)$$

In general this function cannot be presented in a closed form,but rather the relation (3.1) appears in form of a table.

Table 2

outputs

There are a number of difficulties.
i) There may be an enormous variety of inputs and outputs and it may be difficult to find a general law.
ii) Even one input may cause different outputs. This leads to the hypothesis that internal states of the black box are important.
iii) In important cases outputs may appear without any input, e.g. in the brain, so that the relation (3.1) becomes meaningless.
iiii) Finally by introspection we know that processes may go on in our brain without any inputs.
While in a number of cases the black box approach may be very important, it evidently fails in just the cases with which these proceedings are primarily concerned.

b) reduction
The most common method of studying complex systems consists in decomposing them

2

into their elements. Indeed, in quite a number of systems such decomposition is possible and the elements are more or less well defined.

Table 3

gas   ⟶   molecules

crystal ⟶   atoms

brain   ⟶   neurons

organ   ⟶   cells

In the next step of the analysis the properties of the elements are studied, and it is then hoped that the properties of the total system can be understood by means of the properties of the elements. This method is the basis of a school of thought known as ´reductionism´. Though this method has been quite successful in a number of cases, there are basic difficulties. The most important one is due to the fact that at the macroscopic level new qualities arise which are alien to the elements. We note that even in the simple physical system of a gas new qualities arise at a macroscopic level, such as pressure and temperature. At the microscopic level these concepts are not known, because we can describe atoms, for instance, only by their positions and their velocities.

In addition, in many systems cooperative effects may be decisive, so that the cooperation between the parts is much more important for the macroscopic behavior of the system than the properties of the elements on their own. Therefore, while it is very important to study the properties of the individual parts, in general new concepts are needed in addition to understand the total systems.

c) The statistical approach

Let us consider the example of a gas in more detail, as it was done by Boltzmann. While it is impossible to observe and list all the individual atoms with their velocity and position, it is possible to consider their probability distribution p(v) with respect to their velocity, v. In statistical mechanics it becomes possible to derive such probability distributions explicitly.

The properties of the total gas then can be derived from suitable averages, for instance from the mean kinetic energy

$$\frac{m}{2} \langle v^2 \rangle = \int p(v) \frac{m}{2} v^2 \, dv, \tag{3.2}$$

where m is the mass of an atom of the gas. As is known to physicists, at the macroscopic level the corresponding observable is the temperature, or more precisely speaking,

$$(3.2) = \frac{1}{2} kT, \tag{3.3}$$

where T is the absolute temperature and k Boltzmann´s constant. In this way it becomes possible to relate the microscopic quantities, i.e. the velocity of the particles with the macroscopic observable, i.e. the temperature. In the case of the gas and in a number of other physical systems, it is possible to derive distribution functions from fundamental physical laws at the <u>microscopic</u> level.

One may ask whether it is possible to directly derive probability distributions for the microscopic elements from <u>macroscopic</u> observables. For instance, let the temperature of the gas be known or given, can p(v) be determined without a microscopic theory? The basic idea is to give "unbiased estimates" on this probability function . This is achieved by the <u>maximum entropy principle</u> due to Jaynes [1].

Let us denote the still unknown probability distribution by p_j, where the index j may denote specific microscopic states, e.g. the velocity v of an atom. Then the information entropy is defined by

$$S = -\sum p_j \ln p_j.\qquad\qquad(3.4)$$

The relation between the macroscopic observable called $\bar{f}^{(k)}$ with the microscopic probability distributions p_j is given by

$$\sum p_j\, f_j^{(k)} = \bar{f}^{(k)}.\qquad\qquad(3.5)$$

where $f_j^{(k)}$ are properties of the elements at the microscopic level, for instance the kinetic energy of particles in state j. The p_j are then determined by maximizing S under the given constraints (3.5) and the normalization condition

$$\sum p_j = 1.\qquad\qquad(3.6)$$

This principle is so generally formulated that it can be applied to problems outside of physics, e.g. to economic processes or to the reconstruction of pictures as shown by Jaynes in his contribution to these proceedings. There it is also shown how this principle can be extended to time-dependent processes.

In practical applications, the crucial problem consists in finding the appropriate functions $f_j^{(k)}$. For instance, in physical systems driven far from thermal equilibrium, e.g. the laser, the $f^{(k)}$ differ entirely from those known from physical systems in thermal equilibrium. For instance, using for the f's the field mode intensities and the fourth order correlation functions of the field amplitudes, we may recover or in some cases even generalize the probability distribution functions p_j known from the microscopic laser theory, the convection instability etc. [2].

d) Synergetics
In this kind of approach [3] the relation between the microscopic and macroscopic level is studied. In numerous cases of natural systems, but also in a number of man-made systems, the macroscopic state is acquired by means of selforganization of the microscopic elements, i.e. the systems acquire a specific spatial, temporal or functional structure without specific interference from the outside. In synergetics we ask the question whether there are general principles which govern selforganization irrespective of the nature of the subsystems. Such principles could indeed be found, provided the system undergoes qualitative changes at the macroscopic level. Incidentally, quite often the qualitative changes are accompanied by the emergence of new qualities of the macroscopic system, though the microscopic elements are the same. Simple examples are provided in physics by fluids or by the laser. In a fluid (Fig. 1a) a macroscopic spatial structure arises when the system is subjected to a new constraint namely a homogeneous heating, i.e. the fluid acquires a specific macroscopic ordered state which is not imposed from the outside but rather triggered indirectly (Fig.1b). In a laser the incoherent emission acts of the individual atoms (Fig.2a) become correlated, and give rise to a structure ordered in time (Fig.2b).

The specific relation between the macroscopic and microscopic level is established and derived in synergetics by means of two concepts, namely that of the <u>order parameters</u> and the <u>slaving principle</u>. The order parameters are the macroscopic

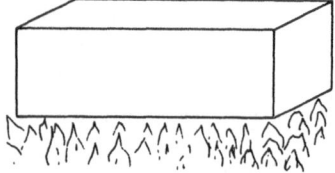

Fig. 1a   Compare text

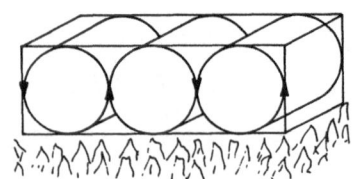

Fig. 1b   Compare text

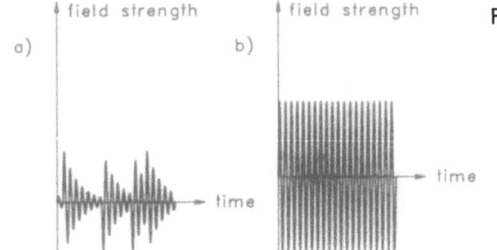

field strength a)

field strength b)

time

time

Fig. 2 a, b   Compare text

observables,which describe the macroscopic behavior of the system. According to the slaving principle,once the macroscopic observables are given the behavior of the microscopic elements is determined (Fig.3). In this way an enormous reduction of the degrees of freedom is achieved, namely in a laser there may be 10^{16} degrees of freedom of the atoms and 1 degree of freedom of the field mode. Due to the slaving principle at the lasing threshold,the whole system is governed by a single degree of freedom. (We mention for the mathematical experts that the slaving principle contains a number of mathematical theorems as special cases, e.g. the center manifold theorem and the slow manifold theorem, cf. [3]).

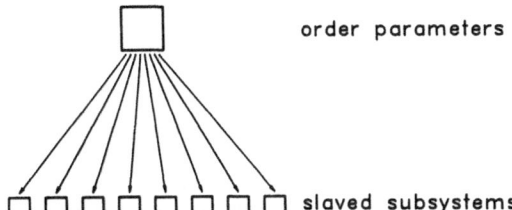

order parameters

slaved subsystems

Fig. 3   Visualization of the slaving principle

In many cases the systems studied in synergetics undergo a series of qualitative changes once a control parameter is changed, for instance the energy input into the total system. More generally speaking,the same elements may show at the macroscopic level quite different patterns of behavior. The laser provides us again with a lucid example [4] (compare Fig.4). With low pump power it can exhibit random emission. At increasing pump power,the emission pattern becomes coherent. At a still higher level regular pulses (quasiperiodic motion) occurs. If another parameter, such as the cavity loss, is changed, the coherent wave can decay into deterministic chaos. Here again various routes from coherent to chaotic motion can be distinguished, for instance by intermittency, where periods of coherent laser emission alternate with chaotic outbursts.

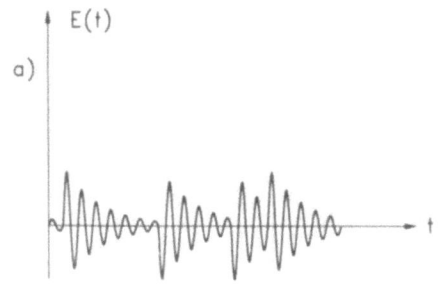

E(t) a)

t

E(t) b)

t

Fig. 4 Various emission patterns of a laser
   a) incoherent

Fig. 4 b) coherent

5

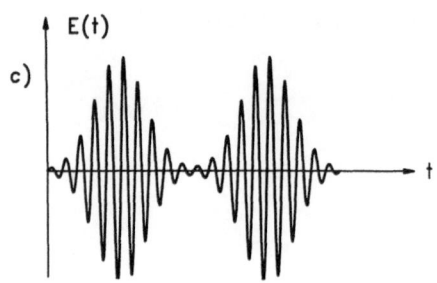

c)

d)

Fig. 4   Various emission patterns of a laser
         c) pulses   d) deterministic chaos

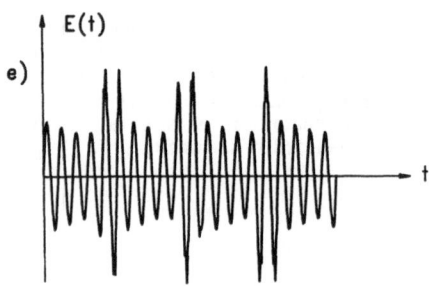

e)

Fig. 4   e) intermittency

I personally feel that the laser can serve as a paradigm of action of neuronal networks. Spontaneous quantum emission effects can become correlated at a macroscopic level and here show quite different patterns of behavior. Changes of patterns of behavior in living beings are well known. We just quote the different gaits of horses or the experiments by Kelso [5] (cf. these proceedings) on hand movements, where even simple hand movements undergo qualitative changes when a single parameter, namely the frequency of hand movement is changed. This phenomenon was recently modelled in terms of concepts of synergetics [6]. I personally believe that speech can be understood as a sequence of behavioral patterns in which each vocal or consonant is an individual pattern governed by only few order parameters, and the problem mastered by nature consists in matching these behavioral patterns so that the movement of muscles can proceed continuous-ly.

As it may transpire from these remarks, synergetics is in a way complimentary to reductionism or, more specifically, to the study of elements. It rather stresses systemic properties irrespective of the nature of subsystems. In this way synergetics establishes profound analogies between the macroscopic behavior of quite different systems.

e) Information theory
When we consider complex systems composed of many subsystems, all the processes going on can be considered as being based on the exchange of information by the individual parts of the system. Therefore, one may hope that information theory may give us insight into the functioning in complex systems. In this strict scientific sense information is defined according to Shannon [7] by

$$I = - K \sum_j p_j \ln p_j, \qquad (3.7)$$

where K is a constant and p_j is the relative frequency of occurrence of the symbol indicated by the index j. Equally well, p_j can be interpreted as probability to find a system in state j.

In spite of the fact that the Shannon information has found important and widespread applications, e.g. in transmission networks, it becomes more and more evident that new concepts are needed. While one of the reasons of the generality of this concept lies in the fact that it does not evaluate the meaning, say, of messages, in many practical cases the semantics is important. Therefore a new kind of information theory will be needed on "semantic information". One possibility is to define a value (or importance) of information by studying the impact of the information on the receiver [8]. For instance, if somebody does not understand Chinese, what a Chinese scientist told him will not cause meaningful reactions of that person. Similarly, when the RNA of phages is put into sand there will be no reaction at all, whereas if it is put into bacteria well defined processes will be triggered. How the value (or importance) of information can be defined or measured can be explained in DNA. Clearly the multiplication rate of species depends on their DNA so that a value of information can be attributed to DNA by the "value landscape" of species. Such values can be introduced by means of the theory of Eigen and Schuster [9] and the value landscape has been recently discussed in more detail by Ebeling and Feistel [10]. Other questions which have not been treated in detail are the generation and transportation of information where in our opinion the slaving principle will play an important role,because it allows for information compression. Just to quote some very simple examples: the meaning of a word like "girl" acts as the order parameter,whereas the individual letters g, i, r, l are slaved elements. Sentences act as order parameters on order parameters and allow for corrections of misspellings, e.g. gill $\longrightarrow$ girl [11] . Quite generally speaking, what is needed in our opinion is a dynamic information theory.

I wish to thank H. Shimizu for profound and exciting discussions we had this year in March on this subject,where we were both put into an "excited, coherent state". According to Shimizu´s suggestion "holonics" may be a new field of research studying "relationships". At the same time,one may try to explore Shannon information in the context of self-organizing systems. Here it turns out that close to instability points Shannon information (also called Shannon entropy) can be decomposed into that of the order parameters and that of the slaved systems (averaged over the order parameter distribution) [12]. In the vicinity of the instability point the information of the order parameter changes, as does the efficiency, whereas the information of the slaved mode does practically not.

f) Evolution

A further way of approaching the problem of complex systems consists in the following. At least in nature,complex systems have evolved from systems of lower complexity. The mechanism by which the evolution to higher complexity proceeds is well described by Darwin as a three-step process:

> 1) mutations
> 2) multiplication
> 3) selection.

Eigen and Schuster [9] succeeded in presenting a mathematical theory of the evolution of biomolecules,where the selection through competition is achieved by the specific multiplication rate. This theory reveals also the connection between multiplication and reliability in reproduction. (Some recent work is reported by P. Schuster in these proceedings,where further references can be found).

One may ask what the purpose of nature is in increasing complexity. Of course, the concept of "purpose" is alien to nature and has an anthropomorphic origin. Nevertheless, one may arrive at the hypothesis that energy consumption and risks to life are reduced by increased information which requires increased complexity of living "structures", but to my knowledge a corresponding theory is still lacking.

g) The study of model systems

Before I shall discuss some specific systems, I wish to discuss some more operational approaches. Since this discussion is somewhat more formal, readers not so much interested in formal ideas are invited to skip this and the following section.

1) Forward modelling

Here we start from the properties of the individual elements of the system and the rules of the combination of these elements or of their interaction. Then we wish to study the time evolution of the total system. An example is provided by cellular automata [13] (Fig.5). Here at time t_1 a state is given, for instance that certain cells of the cellular automaton are occupied, and additional rules, how the state at the next time t_2 evolves by an interaction between elements. For instance, one may set the rules that if two next nearest neighbour elements are occupied and have an unoccupied cell inbetween, then they will become unoccupied while the originally unoccupied cell becomes occupied. Otherwise the cells will become unoccupied or will remain unoccupied.

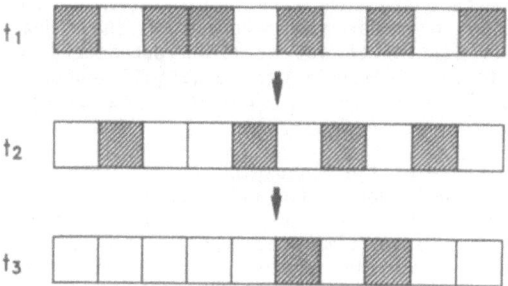

Fig. 5   Example of a cellular automaton at consecutive times t_1, t_2 t_3

In the literature a number of cellular automata have been studied, and various patterns of the evolving states were identified [13]. For instance, there may be processes which tend to a final steady state or where oscillations occur. There is a basic difficulty, namely one can let cellular automata run only for finite times, and therefore the asymptotic behavior cannot be predicted, at least in many cases.

Another example for forward modelling is provided by the theoretical description of physical systems, which consist of many particles and possess interactions between the particles. An example is fluids, where one can study at least to some extent the evolving structures. (See for instance the contribution by Bestehorn, Friedrich and Haken to these proceedings). In all these systems, first the properties of the elements and the rules of their combinations are given, and then the predictions are made.

2) backward modelling

Here the time evolution of a macroscopic system is given. Then we want to draw conclusions on the basic elements and their interaction or, in other words, we want to make a microscopic model of the system.

In recent years some progress could be made, for instance in physical or chemical systems, showing socalled deterministic chaos. In a number of cases (e.g. [14]) it has become possible to reconstruct the attractors so that at least the dimensionality of the underlying processes can be determined, and one may hope for constructing explicit models for the time evolution at microscopic or mesoscopic levels (Fig. 6). A beautiful example for the reconstruction of attractors, even

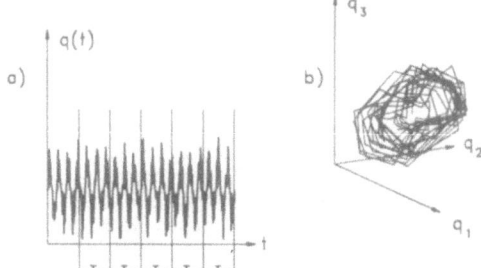

Fig. 6   Scheme of a reconstruction of the trajectories of a
          chaotic attractor (b) from the time series q(t) (a).
          The variables q_1, q_2, q_3 are defined by $q_1 = q(t)$,
          $q_2 = q(t+T)$, $q_3 = q(t+2T)$, where T is an appropriately
          chosen time delay

for brain action (in deep sleep), is provided by Babloyantz in these proceedings.
There are a number of difficulties connected with such approaches. First of all,
the models on the microscopic processes need not be unique. In fact, quite
different microscopic models may lead to the same macroscopic behavior, which has
been   particularly revealed by the field of synergetics. There is a still more
fundamental difficulty which was revealed in mathematics. Let us consider the
Turing machine. Let a specific string of numbers be given. Then one may ask
whether the specific string of numbers can be reproduced by an initial string of
numbers and a program of the Turing machine so that the initial string and the
program is shorter than the initially given number. This is the problem of
reducibility. As can be shown by means of Gödel´s theorem [15], there exists no
general algorithm by which we can decide whether such a reduction can be made,
except maybe for trivial   statements. This has a profound consequence on
modelling in all disciplines of science. Even if a fact can be cast into the form
of a string of numbers, we don't have any general algorithm to check or perform its
reduction. Rather we have to study in each specific case whether and how such a
reduction can be achieved. There may indeed be problems, e.g. in physics which
cannot be reduced by specific models [16].

h) Mathematical modelling
Let us continue section d) Study of model systems. From the mathematical point of
view, the first step in the study of model systems consists in the formulation of
equations and then searching for solutions in closed form or in form of a table.
Of course, the closed form is preferable, because it gives us an insight into the
dependence of the solution on parameters or other constraints. In a way, a
solution of an equation in closed form is just some kind of an algorithm which
tells us which symbols must be replaced by numbers and which processes are then
to be done with these numbers to get a result again in form of a number. With the
advent of computers, things may change entirely. Instead of directly searching for
solutions of given equations, we rather have to develop algorithms to be employed
by the computer. The next step consists in the development of algorithms which
construct algorithms and so on. This leads eventually to the problem of
selfprogramming of computers. There are some difficulties which one should be
aware of. In many cases, an algorithm developed for the solution of a problem is
confronted with the halt problem, i.e. the question, whether one can decide if
the procedure will end after a finite number of steps. For practical applications
the requirements are still more stringent, because only a given time will be
available. On the other hand, only a given accuracy will be needed, so that the
problem need not be solved exactly. This may simplify the halt problem but need
not solve it entirely.

The "algorithmic approach" will eventually require a new kind of human thinking
because what we now understand by "understanding the mechanisms" must change. The
effects of changes of control parameters on the output can become very complex

9

and only indirectly visible by the eventual results of the computer calculation. In this way, by comparison with experimental facts, we can verify that a model is compatible with the properties of the system, but we do not "understand" in which way the result is brought about.

Let us make a final remark in this section on the problem of reducibility. It may well be that for instance processes in the physical world cannot be reduced to a shorter model, i.e. that the computation of that process will require at least the same efforts as the direct study of that process [16]. Nevertheless, we may gain an advantage by modelling, because at least in a number of cases we might be interested only in some relevant features, so that the model then requires less effort for its solution than the study of the total real process. Of course, what is meant by ˉrelevant featuresˉ will depend on the context.

4) Application of concepts of synergetics to computers and life phenomena

a) Computers

Some years ago I suggested reinterpreting self-organizing systems, which produce macroscopic patterns, as parallel computers in the form of continuously distributed elements [17]. Let us consider two typical examples:
i) a chemical reaction developing a macroscopic pattern. If we divide the system into small volume elements, two types of processes can be distinguished:

1) chemical reactions within each cell,
2) transport of matter across the cells.

The processes 1) can be interpreted as computational processes (such as multiplication etc.). This analogy can be easily cast into a rigorous mathematical form. The processes 2) can be interpreted as information flux from one (elementary) computer to a neighboring one. Quite evidently, the result of this computation is the macroscopic pattern.
Similarly, a fluid producing a convection pattern can be interpreted as a parallel computer. Measuring the patterns means to read off the result of the computation.
ii) the laser can also be interpreted as a computer network. Each atom is an individual computer, which via the light field exchanges information with all the other atoms and produces as a result the macroscopic coherent field as an output (= result of the computation).
A few days before the beginning of this meeting I was delighted to learn that Shimizu and his coworkers succeeded in devising a "synergetic computer" (as called so by Shimizu) for pattern recognition. Indeed, this computer has typical properties of synergetic systems, but, of course, much ingenuity had to be put by Shimizu and his coworkers into devising this computer, for instance frequency locking as a tool for pattern recognition, the knowledge of columns in the visual cortex etc. (For more details cf. his contribution to these proceedings). As it seems to me, this computer can be generalized in various ways. Since in its present form it does not (at least in general) distinguish between lines shifted in parallel, one may introduce the counting of knots of two, three, or more lines in order to fix the relative position of lines. In its present form, neighboring cells resonate best if all of them are on a line. It is not difficult, however, to prescribe specific angles of orientation between neighboring cells, so that specific curves can be recognized by specific resonances.

b) life phenomena

In this section I wish to briefly discuss what general conclusions on life phenomena can be drawn from the study of model systems, in particular in the field of synergetics. As already mentioned, physical systems such as lasers and fluids can show pattern formation through self-organisation. Here the evolving structures show a pronounced degree of coherence in space and time. The evolving order is described by order parameters, which slave the subsystems so that coherence is established. It is tempting to apply these concepts to life phenomena. In this interpretation,life phenomena consist in the upconversion of order from the microscopic to the macroscopic level,or, put somewhat differently, an enormous number of microscopic processes are governed by macroscopic coherence. Let us quote some examples:

The microscopic order of DNA is upconverted into living systems.

While this example is still very vague, the following ones are far more concrete.

In morphogenesis the interplay between DNA and positional information leads to cell differentiation , so that the formation of organs becomes possible.

In locomotion,macroscopic movement is achieved by the interaction of many parts of the system,starting e.g. from the neuronal level over muscles and bones. The same is true for the motion of hands etc.

Speech and perception are further examples and,as I believe, thinking must be considered as a process going on on macroscopic scales in the brain.

5) Brain
To our knowledge the brain is the most complex system nature has produced. Of course, it will be far beyond the scope of this article to discuss the brain in any detail. However, we may make a few quite general remarks.

There has been a trend for a while, especially in the field of biology, to search for localized centers for perception, e.g. for the grandmother cell. This means that the final recognition process, for instance the identification of a specific face, is done by an individual cell. The opposite approach,of which I am one of the supporters,is to consider the activity of the brain as that of a coherent field. Very many neurons participate in such an activity. Here I want to make a few further remarks,which are triggered by a preprint I received from Sir John Eccles, (cf. also his contribution to these proceedings), namely the quantum nature of mental processes. It is well known that in perception, such as vision,quantum processes lie at the origin. For instance,in vision single quanta of light can be detected. Recent results on the hearing of cats indicate that there is a very low noise level,which suggests that quantum coherence effects may play a basic role. In his article,Sir John Eccles shows that quantum effects may play a decisive role in the brain,and he has suggested to consider the probability field of quantum mechanics. This will lead to the fascinating problem of developing a quantum theory of the brain. I personally believe that the laser might serve as a nice paradigm for such a quantum theory, because the laser has taught us how quantum coherence on macroscopic scales can evolve from noisy quantum elements. On the other hand,this laser analogy is most probably far too naive to tackle the mind-body problem. I fully appreciate the fact that Sir John Eccles sees in the introduction of the probability field a possibility of solving the mind-body problem,and a possible interpretation of the probability field might be that of an interpreter between mind and body. Here, of course, deep philosophical questions will arise, for instance on whether the probability field must be given the attribute of being material or immaterial. I do not want to enter this discussion here,but I think it will be important to look in an unbiased manner at these problems,which really open new vistas. It may well be that the mind-body problem belongs to a class of problems which are undecidable, but taking time and again a fresh look at this problem will certainly provide us with much deeper insights into the incredibly high levels of abstractions at which our brain works.

6) Concluding remarks
Let us briefly discuss in conclusion of this paper why complex systems present such a fascinating field of research. First of all,I think the understanding of complexity represents an enormous challenge to the human mind; one complex system, the brain, trying to understand other ones or even itself. Complexity

seems to be an essential tool of nature by which it equips living beings to survive,and in this way to develop new qualities, eventually even new qualities which are not needed for the survival of the species, such as music, arts, etc.

On the other hand, the understanding of complex systems becomes of utmost importance for the survival of society. Man must learn how to steer (or not to steer!) complex systems, such as society, or economy. Also,the individual human being may be considered as a highly complex system,so that we may hope for new insights into psychology, psychiatry, and medicine. While society and economy are in a way more or less selforganizing systems (in spite of the opinion of quite a number of scientists),man is now beginning to devise complex systems in form of computers. I think here, we are,in spite of the great success of computer technology,just at the beginning,and knowing more about complex systems will certainly help us in constructing in particular selforganizing computers.

References

[1] E.T.Jaynes, Phys.Rev. $\underline{106}$,4,620 (1957); Phys.Rev. $\underline{108}$, 171 (1957)

[2] H. Haken, Application of the maximum information entropy principle to selforganizing systems, to be published

[3] H. Haken, SYNERGETICS.AN INTRODUCTION, 3rd edition Springer Verlag, Berlin, Heidelberg, New York 1983

H. Haken, ADVANCED SYNERGETICS, Springer Verlag, Berlin, Heidelberg, New York 1983

[4] H. Haken, LIGHT, VOL. 2, LASER LIGHT DYNAMICS North Holland Publ.Comp.,Amsterdam 1985

[5] J.A.S. Kelso, Bull.Psychon.Soc. $\underline{18}$, 63 (1981a)

[6] H. Haken, J.A.S. Kelso, H. Bunz, Biol.Cybern. $\underline{51}$, 347 (1985)

[7] C.E.Shannon, A mathematical theory of communication, Bell System Techn. J. (1948) 27, v. 370-423, p.623-656 C.E.Shannon, W. Weaver, THE MATHEMATICAL THEORY OF COMMUNICATION, University of Illinois Press, Urbana 1949

[8] H. Haken, Some basic ideas on a dynamic information theory in Vol. 21 Springer Series in Synergetics, ed.P.Schuster STOCHASTIC PHENOMENA AND CHAOTIC BEHAVIOUR IN COMPLEX SYSTEMS

[9] M. Eigen, P. Schuster, Naturwissenschaften $\underline{64}$ 541 (1977); $\underline{65}$, 7 (1978), $\underline{65}$ 341 (1978) M. Eigen, Ursprung und Evolution des Lebens auf molekularer Ebene, in Vol. 17 Springer Series in Synergetics, ed. H.Haken EVOLUTION OF ORDER AND CHAOS, 1982

[10] W. Ebeling, R. Feistel, Physical models of evolution processes, in Vol. 28 Springer series in Synergetics, ed. V.I.Krinsky, SELFORGANIZATION, 1984

[11] H. Haken, THE SCIENCE OF STRUCTURE. SYNERGETICS Van Nostrand Reinhold, New York, Cincinnati, Toronto, London, Melbourne 1984

[12] H. Haken, Information, information gain, and efficiency of selforganizing systems close to instability points, to be published

[13] D. Farmer, T. Tiffoli, S. Wolfram, eds.
      Cellular Automata, Physica D <u>10</u> (1984)

[14] J.C. Roux,R.H.Simoni, H.L.Swinney, Physica <u>8D</u>, 257 (1983)

[15] K. Gödel, Monathsh.Math.Phys. <u>38</u>, 173 (1931)

[16] S. Wolfram, Phys.Rev.Lett. <u>54</u>, 735 (1985)

[17] H. Haken, Europhysics News <u>7</u>, July (1976)

Part I

Evolution

Effects of Finite Population Size
and Other Stochastic Phenomena in Molecular Evolution

P. Schuster

Institut für Theoretische Chemie und Strahlenchemie, Universität Wien,
Währingerstraße 17, A-1090 Wien, Austria

Polynucleotide replication is visualized as a stochastic process.
Adaptive selection leads to optimization of mean replication rates
in ensembles of molecules. Due to the intrinsic dynamics of replication, populations become uniform also in the absence of differences
in rate constants. We called this process "random selection" in order
to distinguish from adaption. Random selection leads to random drift
particularly in small populations. Sequence information is stable in
replicating systems only if the process of replication is sufficiently accurate. We apply the theory of multitype branching processes
and derive a stochastic error threshold for replication. In addition,
this theory allows to study accumulation of fluctuations. The total
population size may fluctuate strongly in an ensemble of replicating
molecules, but relative frequencies of individual molecular species
approach a "law of large numbers" in sufficiently large populations.

1. Introduction

Biological evolution is based on two principles: the creation of
variation through erroneous copying processes, commonly called mutations, and selection of the variants which replicate fastest.
Combination of the two principles leads to evolutionary optimization
of properties which are relevant for replication. It is a subtle
combination of a stochastic element and a deterministic process
which makes evolution the powerful driving force of nature. The process of selection has been studied in great detail by the famous
scholars of population genetics. Replication of entire organisms is
an enormously complicated process and hence, the theory of evolution
got more or less stuck in a phenomenological description. A new
approach started out from the biochemistry of nucleic acids. Refinement of experimental techniques made it possible to study replication and adaption of RNA molecules in the test tube [1]. Many
kinetic studies were undertaken since the pioneering works of
SPIEGELMAN and his coworkers, and nowadays the molecular mechanism of
RNA replication is well understood [2,3].

EIGEN [4] proposed a theory of the evolution of molecules which is
based on the equations of biochemical kinetics of replication. Later on
this concept has been worked out in detail [5]. Within the frame of
this theory, molecular evolution is visualized as an optimization pro-

cess of the net rate of RNA production in an ensemble of molecules. Recently, algebraic expressions were derived for potential functions of this process under different environmental conditions [6]. Then, the evolution of the ensemble of RNA molecules follows the gradient of the potential function,and complex behaviour of the system,such as oscillations of individual concentrations or formation of spatial dissipative structures can be excluded (Fig.1).

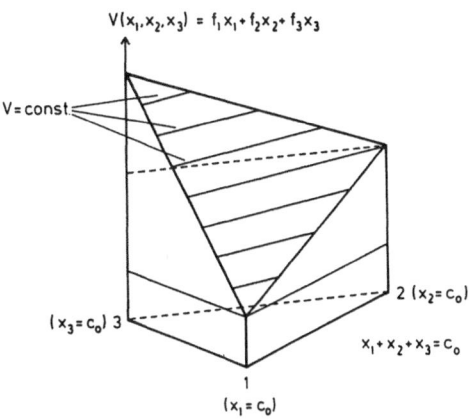

Figure 1: A potential function for the optimization of replication rates in an ensemble of replicating molecules. $V(x_1,...,x_n)$ is a quadratic function which has its maximum at the most efficiently replicating distribution of molecules. The example shown represents the three-dimensional case n=3 with $d_1=d_2=d_3$ and $f_1=1$, $f_2=2$ and $f_3=3$ (For further details see equations (3-5) and [6], [29]).

Conventional chemical kinetics is based on the deterministic description of concentrations by differential equations. Strictly, this approach applies to infinite populations only. In most chemical systems, nevertheless, finite population sizes introduce no uncertainties,because a $\sqrt{N}$ law applies to the fluctuations in concentrations (N, the number of particles, is commonly larger than 10^{18} in chemical kinetics. Then,$\sqrt{N}$ is 10^9 or the standard deviation is 10^{-9} times the expectation value. Fluctuations are hardly detectable). Autocatalytic reactions, in particular replication processes of polynucleotides,represent an exception in this respect. In this contribution, we shall be concerned with stochasticity in finite populations of replicating RNA molecules which results from internal fluctuations.

Polynucleotides, like other heteropolymers,provide a basis for enormous variability,since they can exist in "hyperastronomically" large numbers of isomers. This fact has important consequences for RNA replication. Ordinary chemical reactions commonly involve few molecular species,each of which is present in a very large number

17

of copies. The converse situation is the rule for polynucleotide replication: the number of sequences which are interconverted through
replication, mutation and recombination exceeds by far the number of
molecules actually present. A polynucleotide of a chain length
$\nu=100$ can exist in 10^{60} different sequences, whereas the number of
molecules lies well below 10^{15} in a typical experiment. This lack
of balance between possible realizations and actual population sizes
is one characteristic property common to all biological systems. It
represents a fundamental prerequisite of evolutionary optimization
and adaption.

In section 2 we present some definitions which will facilitate the
forthcoming discussion. Then, we analyse the two most important
effects of population size on ensembles of replicating molecules.

2. Replication, selection and optimization

Let us consider a population of N molecules distributed on n molecular species of types I_1, I_2,...,I_n. We denote the numbers of molecules present at time t by

$$[I_k] = X_k(t); \quad k = 1,2,...,n \tag{1}$$

$$\text{with } \sum_k X_k(t) = N(t) \tag{2}.$$

Changes in the distribution of molecules are caused by three classes
of chemical reactions:

1. error-free replication

$$(A) + I_k \xrightarrow{f_k Q_{kk}} 2 I_k; \quad k=1,2,...,n \tag{3},$$

2. mutation

$$(A) + I_k \xrightarrow{f_k Q_{jk}} I_j + I_k; \quad k,j=1,2,...,n; \quad j \neq k \tag{4}$$

and 3. degradation

$$I_k \xrightarrow{d_k} (B) \tag{5}.$$

The reaction constants f_k contain implicitly the concentrations of
the energy-rich building blocks of the polymers:
A = (GTP, ATP, CTP and UTP). These concentrations are assumed to be
constant as a result of some external control. The accuracy of the

18

replication process is described by the nxn matrix $Q = \{Q_{jk}\}$. Its (dimensionless) elements Q_{jk} represent the probabilities of individual mutations and correct replication respectively. Consequently, we have

$$\sum_{j=1}^{n} Q_{jk} = 1; \quad k=1,2,\ldots,n \tag{6}.$$

In section 3 we shall consider the simple special case of ultimate accuracy of replication. Then, the matrix Q is simply given by the unit matrix of dimension n: $Q_{jk} = \delta_{jk}$.

Degradation of RNA molecules is determined by the rate constants d_k. It leads to energy-poor monomers denoted schematically by B=(GMP, AMP, CMP and UMP)

Natural selection is one basic concept of Darwinian evolution. In order to be able to provide a coherent picture of the evolutionary process,we have to generalize the notion of selection. In our definition of selection we refer exclusively to the distribution of types in the population,and neglect at first the mechanistic aspects. Selection has occurred in a population if all but one types have died out. For example, we say that the molecular species I_m has been selected at time $t=T_m$ if

$$X_m(t)>0 \text{ and } X_k(t)=0 \text{ for all } k\neq m \text{ and } t>T_m \tag{7}.$$

In general, selection occurs only in populations which are subjected to some selection constraint. Several examples of constraints were discussed in previous publications [5-7]. Polynucleotide replication according to equations (3), (4) and (5) is remarkably insensitive to the particular constraint applied [5,8]. In this contribution we shall not study particular constraints explicitly, but we shall assume that the populations considered are finite.

Let us now specify the notion of selection according to the underlying mechanisms. We distinguish two limiting cases: (1) adaptive selection and (2) random selection. Adaptive selection occurs in large (in the limit infinitely large) populations provided the rate constants are sufficiently different (f_k, $k=1,\ldots,n$ and/or d_k, $k=1,\ldots,n$). Random selection dominates in small populations and becomes the exclusive mechanism in the limit of kinetic degeneracy ($f_1=f_2=\ldots=f_n$ and $d_1=d_2=\ldots=d_n$).

In order to illustrate the two idealized processes,we consider error-free replication ($Q_{jk}=\delta_{jk}$). An example of adaptive selection

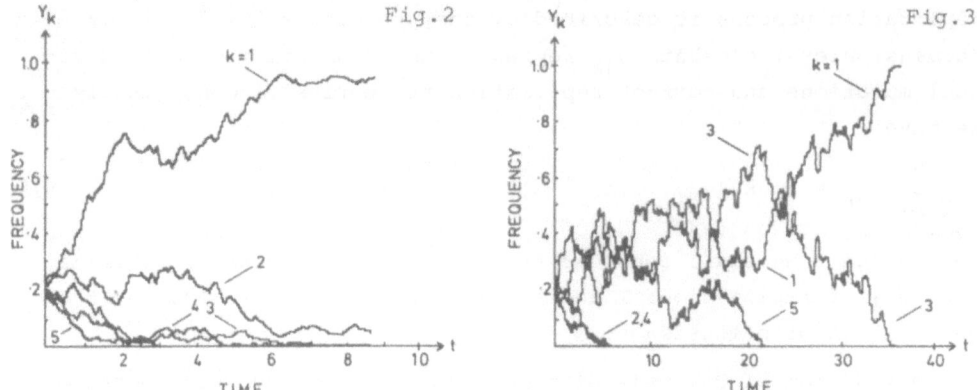

Figure 2: A stochastic simulation of adaptive selection in the replication system (3,5). The rate-constants f_1=4.4, f_2=4.0, f_3=3.6
f_4=3.2, f_5=2.8 and d_1=d_2=...=d_5=4.0 and initial conditions
$X_1(0)$=$X_2(0)$=...=$X_5(0)$=60 were used. The simulation is based on
Gillespie's "Direct Method" [28]. We plot frequencies Y_k(k=1,...,5)
as functions of time. Despite large fluctuations,the molecular species die out in the sequence expected from their rate-constants:
I_5, I_4, etc.

Figure 3: A stochastic simulation of random selection in the replication system (3,5). The rate-constants f_1=f_2=...=f_5=d_1=d_2=...=
=d_5=4.0 meet the condition of complete kinetic degeneracy. The
initial conditions $X_1(0)$=$X_2(0)$=...=$X_5(0)$=60 were used. The simulation is based on Gillespie's "Direct Method" [28]. We plot frequencies Y_k(k=1,...,5) as functions of time.

is shown in figure 2. In the limit of infinite populations we replace
the (discrete) number of molecules by their (continuous) expectation
values: $X_k(t) \longrightarrow x_k(t)$. It is straightforward to verify that the mean
net rate of polynucleotide synthesis increases monotonously [4,9]:

$$\frac{d\bar{E}(t)}{dt} = \frac{d}{dt}\left\{\frac{\sum\limits_k (f_k-d_k)x_k(t)}{\sum\limits_k x_k(t)}\right\} \geqslant 0 \tag{8}.$$

Hence, we are dealing with an optimization process and $\bar{E}(t)$ is the
quantity which is optimized during adaptive selection. This increase
in $\bar{E}(t)$ is a result of systematic changes in the distribution of
molecular species. Those which replicate less efficient than the
mean, decrease in number and ultimately die out. The mean value
($\bar{E}(t)$) increases and more and more species fall below average until
the population becomes uniform,and contains the most efficiently replicating molecular species, say I_m, exclusively:

$$I_m: \quad f_m-d_m = \max(f_k-d_k; \ k=1,...,n) \tag{9}$$

Clearly, adaptive selection is the process Charles DARWIN had in mind when he christened the notion of "natural selection". It is worth noticing, however, that he had foreseen also the role of fluctuations in cases of kinetic degeneracy [10].

In figure 3 we present a typical example of random selection. Other conditions except the differences in rate-constants being equal, random selection is a much slower process than adaptive selection. The type which is ultimately selected cannot be predicted. It will be different in different runs of the experiment. The net rate of polynucleotide synthesis is essentially constant (f_k=f; d_k=d; k=1,...,n):

$$\bar{E} = f - d \tag{10}$$

In the forthcoming section we shall investigate the process of random selection in more detail. Selection in real systems is always a superposition of both processes. Nevertheless, we can have situations which come very close to one of the two limiting cases. It is worth trying to estimate the conditions under which adaptive selection is much faster than random selection [11]. We assume equal rate constants of degradation for this estimate (d_1=d_2=...=d_n) and find that adaptive selection prevails if

$$\frac{\Delta f}{f} > \frac{2 \ln N}{N} \tag{11}.$$

Random selection, as initially stated, dominates in cases of small populations and small differences in rate-constants.

3. Random selection and random drift

In the previous section we pointed out that selection may occur also in absence of any differences in the rate-constants. This is a peculiarity of replication in finite populations which we want to study in more detail now. In this section we consider error-free replication. Then, the analysis of the stochastic process is simplified considerably because the individual stochastic variables $X_k(t)$ are independent and the probability distribution can be factorized

$$\text{Prob } \{X_1(t) = x_1, \, X_2(t) = x_2, \ldots, X_n(t) = x_n\} =$$

$$= \text{Prob}\{X_1(t)=x_1\} \cdot \text{Prob}\{X_2(t)=x_2\} \cdot \ldots \cdot \text{Prob}\{X_n(t)=x_n\} \tag{12}.$$

BARTHOLOMAY [12] modelled the stochastic process in a single variable - equations (3) and (5) with n=1 - as a linear birth and death process,and was able to derive analytic expressions for the probability distributions. The probability of extinction of a given polynucleotide sequence I_k,

$$\text{Prob}\{X_k(t) = 0\} = P_o^{(k)}(t),$$

is particularly important for the forthcoming discussion. We have to distinguish two cases:

$$(1) \quad f_k=d_k: \quad P_o^{(k)}(t) = \left\{\frac{f_k t}{1+f_k t}\right\}^{x_k(0)} \tag{13a}$$

$$(2) \quad f_k \neq d_k: \quad P_o^{(k)}(t) = \left\{1 + \frac{f_k-d_k}{d_k \exp[-(f_k-d_k)t]-f_k}\right\}^{x_k(0)} \tag{13b}$$

Note, that the sequence I_k goes extinct with probability one if $f_k \leqslant d_k$:

$$\lim_{t \to \infty} P_o^{(k)}(t) = \begin{cases} 1 & \text{if } f_k \leqslant d_k \\ \left(\frac{d_k}{f_k}\right)^{x_k(0)} & \text{if } f_k > d_k \end{cases} \tag{14}$$

We define a stochastic variable $T_o^{(k)}$ which describes the event of extinction of I_k. The expectation value $E\{T_o^{(k)}\}$ is finite for $f_k < d_k$ and diverges in the limit $f \to d$ (see figure 4 and [13]). Let us consider now the case of complete kinetic degeneracy: $f_1=f_2=\ldots=f_n=f$ and $d_1=d_2=\ldots=d_n=d$. Then, we are dealing in essence with a single type stochastic process. The rate-constants being equal,

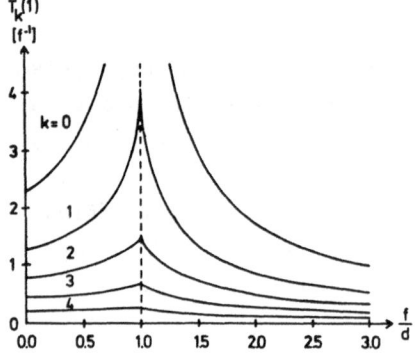

Figure 4: Expectation values of the sequential extinction times $T_k(1)$ as a function of the ratio of rate-constants f/d for n=5. Note, that T_o diverges at f=d. Since we have non-zero probabilities of survival for f>d, the expectation values are based on conditional probabilities in this case. They are normalized to extinction with probability one [13].

we consider the differences in the sequences only for the "book-keeping" of particle numbers in the population. The probability distributions are the same for all n molecular species if we start out from a uniform initial distribution: $x_1(0) = x_2(0) = \ldots = x_n(0) = m$. In order to describe the evolution of the population, we define a sequence of stochastic variables which describes the events at which molecular species become extinct. We have called this sequence "sequential extinction times"[14]. Clearly, these events form a time ordered set:

$$0 = T_n < T_{n-1} < T_{n-2} < \ldots < T_2 < T_1 < T_0.$$

Thus, we have n molecular species present between T_n and T_{n-1}, n-1 molecular species between T_{n-1} and T_{n-2} etc. Random selection is complete at the time T_1. From this event on the population is homogeneous until it becomes ultimately extinct at time T_0.

Expectation values of sequential extinction times were calculated from probability distributions [13-15]. For f=d and $X_k(0) = m$ (k=1,2,....,n) we obtain

$$E\{T_k(m)\} = \frac{1}{f} m(n-k) \binom{n}{k} \sum_{j=0}^{k} \binom{k}{j} (-1)^{j+1} \sum_{l=1}^{m(n-k+j)} \frac{1}{l} \tag{15}.$$

This expression can be largely simplified for the case m=1, i.e. for the initial condition of one copy for each species [14]:

$$E\{T_k(1)\} = \frac{1}{f} \frac{n-k}{k} \tag{16}.$$

The expectation value of T_0 diverges for all m as it did in the single-type problem mentioned before. The expectation values of all other extinction times are finite. The variances, however, become very large in the case of the "late" extinction times (small k values).

The distribution of the expectation values of sequential extinction times is very similar for f<d. In this case we find for $X_k(0) = m$ (k=1,2,...,n):

$$E\{T_k(m)\} = \frac{d}{f} m(n-k) \binom{n}{k} \cdot \tag{17}$$

$$\cdot \sum_{j=0}^{k} \binom{k}{j} (-1)^{j+1} \sum_{l=1}^{m(n-k+j)} \binom{m(n-k+j)-1}{l-1} \left(\frac{d}{f}-1\right)^{l-1} \frac{1}{l}\left\{\ln\left(1-\frac{f}{d}\right) - \sum_{i=1}^{l-1} \frac{1}{i}\left(1-\frac{d}{f}\right)^{-i}\right\}$$

This exceedingly clumsy expression becomes somewhat simpler for m=1:

$$E\{T_k(1)\} = \frac{d}{f}(n-k)\binom{n}{k}(\frac{d}{f}-1)^k(-1)^{k+1} \cdot$$

$$\cdot \sum_{l=1}^{n-k}\binom{n-k-1}{l-1}(\frac{d}{f}-1)^{l-1}\frac{1}{1+k}\{\ln(1-\frac{f}{d}) - \sum_{i=1}^{1+k-1}\frac{1}{i}(1-\frac{d}{f})^{-i}\}$$

(18).

Despite differences in detail, the series of sequential extinction times share some common features. Most molecular species die out within the first phase of random selection. Then, the time intervals between successive extinction times become larger and larger, and eventually the population approaches a homogeneous state with a surviving species chosen at random from the set of initially present competitors. In case the initial distribution of sequences was uniform $(x_1(0)=x_2(0)=\ldots=x_n(0))$ the probability of survival is the same for all species. It amounts $1/n$. Random selection is an intrinsic feature of replication. It will take place almost certainly, by this we mean with probability one, no matter what the initial conditions or the rate-constants $(f\leqslant d)$ were. For the purpose of illustration we show two characteristic examples of sequential extinction times in figure 5.

Random selection leads to random drift in real populations. By random drift we characterize a process going on in populations consisting of selectively neutral types. The distribution of types changes although adaptive selection is absent. In order to illustrate random drift, we consider a simple model based on the previous calculations. We shall assume that error copies occur with a certain low frequency $v=1-Q$. Moreover, we neglect the effect of back mutations

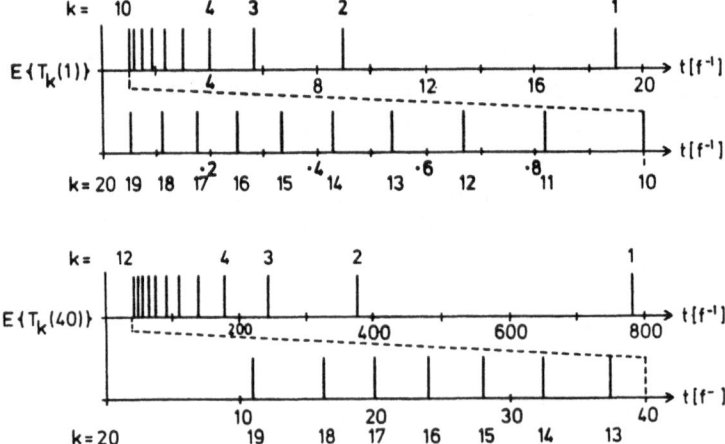

Figure 5: Sequential extinction times. We show two sequences of expectation values $E\{T_k(m)\}$ for f=d and m=1 and 40.

for the moment. The probability for a given mutant to become fixed
in the population is simply 1/N for the case of kinetic degeneracy,
where N is the population size. Spreading of mutants in the population is characterized by two stochastic variables: (1) the time of
fixation (ΔT_f) of a selectively neutral mutant after it appeared at
time t=0 in a population of N individuals and (2) the time of replacement (ΔT_r) of one molecular species by another species in a
population of N individuals (figure 6).

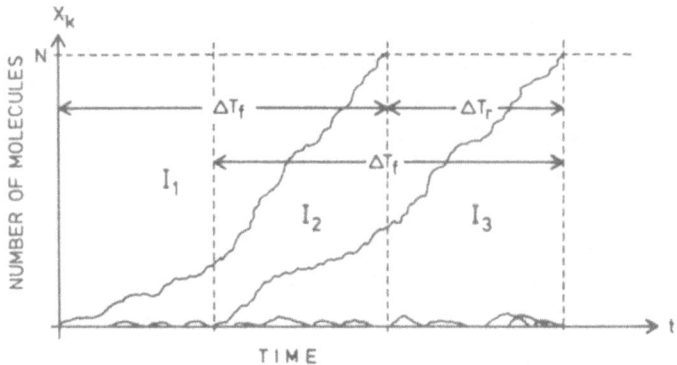

Figure 6: Two stochastic variables characterizing the spreading of
selectively neutral mutants in a population of constant size N: the
time of fixation (ΔT_f) and the time of replacement (ΔT_r).

Both quantities can be calculated by straightforward algebra
from our simple model. Expectation values of both variables are (for
details see [11,16]):

$$E\{\Delta T_f\} = (N-1)\frac{1}{f} \simeq \frac{N}{f} \tag{19}$$

and

$$E\{\Delta T_r\} = \frac{1}{v} \cdot \frac{1}{N} \cdot \frac{N}{f} = \frac{1}{f \cdot v} \tag{20}.$$

Clearly, random drift is a property of finite population size. The
time of fixation of a mutant becomes infinite in infinitely large
populations. Interestingly, the time of replacement of one mutant by
another becomes independent of population size within this model.

The results obtained here are in close analogy to those derived
from population genetics [17]. The basic difference concerns the
mechanism of replication: sexual replication and genetic recombination make the stochastic models in population genetics somewhat
more involved. So far we have treated the mutation as some kind of
"deus ex machina". The mechanism of mutation was not part of the

model. Such a procedure appears to be justified when mutations are extremely rare events, as is the case with higher organisms. In molecular systems mutations are frequent,and the model as described above is no longer adequate.

4. Stochastic error threshold and random replication

Due to the conscious neglect of mutations we were able to separate the multivariable stochastic process corresponding to equations (3-5). Therefore, we could factorize the probability distributions. In this section we shall account for the mutations explicitly and hence we have to deal with the true multivariable process. The probability distributions are no longer separable,and very difficult to calculate. Attempts to derive solutions of the master equation describing the evolution of the probability distribution were successful only if physically unrealistic assumptions on the mechanism of mutations were made (see e.g. [18-20]). The same is true for an investigation based on a Langevin-type equation [21]. Recently, we made an attempt to study polynucleotide replication as a multitype branching process [22]. Here, we shall only present some of the results which are relevant to our discussions.

Replication is considered as a multitype branching process (figure 7). For the purpose of illustration we start with a process discrete in time. Each polymer of type I_k can generate polymers of the same type ($I_k \to 2I_k$) by faithful replication or different types ($I_k \to I_j + I_k$) by erroneous replication or mutation. In each generation,each polymer of type I_k produces r_1 polymers of type I_1, r_2 polymers of type $I_2,\ldots,$ and r_n polymers of type I_n with probability $P_k(r_1,r_2,\ldots,r_n)$.

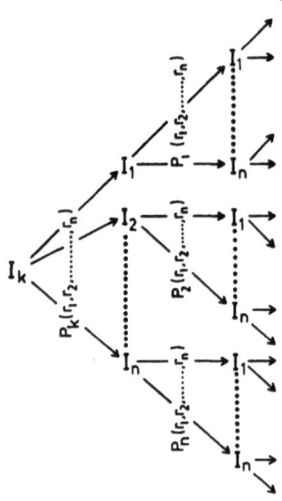

Figure 7: Polynucleotide replication as a multitype branching process. Each type I_k leads to an analogous tree of replication products (k=1,2,...,n)

Replication of an ensemble of molecules is viewed as a superposition of independent simultaneous branching processes of the kind shown in figure 7. The distribution of molecular species in the population at generation "1" is given by the random vector

$$\vec{X}(1) = \{X_1(1), X_2(1), \ldots, X_n(1)\} \tag{21},$$

whose elements are the numbers of molecules of the individual species. The transition law can be illustrated best by means of a simple example: we choose a single molecule of one type, say I_k, as initial conditions at generation $l=0$:

$$\vec{X}(0) = \vec{e}_k = \{X_1(0)=0, \ldots, X_k(0)=1, \ldots, X_n(0)=0\} \tag{22}.$$

Then, the probability generating function for the first generation is of the simple form

$$F^{(1)}(s_1, \ldots, s_n) = f_k(s_1, \ldots, s_n) = \sum_{r_1, \ldots, r_n \geqslant 0} P_k(r_1, \ldots, r_n) s_1^{r_1} \cdot \ldots \cdot s_n^{r_n} \tag{23}.$$

Herein we define the probability generating function for the l-th generation in the conventional manner:

$$F^{(l)}(s_1, \ldots, s_n) = \sum_{x_1, \ldots, x_n \geqslant 0} P^{(l)}(x_1, \ldots, x_n) s_1^{x_1} \cdot \ldots \cdot s_n^{x_n} \tag{24}$$

with

$$P^{(l)}(x_1, \ldots, x_n) = \text{Prob}\{X_1(1)=x_1, \ldots, X_n(1) = x_n\} \tag{25}.$$

The most important properties of the system can be derived from the so-called "mean matrix" M of the branching process. The elements of M are the expectation values of the numbers of molecules in generation $l=1$ derived from single molecules in generation $l=0$:

$$M = \{m_{ik}\}; \quad m_{ik} = E\{X_i(1) \,|\, \vec{X}(0) = \vec{e}_k\} = \left(\frac{\partial f_k}{\partial s_i}\right)_{s_i = \ldots = s_n = 1} \tag{26}$$

The total population size $Z(1) = \Sigma X_k(1)$ as well as the numbers of molecules $X_k(1)$ show time-dependencies which are substantially different from those of the relative variables or frequencies

$$Y_k(1) = \frac{X_k(1)}{Z(1)} \;; \quad \sum_{k=1}^{n} Y_k(1) = 1 \tag{27}$$

The total population size fluctuates strongly in unconstrained auto-catalytic reaction networks [23]. Our system is no exception in this respect. Results derived from the single type process in Z(1) may serve as an illustrative example here. If the mean and the variance of Z in the first generation are

$$E\{Z(1)\} = \bar{m} \quad \text{and} \quad var\{Z(1)\} = \sigma^2$$

and we had a single molecule in the generation $l=0$ $(Z(0)=1)$, then the two quantities grow in the supercritical case (f>d) according to

$$E\{Z(1)\} = \bar{m}^1 \quad \text{and} \quad var\{Z(1)\} = \sigma^2 \frac{\bar{m}^1 (\bar{m}^1 - 1)}{\bar{m} (\bar{m}-1)}$$

Hence, the "window" of probable values of the random variable becomes very large in late generation. For the critical case (f=d) the situation is still worse: the mean is constant whereas the variance grows to infinity. In the multitype case the situation is qualitatively similar, but the expressions for the variance are rather complicated [24].

The behaviour of the relative variables or frequencies $Y_k(1)$ for long times is rather different from that of $Z(1)$. It can be described by means of the largest eigenvalue λ of the mean matrix M and its corresponding right-hand eigenvector u:

$$M\vec{u}^{(k)} = \lambda_k \vec{u}^{(k)}; \quad \vec{u}^{(k)} = (u_1^{(k)}, u_2^{(k)}, \ldots, u_n^{(k)})$$

and $\lambda = \max\{\lambda_k; k=1,2,\ldots,n\}$ with $M\vec{u} = \lambda\vec{u}$.

The eigenvalue determines the qualitative behaviour, which can be characterized best in terms of probabilities of extinction

$$P_o^{(k)} = \text{Prob}\{\exists 1 \text{ such that } \vec{X}(1)=0 | \vec{X}(0)=\vec{e}_k\} \tag{28}$$

From the theory of branching processes follows [24]:

(1) If $\lambda \leqslant 1$ then $P_o^{(k)} = 1$ for all k and extinction occurs with probability one, and

(2) if $\lambda > 1$ then $P_o^{(k)} < 1$ for all k and there is a positive probability of survival to infinite time.

Whenever the population does not become extinct the frequencies of the individual molecular species converge almost certainly, i.e. with probability one, towards a constant value, which is determined by the elements of the largest eigenvector of the mean matrix M:

$$\lim_{1 \to \infty} Y_k(1) = \frac{u_k}{u_1 + u_2 + \ldots + u_n} \tag{29}.$$

Asymptotic convergence of a stochastic variable towards a constant implies that its expectation value converges towards the constant as well,and that the variance approaches zero.

Let us now consider the mean matrix M in more detail. Its elements are derived from the reaction probabilities given in the mechanism (3-5) by comparison with multitype branching processes shown in figure 7:

$$m_{ik} = \left(\frac{\partial f_k}{\partial s_i}\right)_{s_i=\ldots=s_n=1} = f_k Q_{ik} - (d_i - 1)\delta_{ik} \tag{30}$$

with δ_{ik} being Kronecker's delta. In general,we have to know not only the reaction rate-constants (f_f, d_k) but also the mutation frequences Q_{ik} in order to be able to describe the long-time behaviour of the system. A simplified but nevertheless representative example is re-stricted to point mutations and two digits (0,1) instead of the four bases of naturally occurring polynucleotides(G, A, C, U) [25]. These two assumptions allow to compute polynucleotide sequences up to chain length of $\nu \approx 500$. The mutation frequency is related to a mean single digit accuracy $\bar{q}$ and to the Hamming distance D between the two sequences: $\mu_{ik} = D(I_i, I_k)$. The quantity $(1-\bar{q})$ is a measure of the error rate, e.g. $\bar{q}=1$ means ultimately correct replication or no errors. Then, we have

$$Q_{ik} = \bar{q}^{\nu-\mu_{ik}}(1-\bar{q})^{\mu_{ik}} \tag{31}.$$

The stationary sequence distribution is determined by the largest eigenvector $(\vec{u})$ of the mean matrix M, as before. The result is iden-tical with the long-time solution of the corresponding deterministic differential equation,which converges to the eigenvector $\vec{u}$ too.

It is worth to consider the results of the deterministic diffe-rential equation in more detail. The mutant distribution depends on the mean single digit accuracy $\bar{q}$ (figure 8). It shows a rather dra-matic change at a certain critical single digit accuracy, $\bar{q}=q_{cr}$. The transition, which we have called the "error threshold"[4,5], becomes sharper and sharper with increasing chain length ν. In addition, we observe some other analogies to a higher order phase-transition in replication mutation dynamics. Above threshold $(\bar{q}>q_{cr})$ the statio-nary distribution of molecular types is characterized by a "master sequence" I_m and the mutant distribution accompanying it. Such a distribution resembles a biological species in a way,and therefore has been called "quasispecies". The master sequence is characterized by the maximum diagonal element of the mean matrix M. Because of its decisive role in the selection process,this diagonal element is the

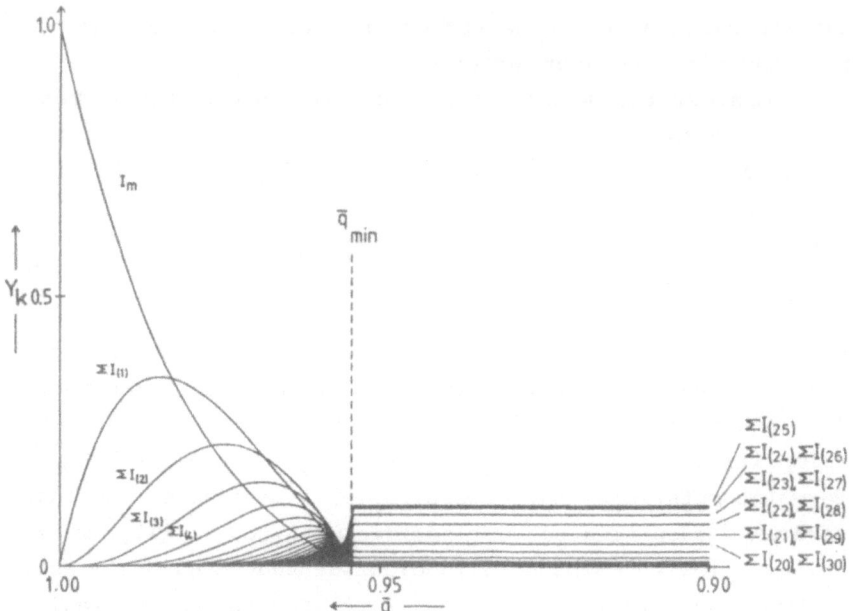

Figure 8: The stationary mutant distribution for the prevailing conditions in a replication-mutation ensemble of polynucleotides, called
"quasispecies". The curves represent the relative concentrations of
the master sequence (I_m), the sum of the relative concentrations of
all one-error mutants ($\Sigma I_{(1)}$), of all two-error mutants ($\Sigma I_{(2)}$) etc.
as functions of the mean single digit replication accuracy   ($\bar{q}$). In
the numerical example presented here the polynucleotide length is
ν=50. The replication rate-constant for the master sequence is
chosen to be f_m=10, for all other sequences $f_1=f_2=\ldots=1$ in arbitrary
time and concentration units. From these parameter values we calculate a minimum accuracy Q_{min}=0.1 (for details see [25]). Thus, we
have a critical single digit accuracy of $\bar{q}_{min}=(Q_{min})^{1/\nu}$=0.945.
Starting from pure master sequence at error-free replication,we observe a pronounced decrease in the relative concentration of the
master sequence with decreasing replication accuracy $\bar{q}$.In sequence,
one-error, two-error, three-error,...,n-error mutants dominate the
quasispecies. We observe a sharp transition in the structure of
the quasispecies at the critical value of the single digit accuracy,
$\bar{q}_{min}$. Below the critical value,the concentrations are exclusively
determined by the statistical weigths of the corresponding sequences.
All single polynucleotides sequences are equally frequent. Hence,
all the sum of the concentrations of 25-error mutants ($\Sigma I_{(25)}$ is
largest, followed by 24- and 26-error mutants ($\Sigma I_{(24)},\Sigma I_{(26)}$) etc.

"selective value" of the corresponding sequence I_k, $w_k=f_k Q_{kk}-d_k$:

I_m: w_m = max{w_k; k=1,...,n}   (32).

At accuracies below the critical value $\bar{q}<q_{cr}$ the replication-mutation
system approaches a uniform distribution of sequences. There is no
master sequence. Replication errors are so frequent that the selective values w_k have no influence on the stationary mutant distri-

bution. Indeed, the mechanism of inheritance breaks down below the critical accuracy. Within the frame of the deterministic approach the error threshold can be cast in quantitative terms. Sufficiently accurate replication requires

$$Q_{mm} = \bar{q}^{\nu_m} > Q_{min} = \bar{q}_{min}^{\nu_m} = \sigma_m^{-1} \qquad (33).$$

Herein, ν_m is the chain length of the master sequence I_m and σ_m is its superiority. This is a combination of rate-constants which expresses the higher efficiency of replication of the master sequence in comparison to its mutant distribution (for further details see [4,5,25]):

$$\sigma_m = \frac{f_m}{d_m - \bar{E}_{-m}} \qquad \text{with} \qquad (34)$$

$$\bar{E}_{-m} = \frac{\sum_{k \neq m} (f_k - d_k) x_k}{\sum_{k \neq m} x_k} \qquad (35).$$

For a given constant mean single-digit accuracy $\bar{q}$ the minimum accuracy of replication implies a maximum chain length of the master sequence:

$$\nu_m < \nu_{max} = -\frac{\ln \sigma_m}{\ln \bar{q}} \simeq \frac{\ln \sigma_m}{1 - q} \qquad (36).$$

The deterministic model of the polynucleotide replication-mutation system was found to be very useful in the prediction of maximum chain lengths of real systems [5,26].

There is, however, a fundamental conceptual problem with the deterministic error threshold. Below error threshold the deterministic model predicts a uniform stationary distribution of sequences. Such a distribution cannot be realized in a finite population. As we pointed out in the introduction, we are dealing with many orders of magnitude more possible polynucleotide sequences than we have molecules in any real population. A uniform distribution of polynucleotide sequences, thus, is far away from reality.

The stochastic approach towards the replication-mutation system, nevertheless, allows a physically meaningful analysis of the error propagation problem in finite populations. Instead of characterizing the details of the mutant distribution in the replication-mutation system, we ask for the probability of survival of the master sequence (I_m) to infinite time: $1 - P_o^{(m)}$, according to the definition given in equation (28). A non-zero probability of survival to infinite time is the stochastic equivalent to a stable quasispecies. Zero probability of survival of the master sequence implies zero probability of survival for all sequences. Thus, we are dealing with a steadily changing ensemble of polynucleotide sequences below the error threshold. The quantitative expression for the stochastic error threshold as derived from the mean matrix M of the multitype branching process is identical with the expression derived from the same matrix of rate-constants and mutation frequencies according to equations (33-36). The stochastic treatment, however, provides also information on the reliability of the results from the mean matrix M by means of the second and higher moments of the probability distribution.

For the purpose of illustration, we used discrete-time multitype branching processes here. In order to derive results which are directly comparable to those obtained from the deterministic differential equations, we have to proceed to continuous-time stochastic processes which are much more difficult to study. Apart from a minor change in the definition of the elements of the mean matrix

$$m_{ik} \longrightarrow a_{ik} = f_k Q_{ik} - d_k \delta_{ik} \tag{30'}$$

the results obtained and the conclusions derived from them are essentially the same for the discrete and continuous process. For details, the interested reader is referred to [22].

The interpretation of the error threshold by means of the stochastic analysis shows some analogy to the "localization threshold" of the mutant distribution in sequence space which has been derived recently by McCaskill [27].

5. Conclusion

In the last section we showed that the replication-mutation system approaches a stationary distribution of polynucleotide sequences provided the accuracy of replication is above the error threshold. Hence, we do not observe selection in the sense of equation (7). We may, nevertheless, perform a linear transformation of variables and use the eigenvectors of the mean matrix, $\vec{u}^{(k)}$ (k=1,...,n), as basis of our coordinate system:

$$\vec{X}(t) = \{X_1(t),...,X_n(t)\} = \sum_k X_k(t) \vec{e}_k = \sum_k U_k(t) \vec{u}^{(k)} \tag{37}$$

Consider now the evolution of the replication-mutation system in the new coordinates $U_k(t)$. In general we start with a mixture of eigenvectors: $U_k(0) \neq 0$; k=1,...,n. After a sufficiently long time the system converges to the largest eigenvector $\vec{u}$ and hence all coefficients $U_k(t)$ except óne become zero. In a "population space" whose coordinate axes are spanned by the eigenvectors of the mean matrix, we observe selection in the sense of equation (7): The system approaches a homogeneous distribution which corresponds to the largest eigenvector of M.

Below error threshold the system behaves differently: it does not approach a stationary state and its behaviour may be characterized as a kind of random walk in population space.

The stochastic treatment of polynucleotide replication led to three results:

(1) Essentially stochastic phenomena like random selection or random drift were introduced into the mathematical model. We studied unconstrained systems in which these stochastic phenomena are largest. Random selection and ramdom drift do, nevertheless, play a role also under other environmental conditions, particularly in cases of kinetic degeneracy and in small populations.

(2) We observed an interesting difference in the general long-time behaviour between total numbers of molecules and relative population numbers or frequencies in the replication-mutation system. Total numbers fluctuate strongly in agreement with the behaviour of other unconstrained autocatalytic systems. The frequencies, however, converge asymptotically to constant values. They obey a $\sqrt{N}$ law and their behaviour justifies the use of differential equations to describe the evolution of the frequencies in systems without constraints.

(3) The stochastic treatment provides a reinterpretation of the error threshold relation derived previously from the deterministic equations. Zero or non-zero probability of survival to infinite time of the master sequence becomes the decisive quantity in the derivation of a stochastic error threshold. By this definition,the behaviour of the system becomes compatible with the underlying model on both sides of the critical accuracy of replication.

Acknowledgements: The work reported here has been supported financially by the Austrian "Fonds zur Förderung der wissenschaftlichen Forschung" (Project.No.5286) and by the "Hochschuljubiläumsstiftung der Stadt Wien". Technical assistance in the preparation of the manuscript by Mrs.J.Jakubetz and Mr.J.König is gratefully acknowledged.

References:

1. S.Spiegelman: Quart.Rev.Biophysics $\underline{4}$, 215 (1971)
2. C.K.Biebricher: Evolutionary Biology $\underline{16}$, 1 (1983)
3. C.K.Biebricher, M.Eigen, W.C.Gardiner,jr.: Biochemistry $\underline{22}$, 2544 (1983)
4. M.Eigen: Naturwissenschaften $\underline{58}$, 465 (1971)
5. M.Eigen, P.Schuster: "The Hypercycle - a Principle of Natural Self-Organization" Springer-Verlag Berlin 1979. The booklet is a combined reprint of three papers: Naturwissenschaften $\underline{64}$, 544 (1977) and $\underline{65}$, 7 and 341 (1978)
6. P.Schuster, K.Sigmund: Ber.Bunsenges.Physikal.Chemie $\underline{89}$ (6), in press (1985)
7. J.Hofbauer, P.Schuster: "Dynamics of Linear and Nonlinear Auto-catalysis and Competition", in P.Schuster, ed.: Stochastic Phenomena and Chaotic Behaviour in Complex Systems, pp.160-172. Springer-Verlag, Berlin 1984
8. P.Schuster, K.Sigmund: Selection and Evolution in General Open Systems, Preprint 1985
9. B.L.Jones: J.Math.Biol. $\underline{6}$, 169 (1978)
10. C.Darwin: "On the Origin of Species by Means of Natural Selection, or the Preservation of Favoured Races in the Struggle for Life". J.Murray, London 1859
11. P.Schuster: "Polynucleotide Replication and Biological Evolution" in E.Frehland, ed.: Synergetics - from Microscopic to Macroscopic Order, pp.106-121, Springer-Verlag, Berlin 1984
12. A.F.Bartholomay: Bull.Math.Biophysics $\underline{20}$, 97 (1958)
13. W.Fontana: Diplomarbeit, Univ.Wien 1984
14. P.Schuster, K.Sigmund: Bull.Math.Biol. $\underline{46}$, 11 (1984)
15. W.Fontana, P.Schuster, K.Sigmund: Random Selection and Sequential Extinction Times,Preprint 1985
16. P.Schuster, K.Sigmund: "Random Selection and the Neutral Theory - Sources of Stochasticity in Replication" in, P.Schuster, ed.: Stochastic Phenomena and Chaotic Behaviour in Complex Systems pp.186-207. Springer-Verlag Berlin 1984
17. M.Kimura: "The Neutral Theory of Molecular Evolution", Cambridge Univ.Press, Cambridge, U.K. 1983
18. B.L.Jones, K.H.Leung: Bull.Math.Biol. $\underline{43}$, 665 (1981)
19. R.Heinrich, I.Sonntag: J.Theor.Biol. $\underline{93}$, 325 (1981)
20. J.S.McCaskill: Biol.Cybernetics $\underline{50}$, 63 (1984)
21. H.Iganaki: Bull.Math.Biol. $\underline{44}$, 7 (1982)
22. L.Demetrius, P.Schuster, K.Sigmund: Bull.Math.Biol. $\underline{47}$, 239-262 (1985)

23. P.Schuster: "Selection and Evolution in Molecular Systems -
    a Combined Approach to Stochastic and Deterministic Chemi-
    cal Kinetics", in B.Gomez, S.M.Moore, A.M.Rodriguez-
    Vargas, A.Rueda, eds.: Stochastic Processes Applied to
    Physics and other Related Fields, pp.134-159. World
    Scientific Publ.Co., Singapore 1983
24. T.E.Harris: "The Theory of Branching Processes", Springer-
    Verlag, Berlin 1963
25. J.Swetina, P.Schuster: Biophys.Chem. 16, 329 (1982)
26. M.Eigen, P.Schuster: J.Mol.Evol. 19, 47 (1982)
27. J.S.McCaskill: J.Chem.Phys. 80, 5194 (1984)
28. D.T.Gillespie: J.Comp.Phys. 22, 403 (1976).
29. K.Sigmund: The Maximum Principle for Replicator Equations,
    IIASA WP 84-56 (1984)

Part II

Functions of the Brain in Man and Animals

Computation of Sensory Information by the Visual System of the Fly (From Behaviour to Neuronal Circuitry)

W. Reichardt

Max-Planck-Institut für Biologische Kybernetik, Spemannstraße 38,
D-7800 Tübingen, Fed. Rep. of Germany

A detailed understanding of the principles of sensory information processing by the nervous system requires investigations with a variety of "model" systems at different levels of complexity. The fly as a model system represents a compromise in the sense that the properties of information processing are taking place in a rather complex nervous system that is amenable to a quantitative analysis at three different levels: At the phenomenological level the overall function of the system, its input-output behaviour, and its logical organization are studied. At the second level, the functional principles of the subsystems, expressed by its algorithmic properties are the object of the analysis. At the third level the investigation focuses on the individual nerve cells and the neuronal circuitry involved in the computations. - The approach, described here, deals with an analysis of the visuo-motor control system of the fly at the phenomenological level, which leads to an algorithmic description of the main subsystems for the extraction of motion- and position-information. The analysis at the third level is connected with the investigation of the so called figure-ground and pattern discrimination problem by the visual system of the fly. In this connection the elements of a new movement detector theory are mentioned.

1. Introduction

Complex systems such as the central nervous system, or even parts of it like the visual system, posses a large number of components. A determination of the detailed function of any individual component seems rather hopeless at the present time, but fortunately is not needed if one is mainly interested in features on a macroscopic scale such as the behaviour of an organism in its environment.

How optical information is processed by the fly's visual system has been explored in great detail. The objective is to unravel the logical organisation of the system and its underlying functional and computational principles. There are three levels of analyzing and "understanding" complex systems. At the first, strictly phenomenological level, one investigates the overall function, the input-output behaviour of the system and its logical organization. At the second level the functional principles of the subsystems are the object of the analysis. At the third level, one studies individual neuronal components and the detailed circuitry.

In this review-like article a phenomenological theory of the fly's visual orientation behaviour is outlined. The theory describes and predicts, at the first level, a rather complex behaviour in terms of some simple computations performed by the neuronal interactions on the visual input. At the second level, functional properties of the interactions underlying the computations are given. These two levels of analysis cannot simply follow from single-cell recordings or from histology. Finally, an outline at the cellular level is described, as it follows from more recent investigations of figure-ground and pattern discrimination.

2. A Smooth Fixation and Tracking Control System

Male and female flies fixate - that is fly towards - small, contrasted patterns and they track moving objects. The theoretical analysis of this visual control system relied almost completely on experiments performed in the laboratory with a flight simulator device shown in Fig. 1, which enables one to simulate the visual input under free-flight conditions in one degree of freedom, rotation around the vertical axis. Results obtained with the device can be extended to free-flight conditions.

Fig. 2 represents the equivalent free-flight situation. In the horizontal plane α_f designates the instantaneous direction of flight with respect to an arbitrary zero direction, α_p the instantaneous angular position of an object.

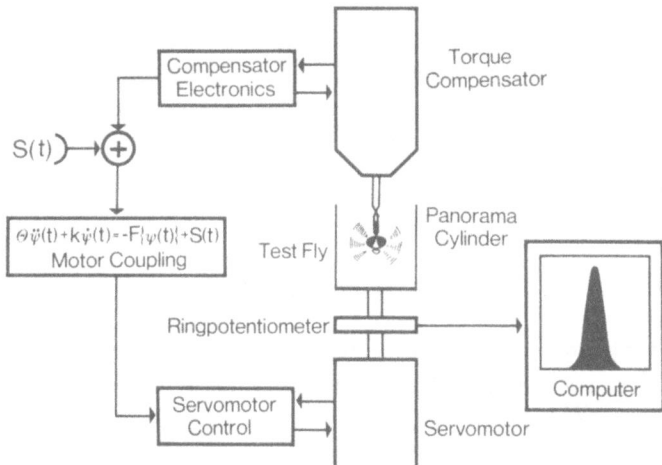

Figure 1 Simplified scheme of the basic closed-loop flight simulator setup developed by REICHARDT and WENKING [1]. A fly, fixed to the torque compensator, controls the velocity of a surrounding panorama by its own torque signal through an analog simulation of the flight dynamic. Redrawn from REICHARDT [2].

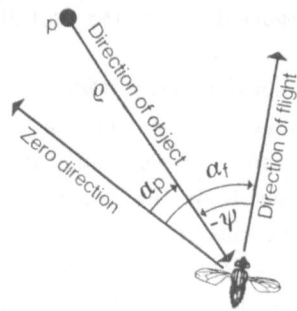

Object in the environment

Figure 2 Polar coordinate system describing the fly's rotational degree of freedom around the vertical axis (on the horizontal plane). $\alpha_p(t)$ designates the angle between an arbitrary zero direction and the direction of the object in the fly's environment; $\alpha_f(t)$ is the angle between the zero direction and the fly's direction of flight; $\psi(t) = \alpha_p(t) - \alpha_f(t)$ is the error angle between the fly's direction of flight and the object. The distance ρ between the object and the fly is usually assumed to be large in order to neglect translation effects. From REICHARDT and POGGIO [3].

The angle $\psi = \alpha_p - \alpha_f$ is referred to as the error angle. If the head is fixed relative to the thorax, $\psi(t)$ represents the location of the image of the object on the retina at instant t. When $\psi = 0$, the projection of the fly's long axis on the horizontal plane points toward the object. Translation effects — the object is far away from the fly — are here neglected, and it is assumed that the flight path of the fly is confined to the horizontal plane.

It was shown in a series of studies (see review REICHARDT and POGGIO [3]) that the flight dynamic is well approximated by the following equation

$$\ddot{\alpha}_f(t) + k\, \dot{\alpha}_f(t) = F(t) \tag{1}$$

where θ and k are the moment of inertia and an aerodynamic friction constant of the fly, respectively, and F(t) is the instantaneous torque produced by the fly's wings. Since θ is very small compared to k, the angular velocity $\dot{\alpha}_f$ is essentially proportional to the torque ($\theta/k = 8 \cdot 10^{-3}$ sec). Equation (1) can be rewritten as

$$\theta\, \ddot{\psi}(t) + k\, \dot{\psi}(t) = -F(t) + S(t) \quad \text{with} \quad S(t) = \theta\, \ddot{\alpha}_p(t) + k\, \dot{\alpha}_p(t) \tag{2}$$

where S(t) reflects the trajectory of the object. Thus the fly controls its angular velocity through its own torque F. The basic problem here was to determine how F depends on the visual input, that is, which control system is used by the fly. A large series of experiments has led to the following conclusions:

a) The observed torque process is stationary under normal, experimental conditions, implying that there is no switching between different control systems.

b) The term F(t) can be approximated as the sum of two terms: a visually evoked response $R_t\{\psi(s)\}$, representing a functional of the error-angle history, and a component that can be characterized stochastically as a Gaussian process N(t).

c) The visually induced response depends smoothly on the error-angle history. Under this condition, rigorous theorems (see for instance COLEMAN [4]) ensure that R_t may be approximated by a function of $\psi(t)$ and its derivatives. The first approximation is given by

$$R_t = D(\psi)+r(\psi)\ \dot{\psi}(t) \tag{3}$$

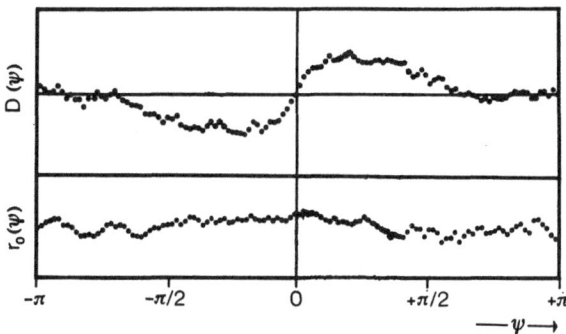

Figure 3 The functions D(ψ) and r(ψ) characterizing the position and the velocity computation elicited by a narrow vertical black stripe segment in Musca females. The stripe segment was rotated with a constant angular speed (8°/sec) and the measured (open-loop) torque was decomposed into the direction-insensitive component D(ψ) and into the direction-insensitive one r(ψ)·$\dot{\psi}$. From REICHARDT and POGGIO [3].

Experimentally it has been verified that (3) is a satisfactory approximation under a wide range of conditions. The terms D(ψ) and r(ψ)·$\dot{\psi}$ with r(ψ) $\approx$ r_o are shown in Fig. 3. In addition, the reaction shows a small delay ($\varepsilon \approx$ 20 msec). Thus (2) becomes

$$\Theta\ \ddot{\psi}(t)+k\ \dot{\psi}(t) = -D[\psi(t-\varepsilon)]\ -r_o\dot{\psi}(t-\varepsilon)+N(t)+S(t) \tag{4}$$

where $r_o \cdot \dot{\psi}$ is the contribution of a velocity computation; the term D(ψ) carries the position information and represents the "attractiveness" profile associated with the specific pattern. The terms have been characterized quantitatively through independent experiments. Through (4) the theory can predict nontrivial behaviour in quantitative detail. An example of the predictive power of (4) is shown in Fig. 4. In a stochastic sense (4) predicts the angular trajectory of a fly fixating or tracking patterns. The experimental data agree with the theoretical predictions.

Although connections with physiological and anatomical data have been established (see HAUSEN [6,7]; WEHRHAHN [8]), the theory and, in particular, the

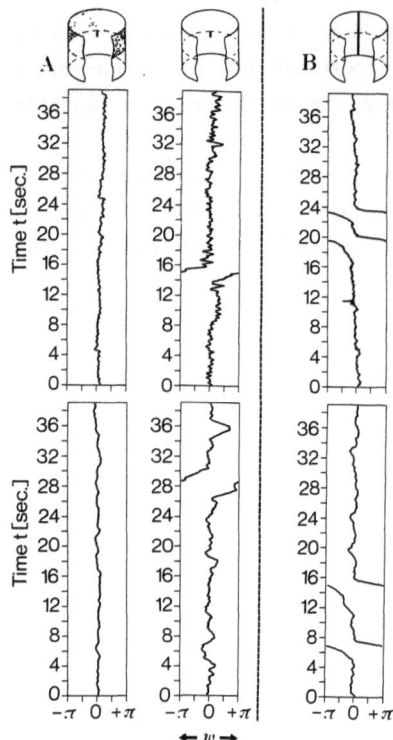

Figure 4 a) Dynamics of fixation of two different patterns by <u>Musca</u> females as measured (upper part) and as predicted (lower part) by (4) with standard values for the parameters. For details see POGGIO and REICHARDT [5]. b) The trajectory $\psi(t)$ for tracking of a black stripe rotating at constant angular velocity as measured (upper part) and predicted (lower part) in a stochastic sense by the same equation.

derivation of (1) are based on behavioural data. Equation (4) describes the basic organization of the fixation and tracking system used by the fly to control its flight. It implies that this control system relies on computations performed on the visual input extracting movement information and position information, and it states how this information is used to control the flight trajectory. A description like (4) could not have been obtained from single—cell recordings or from histology alone. This is an example where there seems to be little predictive extrapolation from the component level to the control system level.

3. Free-Flight Behaviour

A critical validation of the theory is due to free—flight experiments. LAND and COLLETT [9]] were able to film chases in which male flies (<u>Fannia</u> <u>canicularis</u>) pursued other flies. Their data led to the conclusion that the control system used by the chasing fly can be described essentially by (4). In addition, they have shown that such an equation can correctly simulate the trajectory of the chasing fly, given that of the leading fly. A later analysis of the flight behaviour in the hoverfly (COLLETT and LAND, [10,11]) uncovered different control systems used by this fly in various circumstances. One of them is again a smooth system, phenomenologically identical to the control system in Fannia and Musca. Thus the control system described by the Reichardt-Poggio theory is equivalent, from the point of view of logical organization and information processing, to the control system used in free flight by male Fannia and, often, by Syritta.

Collett and Land, however, believe that males but not females can track other flies. It would, in fact, be surprising if females could not track other flies at all, since the control system described by (4) is an effective tracking system up to a certain angular velocity, and since tracking has been experimentally demonstrated in restrained females (VIRSIK and REICHARDT, [12]). It, however, has been shown that the free-flying Musca females do indeed track other flies, although not as often or as effectively as their male companions (WEHRHAHN, [13]; WAGNER, [14]). Since tracking in males is characterized by higher angular velocities, the term "chasing" should be reserved for this behaviour and that both males and females be considered to perform "normal" fixation and tracking.

The difference between chasing in males and tracking in females does not necessarily imply a unique male-specific neural circuit specifically designed for chasing (REICHARDT and POGGIO, [3]), although there must be, at least, trigger, gain control, and velocity-distance neural circuits. Collett and Land's conjecture that there should be male-specific visual interneurons in the optic lobes has been supported by Hausen and Strausfeld (HAUSEN and STRAUSFELD, [15]; STRAUSFELD, [16]). WEHRHAHN's [13] demonstration that during chasing male flies keep the target in the superior frontal part of the visual field, while female flies track with the inferior frontal part of the eye; see also WAGNER [14]. This observation is consistent with the behaviourally-determined function $D(\psi)$ (see (3)), and the equivalent lift response to vertical displacements in ϑ , $L(\vartheta)$ by WEHRHAHN and REICHARDT [17,18]. Furthermore,the fact that males and females use different parts of the eye strongly suggests that the neural circuitry of the male chasing system is distinct from the neural circuitry underlying normal fixation and tracking in females and males.

4. Reconsideration of the Smooth Control System

The phenomenological theory outlined here is restricted to one degree of dynamic freedom, the rotation around the vertical axis of the fly. The vertical degree of freedom, involving the fly's lift, can be described by an equation similar to (4) (WEHRHAHN and REICHARDT, [17,18]). WEHRHAHN [19] has shown that the two degrees of freedom are, with respect to the position response, essentially independent, as hypothesized earlier (REICHARDT and POGGIO, [3]). The results allow a quantitative description of fixation and tracking behaviour in two degrees of freedom. Of course,one would like to extend this analysis to all six degrees of freedom, including roll, pitch etc. For instance, the description of translations involves nonlinear terms, arising from the geometry of the situation, which can lead, in their interplay with the control system of the fly, to a complex series of fixation and tracking behaviours. In this connection,let us consider again Fig. 2. If the distance ρ is not very large, (4) can no longer be applied, and one has to take into account the geometrical effect of the fly's translation on the error angle ψ. The appropriate description is given by adding to (4), where we put $\Theta = 0$, a suitable geometrical term, and by introducing an additional equation that gives the rate of change of ρ :

$$\dot{\rho} = -v_o \cos(\psi) \; ; \quad \dot{\psi} = -\frac{1}{k} D[\psi(t-\varepsilon)] - \frac{r_o}{k} \cdot \dot{\psi}(t-\varepsilon) + \frac{v}{\rho} \cdot \sin(\psi) + \frac{N(t)}{k} \quad (5)$$

In order to solve (5) it is necessary to know how the translation velocity v_o is controlled. A simple mathematical analysis of (5) has been given,together with some possible biological implications; (REICHARDT and POGGIO, [20]).

One of the major tasks for the future is to understand, at the phenomenological level, how the various control systems interact, and how switching among various subroutines of behaviour occurs which are released by rotational and translational motions. Three-dimensional computer reconstruction of flight trajectories and their analysis promises to yield many new data, as for example shown in Fig. 5.

The analysis of visually guided movements in insects offers a good example of the understanding one can achieve at the highest level of behavioural

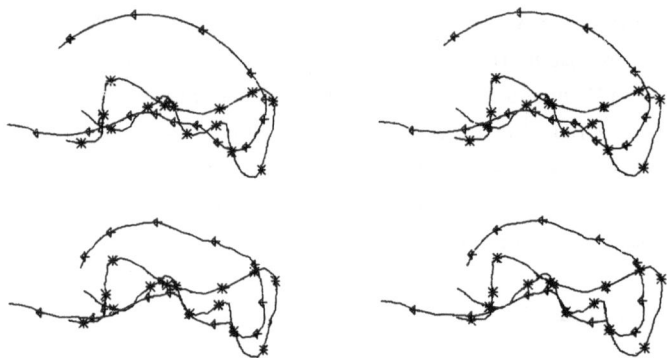

Figure 5 A 1440-msec chase between flies (Musca domestica) flying freely in a cage. The bottom stereo plot shows the 3-dimensional trajectories of the two flies (* leading fly, Δ chasing fly). The top stereo pair shows a computer simulation of the flight of the chasing fly, according to the theory described here. If possible,the reader should look at the figure with a standard stereo viewer. The film was made by H. Wagner.

organization. The analysis considers the logical organization of a specific control system; it neglects explicit computational analysis since the goal of the system is clearly tracking and fixation, provided by modules for movement and position detection.

5. Functional Specification of Neural Interactions at the Algorithmic Level

The phenomenological theory characterizes the basic logical organization of flight behaviour controlled by the optical environment. However, the theory does not allow us to specify the interactions because the variables ψ and $\dot{\psi}$ are only indirectly related to the organization of receptor inputs, and to the interaction processes in the visual nervous system. The phenomenological theory requires the processing of optical information in the neural network between the receptors and the flight muscles to perform two main computations on the visual input. One computation extracts movement information (the term $r(\psi) \cdot \dot{\psi}$ of the phenomenological theory (4)). The other provides position information (the term $D(\psi)$).

In the following the algorithms for movement, position and relative movement computations will be characterized:

a) The Algorithm for Directionally Selective Movement Computation: If a system or an interaction is movement-selective, it is required that it shows an average response that is direction-selective. The computation of movement requires that the system has at least two light receptor inputs for representing the vector of movement. The property of direction selectivity forces to assume that the interaction between signals taken up by the receptors are interacting in an antisymmetric fashion. An actual experiment that determines the interaction properties is outlined in Fig. 6. A test fly is fixed to a flight-torque compensator, and its yaw-torque is measured. Positive (negative) torque implies that their movement-detecting system measures movement to the right (left). The results plotted and described in Fig. 6 show that when a movement to the right is simulated, the average (dc-free) response is positive (negative) and the phase dependence of the time-averaged response in good approximation proportional to sin(∅). This and many other experiments imply that the algorithm for directional movement detection can be characterized in terms of many 2-input systems with multiplication-like, antisymmetric interactions, of which the Hassenstein-Reichardt model is a specific example (see e.g. REICHARDT, [21]); a graph of the interaction is respresented in Fig. 7A.

b) The Algorithm for Position Computation: A system can be said to detect a small contrasted object moving against a white background if its time-averaged output depends on the position of the object in front of the system's photoreceptors. Under the experimental conditions, pictured in Fig. 6, the

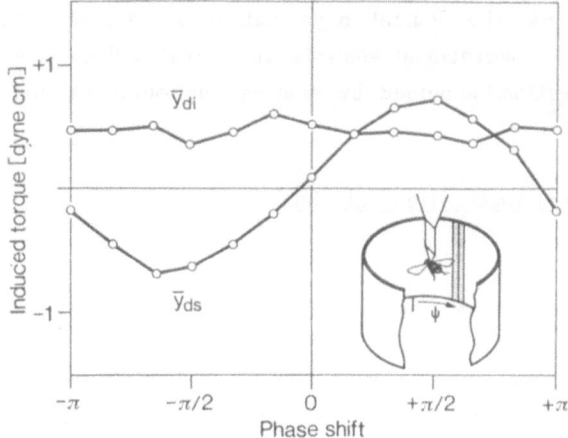

Figure 6 The mean (time) torque response of a test fly elicited by two vertically oriented filament lamps whose intensities are sinusoidally modulated and phase-shifted with respect to one another ($x_1 = \sin(\omega t)$; $x_2 = \sin(\omega t + \emptyset)$). The phase lag $\emptyset$ is defined as positive if the luminance modulation of the right lamp follows that of the left lamp. The half-sum $(\overline{y_{ds}})$ and the half-difference $(\overline{y_{di}})$ of the reactions induced by the two lamps located first in the mean position $\psi = +15°$ and then in $\psi = -15°$. $\overline{y_{ds}}$ represents the direction-sensitive and $\overline{y_{di}}$ the direction-insensitive component of the mean yaw-torque. Redrawn from PICK [22].

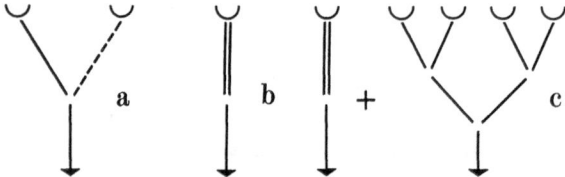

Figure 7 Graphical representations of algorithms for movement, position and relative movement computations. Redrawn from REICHARDT and POGGIO [24].
a) Movement computation. The graphical representation of the algorithm pictures a system (module) with two receptor inputs and an antisymmetric, multiplication-like interaction (expressed by a dotted line). The time-averaged response component $\overline{y_{ds}}$ as a function of the phase $\emptyset$ is given by $R(\emptyset)\alpha \sin(\emptyset)$, as pointed out in connection with the experiment described in the legend of Fig. 6.
b) Position computation. The graph of the algorithm represents a system (module) with one receptor input and a multiplication-like interaction that ensures a time averaged response when an object is moved across the receptor input. The system is parametrized, in the sense that its response is dependent on the location of the position-detector in the compound eye.
c) Relative movement computation. The graphical representation of the system consists of two modules. The one on the left represents a position-detector, the other a combination of three movement detectors with multiplication-like symmetric interactions, measuring the cross-coherency of the directionally independent movement signals. The graph on the right has four receptor inputs and is therefore double orientation specific in accordance with the experiments.

position-dependent part of the time averaged yaw-torque of the fly does not depend on the light phase, indicating that 1-receptor flicker detectors distributed homogeneously in the eye, play a fundamental role in this computation (PICK, [22,23]). A graph for the most probable interaction responsible for position computation is shown in Fig. 7B.

c) The Algorithm for Relative-Movement Computation: Relative movement computation has not been explicitly treated in connection with the phenomenological description, expressed by (4), but should be discussed here at the algorithmic level. A system can be said to detect relative movement if its average output depends on the position of a visual object moving relative to a larger background, irrespective of its texture. An experimental set-up for studying figure-ground discrimination is shown in Fig. 8. A small object oscillates in front of a textured background, that is moving with respect to the fly's array of photoreceptors. In the absence of relative movement, the object disappears in the texture. Relative movement, however, allows a determination of the objects location and, in the fly's case, induces an attempt to fixate it. When a stripe and a random-dot texture are oscillated with slightly different frequencies, the stripe is seen and the fly tries to turn toward it. Fig. 9 shows the results of a paradigmatic experiment, where figure and ground are

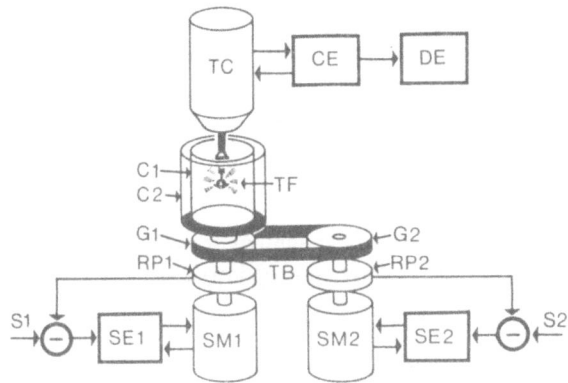

Figure 8 Schematic representation of an experimental set up used in the figure-ground test. A test fly TF is suspended from a torque compensator TC, which in turn is connected with the compensator electronics CE. The flight torque signal, generated by the fly under compensation, is proportional to the voltage output of CE. The CE output is evaluated by a data evaluation box DE. The motions of two cylinders C1 and C2 are controlled by two 400-Hz servomotors SM1 and SM2 whose shafts carry ringpotentiometers RP1 and RP2 and gear wheels G1 and G2. The inner cylinder is connected with servomotor SM1 whereas the outer cylinder is driven by a transmission belt TB from servomotor SM2. The ring potentiometer voltages are fed back to the inputs of the servomotor electronics SE1 and SE2, so that SM1 and SM2 are operated under position control from the two inputs S1 and S2, respectively. Under these conditions an angular displacement of a cylinder is strictly proportional to the voltage applied to an input of the SE electronics. The two cylinders are illuminated by four direct-current-driven fluorescent ring bulbs. Redrawn from REICHARDT and POGGIO [24].

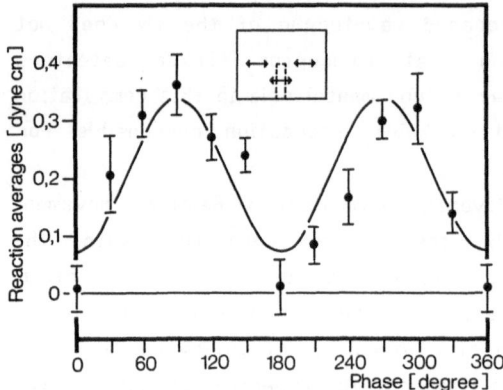

Figure 9 Phase-dependence of figure—ground discrimination. Time-averaged torque responses of ten flies to sinusoidally oscillating foreground and background patterns, measured under conditions described in Fig. 8. The foreground consists of a black vertical stripe, 3° wide, positioned in the lower part of the panorama, oscillated around the mean position $\psi = +30^{\circ}$. The background pattern consists of a random-dot texture that can be moved independently from the foreground. The oscillation amplitude of the stripe and the ground was $\pm 1^{\circ}$ at 2,5 Hz frequency. Redrawn from REICHARDT and POGGIO [24].

oscillated with the same frequency (2,5 Hz). Its outcome depends on the relative phase between the oscillations. The fly, of course, does not see the stripe when $\emptyset = 0^{\circ}$. However, it reacts strongly when $\emptyset = 90^{\circ}$ or 270°. Interestingly, it does not react to the figure when $\emptyset = 180^{\circ}$. The phase portrait, shown in Fig. 9, is well approximated at this oscillation frequency ω by the expression

$$\bar{R} = k_{o}(\omega) - k_{4}(\omega)\cos(2\cdot\emptyset) \tag{6}$$

which speaks in favour of an algorithm whose graphical representation is pictured in Fig. 7c; see in this connection REICHARDT and POGGIO [24].

6. The Cellular Basis of Figure—Ground Discrimination and Pattern Discrimination

Motion of an animal in the visual world generates a distribution of apparent velocities on its eyes. Discontinuities in its optical flow field are a good indication of object boundaries and can be used to segment images into regions that correspond to different objects (HELMHOLTZ, [25]). In particular, the relative movement of an object against a background can be used to reveal its presence and to delineate its boundaries. The human visual system is very efficient in exploiting this fact (JULESZ, [26]; BAKER and BRADDICK, [27]; van DOORN and KOENDERINK, [28]; REGAN and BEVERELY, [29]). Similarly, a fly is able to detect and to discriminate an object or a figure that moves relative to a background texture as has been mentioned before in section 5 (VIRSIK and REICHARDT, [12]; REICHARDT and POGGIO, [24]).

a) Neuronal Circuitry: A model of the neuronal circuitry for figure-ground discrimination via relative movement by the visual system of the fly has been suggested more recently (POGGIO, REICHARDT and HAUSEN, [30], and REICHARDT, POGGIO and HAUSEN, [31]). The design of the theory is based on the analysis of time-averaged behaviour and on the dynamic of the torque response at the onset and in the stationary phase of relative movement. Two main results have been obtained:

1) The characteristic properties of figure-ground discrimination are observed not only under binocular but also under monocular stimulation by the ground texture.

2) The torque response of the fly follows the oscillation of a given pattern, as is well known from studies of the optomotor behaviour. Experiments clearly show that the amplitude of the response increases with an increasing movement amplitude (pattern speed), whereas it is almost independent of the dimensions of the pattern, suggesting a quite specific gain control mechanism. Consequently, the effect of a small figure in contributing to the optomotor response is as large as the ground's, despite their different shapes and areas. These and other findings have led to a proposal for a neuronal circuit obeying the following constraints: The circuit should contain a gain control mechanism for the overall optomotor reaction. The interaction between movement detectors is not realized through a lateral inhibitory network but through large-field movement – selective neurons inhibiting elementary movement detectors. The neuronal theory should reproduce the characteristic dynamics of the response for all phases (relative velocities) between figure and ground and for binocular and monocular stimulation.

A model for the neuronal circuitry satisfying the constraints is outlined in Fig. 10; its detailed operations are described in the legend of the figure. According to the circuitry, the behavioural response is given by the relation

$$R = \frac{Nx^n}{[\beta + (Nx)^q]^n} \tag{7A}$$

except for the running average at the output and provided that all channels N are equally excited by a moving pattern. x designates the output of each of the N movement detectors, β the shunting inhibition coefficient, n the postsynaptic characteristic of the synapses at the inputs to the X-cells, and q a saturation of the S-cells. Experiments require that, for increasing N, the response R should become independent of N (gain control), which means that n·q has to be unity, neglecting β for $(N \cdot x)^q \gg \beta$. Since the response under relative phase 0° conditions between figure and ground is sinusoidal, n = 2 so that q = 1/2. With

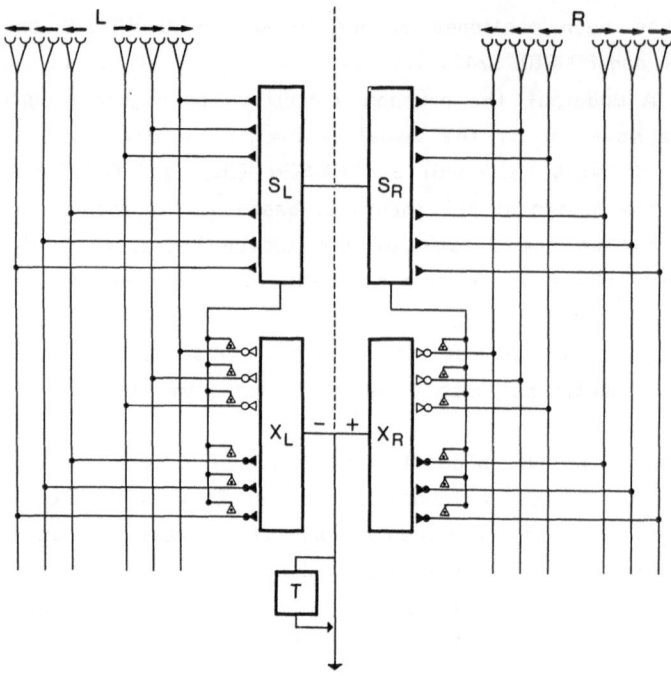

<u>Figure 10</u> Outline of the model circuitry for figure-ground and pattern discrimination for the right and for the left compound eye. Considering the right eye as an example, two retinotopic arrays of elementary movement detectors, responding selectively to progressive (——▶) and regressive (◀——) motion serve as input channels to the neural circuitry. The two arrays share the same field of view. They are drawn apart from each other. A pool neuron (S_R) summates the movement detector outputs (——◀indicate excitatory synapses) as well as the input from its contralateral homologue (S_L). Its output is assumed to undergo a saturation effect (modeled by taking the square root of its overall excitation; q = 0,5) and to shunt inhibit each movement detector channel via presynaptic inhibition. The synapses involved (——◁) should therefore inhibit (opening ionic channels with an equilibrium battery close to the resting potential) the output of each elementary detector channel. The cell X_R summates the progressive (excitatory ——●◀) and regressive (inhibitory ——○◁) detectors. Progressive channels have a higher amplification than regressive ones, possibly because of the different ionic batteries involved. The synapses on the X cell are assumed to operate with a nonlinear input-output characteristic, leading to postsynaptic signals that are approximately the square (n = 2) of the inputs. As far as presently known, the motor output is controlled by the X-cells via a direct channel and a channel T computing the running average of the X-cells output. Redrawn from REICHARDT, POGGIO and HAUSEN [31].

these parameter specifications (7A) reads

$$R = \frac{Nx^2}{[\beta+(Nx)^{0.5}]^2} \approx x \quad if \quad (Nx)^{0.5} >> \beta \tag{7B}$$

Let us now consider the case that figure (F) and ground (G) both excite N_F and N_G detector channels, respectively. With x, y we designate the outputs of the movement detectors which receive their inputs from the moving figure and the moving ground, respectively. Under these conditions (7A) becomes

$$R = \frac{N_F \cdot x^n + N_G \cdot y^n}{[\beta + (N_F x + N_G y)^q]^n} \tag{8}$$

Usually $(N_F x + N_G y)^q \gg \beta$, and with $n = 2$, $q = 1/2$ one arrives at

$$R \approx (a+b)\left[\tilde{x} + \tilde{y} - \frac{2\tilde{x} \cdot \tilde{y}}{\tilde{x} + \tilde{y}}\right] - \frac{b\tilde{x}^2}{\tilde{x} + \tilde{y}} - \frac{a\tilde{y}^2}{\tilde{x} + \tilde{y}} \tag{9}$$

where $\tilde{x} = N_F \cdot x$, $\tilde{y} = N_G \cdot y$, $a = \frac{1}{N_F}$, $b = \frac{1}{N_G}$. The response (except for the running average) of the fly depends linearly on the movement fields x and y which are inhibited (minus sign) by the weighted correlation of x and y. In addition, there are selfinteraction terms in x and y. Equation (9) says that the contributions from the figure (x) and from the ground (y) to the response become independent in the time-average when figure and ground move in statistical independence, a prediction which has been tested and confirmed in a series of experiments (REICHARDT, POGGIO and HAUSEN, [31]).

b) Figure-Ground Experiments and Theoretical Predictions: The theory outlined here has been tested in a series of figure-ground experiments. Only a few tests are shown here, and compared with computer results that are based on the model presented in Fig. 10.

Gain Control for Overall Optomotor Responses: When only a figure is oscillated in front of one of the two compound eyes and the ground is at rest, one observes an oscillatory response of the fly with the same frequency.

The outcome of these experiments is shown in Fig. 11a; parameter is the amplitude of the oscillating figure. The experiments indicate that response amplitudes are relatively independent of the width of the oscillating figure, but increase when the amplitude of the oscillating figure is increased. Testing the gain control mechanism with the neuronal model leads to the plots shown in Fig. 11b. The response amplitude increases with the pattern's angular extent and reaches a constant level. However, it depends strongly on the detector channel output. With increasing output, i.e. with increasing oscillation amplitude of the figure pattern, one gets response curves of different levels. The computer simulation is quite in accordance with the experimental data plotted in Fig. 11a.

A Typical Figure-Ground Experiment: A test fly is stimulated by a small figure (a vertically oriented and textured stripe), which is sinusoidally oscillated in front of one of the fly's two compound eyes. A random-dot textured ground (of the same texture as the figure) is also oscillated with the same frequency and amplitude as the oscillating figure. The response plotted in Fig. 12a consists of an initial part in which the phase between figure and ground oscillation is $0°$, followed by a part in which the phase has been switched to $90°$. The oscillatory response increases after the phase is switched to $90°$ and reaches a positive level around which it oscillates in a rather complicated fashion. In the $90°$

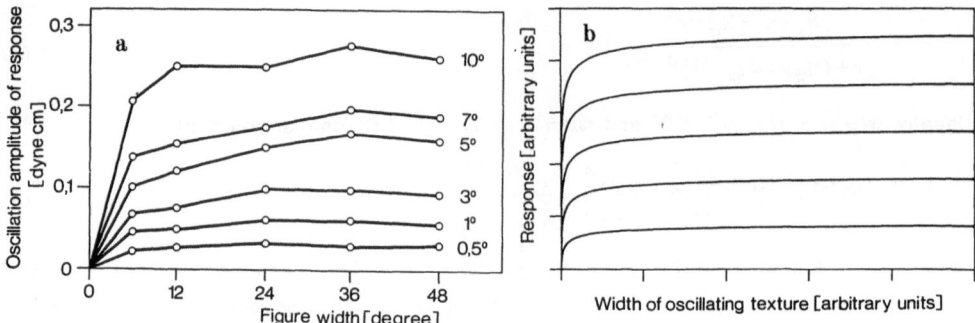

Figure 11 Gain control mechanism that operates on the angular extent of moving patterns. Redrawn from REICHARDT, POGGIO and HAUSEN [31].
a) Oscillation amplitude of the torque response as function of the angular extent of an oscillating figure ($0°$ to $48°$ wide). Parameter is the oscillating amplitude of the figure ($0.5°$ to $±10°$). In the experiments, the ground is not oscillated. For a given amplitude of the oscillating figure, the amplitude of the response increases with increasing figure width, but reaches a fairly constant level as soon as the figure's extent amounts to more than $12°$. Each point derived from 100 sweep averages.
b) Computer plot demonstrating the gain control mechanism. In the simulations, the parameter settings are $n = 2$, $q = 0.5$ and $ß = 0.05$. The diagram shows the response (oscillation amplitude) as a function of the width of the oscillating texture. Parameter of the plot is the oscillation amplitude of the figure pattern.

phase region, the time-average of the response is positive: the fly is attracted by the oscillating figure. This finding agrees with the earlier results (shown in Fig. 9) when the time-average of the response was measured. The result of the corresponding computer simulation, based on the neuronal circuitry, is plotted in Fig. 12b. The response at phase $0°$ is a sinusoid that changes dramatically when the phase changes. Since the running average time amounts to 0.4 sec, the transition phase ends at 0.8 sec. From 0.8 sec on the response is stationary, quite similar to the torque response plotted in Fig. 12a. Many other experimental and computational tests of the model have been carried out and are published in REICHARDT, POGGIO and HAUSEN, [31], and GUO and REICHARDT, [32].

Recent experimental and theoretical investigations by REICHARDT and GUO [33] have shown that pattern discrimination by the fly is a special (degenerate) case of figure-ground discrimination. Applying (8) to the special case that two patterns (F, G) of equal size are oscillated in counter-phase in front of the two compound eyes (each in front of one), we get with $N_F = N_G = N$

$$R = \frac{Nx^n - Ny^n}{[ß + (Nx + Ny)^q]^n} \quad \frac{x^2 - y^2}{x + y} = \frac{(x+y)(x-y)}{x+y} = x - y \tag{10}$$

so that the response R is equivalent to the difference of the detector output signals (x, y) that are elicited by the patterns F and G.

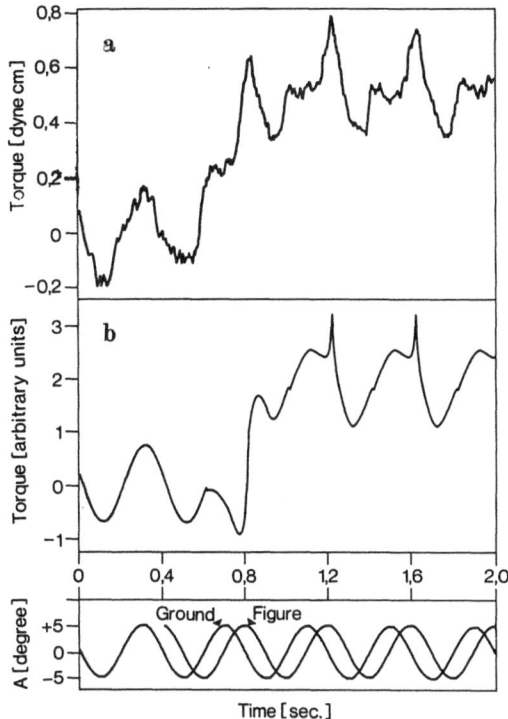

Figure 12 A typical figure-ground discrimination test. The patterns of figure
and ground consist of statistically distributed black and white dots (Julesz-
patterns) and are moved relative to each other. Redrawn from REICHARDT, POGGIO
and HAUSEN [31].
a) A textured stripe (figure) of 12° angular width is sinusoidally oscillated in
front of a 360°-textured background which oscillates with the same amplitude
and the same frequency (2,5 Hz). The stripe oscillates around the mean
position ψ = +30° that is in front of the right eye. Oscillation amplitudes
amount to ±5°. The time courses of the figure and ground oscillations (in terms
of positions) are plotted in the lower part. With regard to the right eye of the
fly, movements from -5° to +5° are progressive movements, whereas movements
from +5° to -5° are regressive movements. At time 0,5 sec the relative phase
between figure and ground switches from 0° to 90°. The response (torque) of
the fly increases after the phase has shifted, and oscillates around a positive
average response level with 2,5 Hz frequency. The increase of the response
means that the fly is attracted by the oscillating figure. The response plotted
is an average of 100 sweeps.
b) Computer simulation of the figure-ground experiment. Parameters of the
simulation (n, q, ß) are the same as specified in Fig. 11b. The figure-ground
phase changes from 0° to 90° at 0,4 sec. The transition phase is completed at
0,8 sec. since the running average time (T-box in Fig. 10) is one period of
response of 0,4 sec.

Under the condition that the torque response R of a fly depends directly on

the motion fields (see equations (7B) and (10)), it is important to clarify the

functional role of the movement detectors whose outputs do not only depend on

pattern velocity, but also on contrast, average brightness and other pattern

parameters. To this purpose, the Hassenstein-Reichardt theory on movement

detection, as sketched in Fig. 13a,b, has been reconsidered in terms of a

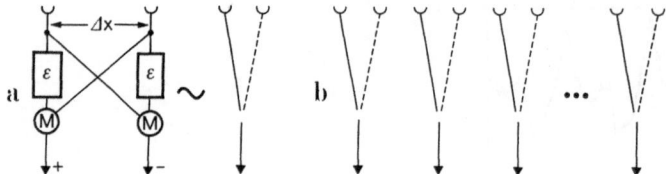

Figure 13 An algorithmic representation of a movement detector and an array of movement detectors.
a) An individual movement detector consists of two light receptors, two lowpass filters which in first approximation delay the input signals by ε and by second order, multiplication-like interactions. A shorthand graphical representation, as already shown in Fig. 7a, is drawn on the right side.
b) A one dimensional array of movement detectors.

continuous approach which covers not only the time-average of the detector response (as the earlier analysis did), but also its dynamic properties when patterns are moved with arbitrary time-dependent velocities. In reality, the number of movement detectors per eye is certainly finite, but in the approximation we assume that this number is infinite. Consequently, the separation angle Δx between adjacent light receptors is approaching the differential dx. If x is a spatial coordinate, F = (x+s(t)) a representation of a one-dimensionally contrasted pattern that is moved according to s(t), we get at the outputs of an array of such movement detectors a response distribution that is given approximately by the density expression

$$\frac{dD(x,t)}{dx} = -\varepsilon \frac{ds(t)}{dt} \left[\left(\frac{\partial F(x+s(t))}{\partial x} \right)^2 - F(x+s(t)) \frac{\partial^2 F(x+s(t))}{\partial x^2} \right] \tag{11}$$

where ε represents a small delay of the filters shown in Fig. 13a. Equation (11) tells us that the detector response is proportional to $\frac{ds(t)}{dt}$, the instantaneous pattern speed, but also depends on F in a way specified by the expression that is given in the bracket.

A first new insight into the detector's operation follows from (11) when the pattern contrast function was assumed to be F = A ± B $e^{-a^2(x-ct)^2}$; a Gaussian contrast superimposed on a constant brightness which is moved with constant velocity $\frac{ds}{dt}$ = −c across a detector array. The two contrast patterns are shown in Fig. 14a,b. Evaluating (11) with the specific contrast functions leads to an expression plotted in Fig. 14c,d. The results, which in the meantime have also been established experimentally, show an interesting phenomenon, namely the existence of a motion contrast. That is to say, when the contrast pattern in Fig. 14a is moved to the right (−c), the detectors receiving inputs from the central part of the pattern are responding in accordance with the direction of motion, whereas the detectors receiving inputs from the peripheral parts of the contrast pattern are responding inversely, so as if the pattern would move to the left. That situation changes when the contrast of the pattern is inversed, but movement is still from left to right, as is shown in Fig. 14d. If the outputs

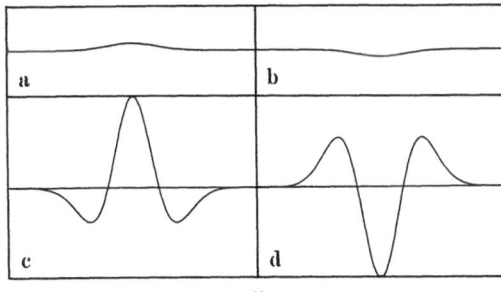

Figure 14 Responses of a linear array of differentially sized movement detectors to moving contrast patterns. Upper figures a and b: Representations of moving contrast functions $F = A \pm Be^{-a^2(x-ct)^2}$, with c the velocity of motion. Lower figures c and d: At the detector output channels one observes a response distribution which moves with the pattern speed c (from left to right). The output distributions, generated by the movement detector array, surprisingly consist of two parts, one which signals the sign of pattern motion in a correct way, whereas the other signals apparent counter motion. When, however, the integral over all the individual motion contributions is taken, the integrated detector outputs correctly reflect the direction of pattern motion.

of movement detector arrays are summated, the total detector response complies with the rule that it signals the direction of motion correctly.

It should be added here that attempts have been made to clarify the movement detector scheme, as specified in Fig. 13a, at the cellular level. In this connection the work of RIEHLE and FRANCESCHINI [34] should be mentioned. These authors have shown by recordings from a movement-sensitive large field neuron (H1-cell) and by stimulating individual light receptors, that the multiplication scheme of the Hassenstein-Reichardt type and not the Barlow-Levick scheme (see BARLOW and LEVICK [35]; TORRE and POGGIO [36]) is implemented by the visual system of insects.

The continuous approach applied here to determine approximately the spatial response distribution at the outputs of an array of movement detectors to a moving pattern (see (11)) has a remarkable advantage over a discontinuous approach or a computer simulation. Namely, it gives us an insight into the rules which govern the nonlinear local transformation carried out by the movement detectors of the fly. The approximation has been treated in mathematical detail and will be published in REICHARDT and GUO [33].

In the meantime, the approach has been extended to two dimensions. The motion field $\vec{v}^*(x,y)$ that is generated by two-dimensional arrays of (orthogonally oriented) movement detectors in response to the velocity field of a moving pattern, is now available. The local, nonlinear transformations, provided by a pair of movement detectors, have tensorial character and are given by the following expression

$$
\begin{bmatrix} v_x^*(x,y) \\[2em] v_y^*(x,y) \end{bmatrix} = -\varepsilon \begin{bmatrix} (\frac{\partial F}{\partial x})^2 - F\frac{\partial^2 F}{\partial x^2} & \frac{\partial F}{\partial x}\cdot\frac{\partial F}{\partial y} - F\frac{\partial^2 F}{\partial x\partial y} \\[2em] \frac{\partial F}{\partial y}\cdot\frac{\partial F}{\partial x} - F\frac{\partial^2 F}{\partial y\partial x} & (\frac{\partial F}{\partial y})^2 - F\frac{\partial^2 F}{\partial y^2} \end{bmatrix} \begin{bmatrix} v_x(x,y) \\[2em] v_y(x,y) \end{bmatrix} \quad (12)
$$

with $F = F[x+s_x(t),\ y+s_y(t)]$, $v_x = \dfrac{ds_x}{dt}$, $v_y = \dfrac{ds_y}{dt}$. The meaning of (12) is that in general the direction of the motion vector $\vec{v}^*$ is different from the direction of the velocity vector $\vec{v}$, except for two "eigen"-directions which entirely depend on the pattern structure. The tensorial character of (12) diminishes when the off-diagonal elements of the symmetrical matrix disappear. Interestingly, this is the case for $F[x+s_x(t),\ y+s_y(t)] = f[x+s_x(t)]*g[y+s_y(t)]$, where f and g are one-dimensional contrast functions. It is expected that the two-dimensional movement detector theory will play a dominant role in formulating a design of a theory on pattern discrimination in insects. The details of the derivation of (12) will be published elsewhere, REICHARDT [37].

The work reported so far was aiming for a design of a cellular topology, derived from quantitative, behavioural experiments. It is necessary, of course, to determine the cellular components directly, by means of histology and electrophysiology, in order to associate real nerve cells to the topological structure outlined in Fig. 10. Extensive investigations have been carried out recently at the level of the third optical ganglion (lobula plate), where large neurons are "integrating" the output signals of large arrays of movement detectors. Specifically, it has been shown by EGELHAAF, [38,39,40] that neuronal correlates of the output-cells of the model circuitry, shown in Fig. 10, are located in the lobula plate. These output cells can be subdivided into two classes with different spatial integration properties. One of them is more sensitive to large-field pattern stimulation, whereas the other responds best to the motion of small patterns (objects).

I thank Martin Egelhaaf for carefully reading the manuscript, Leo Heimburger for drawing the figures, and Inge Geiss for typing the text.

References

1   W. Reichardt and H. Wenking: Naturwiss. 56, 424 (1969)
2   W. Reichardt: Naturwiss. 60, 122 (1973)
3   W. Reichardt and T. Poggio: Quart. Rev. Biophysics 9, 3, 311 (1976)
4   B.D. Coleman: Arch. Ration. Mech. Analysis 43, 1 (1971)
5   T. Poggio and W. Reichardt: Kybnernetik 12, 185 (1973)
6   K. Hausen: "Signal Processing in the Insect Eye", in Function and Formation of Neural Systems, ed G.S. Stent (Dahlem Konferenzen Berlin 1977) p. 81
7   K. Hausen: Verh. Dtsch. Zool. Ges. 1981, 49 (1981)

8   C. Wehrhahn: "Visual Guidance of Flies During Flight", in
     Comprehensive Insect Physiology, Biochemistry and Pharmacology,
     eds. G.A. Kerkut and L.I. Gilbert (Pergamon Press Oxford 1984)
9   M.F. Land and T.S. Collett: J. Comp. Physiol. 89, 331 (1974)
10 T.S. Collett and M. Land: J. Comp. Physiol. 99, 1 (1975a)
11 T.S. Collett and M. Land: J. Comp. Physiol. 100, 59 (1975b)
12 R. Virsik and W. Reichardt: Naturwiss. 61, 132 (1974)
13 C. Wehrhahn: Biol. Cybern. 32, 239 (1979)
14 H. Wagner: Dissertation, University of Tübingen 1985
15 K. Hausen and N.J. Strausfeld: Proc. R. Soc. Lond. B 208, 57
     (1980)
16 N.J. Strausfeld: Nature, Lond 283, 381 (1980)
17 C. Wehrhahn and W. Reichardt: Naturwiss. 60, 203 (1973)
18 C. Wehrhahn and W. Reichardt: Biol. Cybern. 20, 37 (1975)
19 C. Wehrhahn: Biol. Cybern. 29, 237 (1978)
20 W. Reichardt and T. Poggio: "Visual Control of Flight in Flies"
     p. 135. "Appendix 6: On Mathematical Trajectories of Flies" p.
     205, in Theoretical Approaches in Neurobiology, eds. W.E.
     Reichardt and T. Poggio (MIT Press Cambridge Mass. 1981)
21 W. Reichardt: "Autocorrelation, a Principle for Evaluation of
     Sensory Information by the Central Nervous System" in
     Principles of Sensory Communications, ed. W.A. Rosenblith (John
     Wiley New York 1961) p. 303
22 B. Pick: J. Naturforsch. 29c, 310 (1974)
23 B. Pick: Biol. Cybern. 23, 171 (1976)
24 W. Reichardt and T. Poggio: Biol. Cybern. 35, 81 (1979)
25 H. Helmholtz: Handbuch der physiologischen Optik (Voss Hamburg
     Leipzig 1896)
26 B. Julesz: Foundations of Cyclopean Perception (University of
     Chicago Press Chicago 1971)
27 C.L. Baker and O.J. Braddick: Vision Res. 22, 851 (1982)
28 A.J. van Doorn and J.J. Koenderink: Biol. Cybern. 44, 167 (1982)
29 D. Regan and K.I. Beverly: J. Opt. Soc. Am. A1, 433 (1984)
30 T. Poggio, W. Reichardt and K. Hausen: Naturwiss. 68, 443 (1981)
31 W. Reichardt, T. Poggio and K. Hausen: Biol. Cybern. Suppl. 46,
     1 (1983)
32 A. Guo and W. Reichardt: Biol. Cybern. in prep. (1985)
33 W. Reichardt and A. Guo: Biol. Cybern. in prep. (1985)
34 A. Riehle and N. Franceschini: Exp. Brain Res. 54, 390 (1984)
35 H.B. Barlow and W.R. Levick: J. Physiol. 178, 477 (1965)
36 V. Torre and T. Poggio: Proc. R. Soc. Lond B 202, 409 (1978)
37 W. Reichardt: In preparation 1986
38 M. Egelhaaf: Biol. Cybern. in print (1985a)
39 M. Egelhaaf: Biol. Cybern. in print (1985b)
40 M. Egelhaaf: Biol. Cybern. in print (1985c)

Multi-Neuron Experiments:
Observation of State in Neural Nets

G. Gerstein, A. Aertsen, M. Bloom, E. Espinosa, S. Evanczuk,
and M. Turner

Department of Physiology, University of Pennsylvania,
Philadelphia, PA 19104, USA

1. Introduction

During the last twenty years it has become common in the neurophysiology laboratory to record and study the electrical activity of single
neurons, one at a time, and in a wide variety of stimulus and
behavioral conditions in many creatures. Using statistical and computer methods (reviewed in the book by GLASER and RUCHKIN [1]), it is
possible to associate firing patterns of single neurons with transmission of particular sensory or motor information. A striking discovery
of this era has been that neurons with similar stimulus selectivity
(preferences or tuning) lie together in spatially restricted clusters
which may repeat at approximately regular spatial intervals. The best
known examples are the cortical columns first described by MOUNTCASTLE
[2] for somatosensory cortex and by HUBEL and WIESEL [3] for visual
cortex.

Even as single neurons studies gained popularity, it was widely accepted that most of the complex properties and abilities of the nervous system can not be accomplished by single neurons acting alone.
The idea that individual neurons join into varying functional assemblies for special and possibly temporary purposes originated with
SHERRINGTON [4], was refined by HEBB [5], and has been an implicit
part of neurophysiological theory ever since. Unfortunately there is
remarkably little direct experimental knowledge about such neuronal
assemblies and their properties (and hence very unsatisfactory theory)
simply because most studies have involved only sequential observations
of single neurons.

Sequential study of single neuron properties can be used to infer
some static aspects of some assemblies, as for example the shared
stimulus preferences of neurons in a cortical column. However, dynamic
aspects of neural assembly organization can only be studied by simultaneously and separably observing the activities of many individual
neurons. The experimental circumstances must be arranged so that the
observed portion of nervous system is actively engaged in analysis of
stimuli and in production of learning and behavior. In such studies it
becomes possible to examine directly the membership and organization
of neuronal assemblies under shifting stimulus and behavioral conditions, as well as to trace the time-varying flow of information
through the assembly. If the experimental conditions involve learning, it becomes possible to determine when there are effective connectivity changes and whether the flow of information is rerouted during
the acquisition of the task. In other words, with simultaneous observation of neurons, we may seek emergent properties of assemblies that
transcend those of the single neurons of which the assemblies are composed. An important, and as yet unstudied aspect of such experiments
is the sampling problem: what fraction of a neuronal assembly must be
put under observation in order to make significant conclusions about
the entire assembly.

2. Experimental Technology

Multi-neuron experiments unfortunately represent a nonlinear increment in difficulty over standard single neuron studies. Special equipment is required, but far more importantly, it becomes necessary to handle an enormous flow of data, even if only some 20 neurons are under observation. It is essential that interpretation of the data occur rapidly enough so that parameters can be manipulated sensibly during an experiment. The critical factors are a suitable abstraction of the data for presentation to the experimenter, and sufficient computer power to produce the abstractions at near real-time speeds. Matters of technology and interpretation are reviewed in [6].

a. Electrodes. The objectives are to sense the electrical activity of the maximum possible number of single neurons with a minimum of physical damage. For solid brain tisue these objectives are presently best attained with bundles of wires, each being separately insulated and pointed. In some realizations of this methodology, each wire can be individually moved along its axis, using an appropriate (and large) driving device. We have made satisfactory use of simpler bundles of fixed wires which are moved in unison: 12 or 24 micron insulated Tungsten wires with sharpened points are glued into a stainless steel tube some 5 millimeters back from the tips. The front of the electrode structure is shown in Fig. 1. A slightly larger guide tube is implanted on a secant towards the cortical structure of interest, with its end just short of the target. This causes cortical damage only to distant, irrelevant locations; pia over and afferents below the target cortex receive minimal perturbation. The electrode structure is then inserted through this protecting guide tube; the electrodes need only pass through brain tissue during the last five millimeters of movement. The technology can be used both in acute and chronic circumstances.

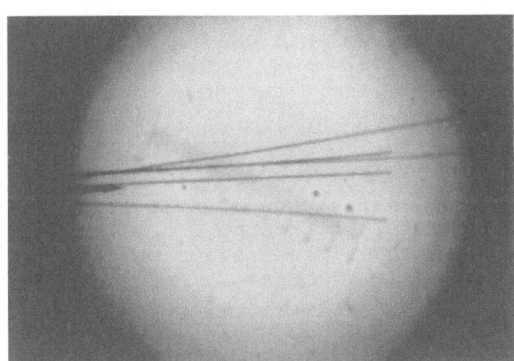

Fig. 1: Multi-wire electrode bundle. Each electrode is an epoxy insulated 12 micron Tungsten wire which has been mechanically ground to a sharp point. Supporting mandrel just visible on left. From [6].

b. Spike Shape Sorters. Wire electrodes of the type just described often will pick up trains of extracellular action potentials from several neurons. The spike train originating from a particular neuron can be distinguished on the basis of its waveform, which is the result of the geometrical and spatial relations between the particular neuron and the electrode. Waveform sorting can be accomplished with varying degrees of success using simple or more complex hardware devices, or by means of software alone. A review of the possibilities is given by SCHMIDT [7]. We have made use of a device based on principal components of a typical set of waveforms, and are finishing development of another scheme based on software only. For examples of this process, see GERSTEIN et al. [6]. The end result of this technology is that we can sort out the activity of several neurons through a single electrode; since we normally use seven wires, a typical recording may have the distinguishable activity of some 20 neurons.

c. Computer Complex. During a typical multi-neuron experiment, we must (1)control stimulus and assess response behavior, (2) sort waveforms and create a time list of when each event (action potential, stimulus, response, etc.) occurred, and (3) analyse and interpret the data stream. We accomplish these several goals by a network of four computers organized as shown in Fig. 2. Computer B, an LSI-11/23, manages stimuli and responses; computer A, an LSI-11/2, manages waveform sorting and preliminaries of data acquisition; computer C, an LSI-11/23 acts as master for computers A and B, and finishes the data acquisition process. Finally, computer D, a Data General MV10000, carries out various types of analysis and subsequent graphical display. The immense power of the last computer (about equal to a VAX 785) is essential for the kind of calculations described at the end of this paper.

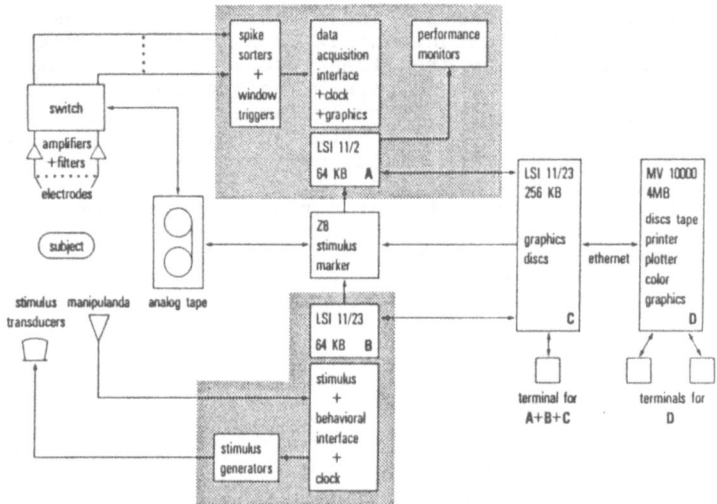

Fig. 2: Diagram of signal flow and of analysis devices in the laboratory. Each of the blocks around computers A through D has separate logical functions concerned with data acquisition, stimulus control, or data analysis. See text.

3. Analysis by Cross-Correlation

Traditionally, analysis of multi-neuron experiments has been carried out with appropriate variants of cross-correlation. Again, [1] offers a good review with many references to the original literature.

The fundamental calculation is cross-correlation of two point processes (each representing one neuronal spike train), and the results are usually presented as a cross-correlation histogram (or cross-correlogram) of some convenient time resolution. There are two possible types of interpretation of this measurement:
  (1) Since a correlogram peak simply indicates (delayed) coincident firing of the two neurons, this can represent a coding scheme for representation of information.

(2) The coincident firing can be taken as a consequence of synaptic (or other) connection between the two neurons. Thus we may infer an "effective logical connectivity" between and to the observed two neurons.

We stress that "effective logical connectivity" is NOT the same as anatomical connectivity. Anatomical connections must of course underlie "effective connections". However, a given anatomical connection may be silent or ineffective at a particular time, hence will have no effect on relative firing of the two neurons, and will not be detected by the cross-correlogram measurement. In general, the cross-correlation measurements have considerably greater sensitivity for excitatory interactions between neurons than for inhibitory interactions [8].

The "effective logical connectivity" between two neurons can be expressed in terms of the connections shown in Fig. 3. The two observed neurons are labelled A and B, U is an unobserved (inferred) source of shared input to the observed neurons, and S is an unobserved stimulus related input to the two neurons. Characteristic features in the cross-correlogram (like width of a peak or valley, or its offset from zero delay) can be associated with each of the pathways of Fig. 3. The stimulus inputs are measured by time-shifting one of the spike trains by one full stimulus period, and recomputing the cross-correlation; this is called the shift-control correlogram. The interested reader is referred to the literature for additional detail.

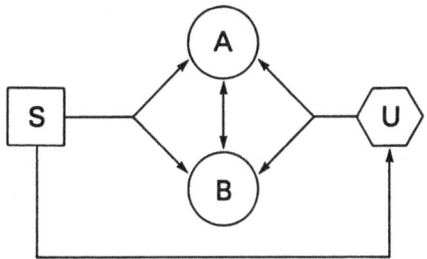

Fig. 3: Diagram of possible types of logical connection between two observed neurons A and B. The existence of S (stimulus) and U (shared) inputs can be inferred from cross-correlograms. See text.

The correlation of firing times, and associated interpretation in terms of "effective connectivity" may be a time-varying quantity. The time-scale of such variations can range from hours (associated with general biological nonstationarities) to 100 msec (associated with presentation of a stimulus. Such dynamic variations in effective connectivity or state are probably the most interesting quantity that has become experimentally accessible with the development of multi-neuron experiments. A simple process of time-editing which identifies, isolates and concatenates selected portions of the multi-neuron data (for example associated with delivery of some particular stimulus condition) suffices to let the ordinary cross-correlation methods be used to study dynamic variations of neural state. Such time-editing methods have been used in all the experiments described below.

So far we have considered cross-correlation applied to a single pair of neurons. In a typical multi-neuron experiment, however, we have many pairs of neurons. Thus we can build up a fairly extensive analysis of the effective connectivity among all the observed neurons, together with some inference about unobserved stimulus and other shared inputs to the network. However, note that there is a combinatorial proliferation of calculation. For a typical experiment which records 10 neurons under 10 different stimulus conditions, we end up with the need to examine some 1800 cross-correlograms alone. (45 neuron pairs, x2 for stimulus shift-control, x2 for two-time resolutions, x10 for the stimulus conditions = 1800). Thus, the experimenter is faced with the need to absorb incredibly large amounts of correlogram information in order to analyze the state of the observed neural assembly; this is both difficult and extremely time consuming. A more global approach to the analysis of a multi-neuron experiment is essential, preferably treating the entire observed neural network as an entity rather than as a combination of pairs. One such calculation has recently been developed in our laboratory, and is reviewed in the last part of this paper.

4. Neural Assembly Properties in Cortex

For the past several years our laboratory has been engaged in multi-neuron experiments with the auditory cortex of the cat. This is a well studied region, so that we were able, on the basis of known results in the literature, to select stimulus conditions that should be particularly favorable for the demonstration of emergent properties of neuronal assemblies. We sought conditions in which destruction of auditory cortex caused severe impairment of behavior based on stimulus interpretation, but yet where there were no reports of cortical single neuron properties that "tuned" for that class of stimuli.

At least two types of stimulus conditions and associated behavior satisfy these requirements:
(1) The task is to identify and walk to one of several sources of brief sound. These sources are arranged at about 10 degree intervals in the horizontal plane as seen from the starting position. A series of experiments by JENKINS and MERZENICH [9] clearly demonstrated that bilateral ablations of auditory cortex impaired execution of this task. If the ablation was limited to iso-pitch stripes (medial to lateral), the behavioral deficit occurred only at a pitch corresponding to that of the ablated portion of auditory cortex. On the other hand, MIDDLEBROOKS and PETTIGREW [10] have examined the sensitivity of cortical single-unit responses to direction of sound source, and have found ONLY neurons that respond to an entire hemifield, or in a more limited way along the spatial axis of the outer ear. No cortical neurons were found to selectively respond to spatial direction in a way corresponding to the behavioral discrimination of direction. (Neurons with spatial auditory tuning properties have been described in the superior colliculus [11]; the relationship to cortical neurons and to cortical ablations is not known).
(2) The task is to discriminate short "melodies" made up of a few tones. NEFF [12] reports that cats with ablation of the auditory cortex can not do such tasks. There are NO reports in the literature that single neuron responses can select particular "melodies", although some single neurons seem to "prefer" ascending or descending pitch tone sequences [13].

Our experiments [14], [15] have made use of both types of stimuli, but have not included active behavioral tasks. Typically, some ten neurons were recorded simultaneously, and the data analyzed by standard cross-correlation methods.

The possibility of stimulus coding by near-coincident firing rather than the firing patterns of the individual single neurons can be examined with both types of stimuli described above, but is conceptually easier to discuss in the case of stimuli that correspond to different direction of the sound source. In this situation, we observed many cases where the spatial "tuning" (i.e. preference) of individual neu-

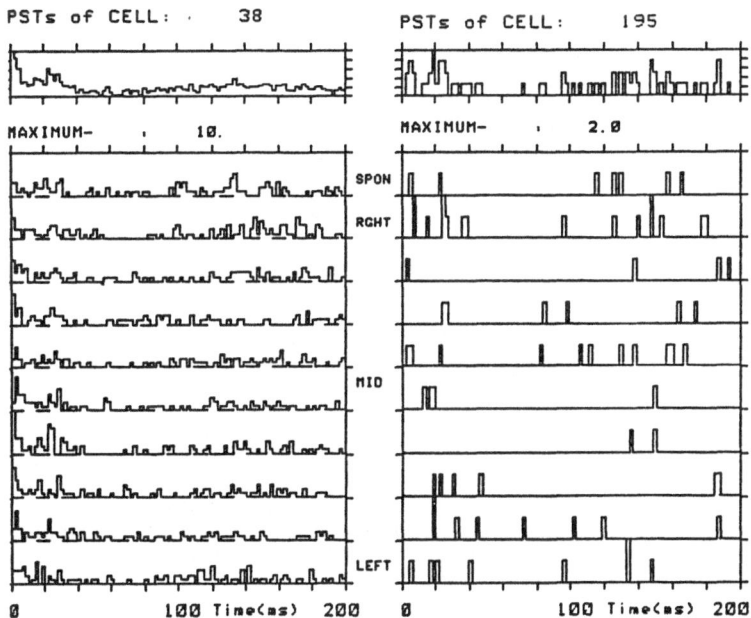

CWMVPA.PØ1

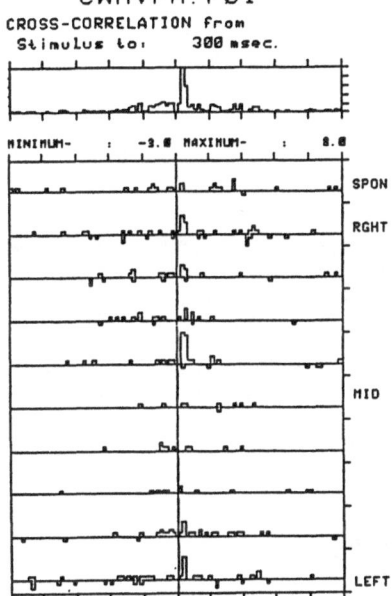

CWMVPA.PØ1

Fig. 4: Spatial tuning shown in cross-correlograms (right column) shown by peak in correlogram just above "MID". Note absence of similar tuning in PST histograms for each neuron individually (left columns).

rons was fairly broad, while the spatial "tuning" of the correlograms for these same neurons showed a peak (or its lack) only at some particular spatial angle. An example of such contrasting spatial tuning is shown in Fig. 4. The vertical dimension represents different spatial angles of the sound source; the first two columns are PST histograms (stimulus - locked averages) for the responses of the individual neurons; the last column shows cross-correlograms for the same two neurons. Each column also includes an entry for spontaneous activity (i.e. no stimulus), and an entry which is the response or correlogram summed over all spatial angles.

Such experimental data demonstrate that information about spatial direction of a sound source is available in the near-coincident firing of neurons even though this information is not present in the individual spike trains of the neurons. This, therefore, is an emergent property of the neural assembly which can not be predicted from the individual properties of its components. Note that although the experimenter in the laboratory can extract information about spatial direction of a sound source from the near-coincident firing of observed neurons, it is not demonstrated that this is a coding mechanism actually used by the nervous system. For the moment, near-coincident firing of these neurons must be considered a putative coding mechanism for spatial direction, available, but not necessarily used.

Changes in "effective connectivity" of an entire assembly of neurons can also be examined with either of the two classes of stimuli defined above. The basic result is that the logical connections among the observed neurons vary with stimulus conditions. (Note that if a sequence of different stimuli are repeatedly presented, it is necessary to select the data at and immediately following the repetions of a particular stimulus for analysis; such selected segments are treated so as to avoid any edge effects from the segment boundaries, and the diagram we obtain represents an average over all repetitions of the particular stimulus.) Examples of stimulus·related changes of effective connectivity are shown in Fig 5. The short tone stimuli were presented to left, both, or right ear, corresponding to extremes of left, center, and right positions of a sound source. The Figure shows only direct interactions between the neurons: shared inputs from unobserved neurons are not shown, and there has been correction for all stimulus effects on each neuron individually. (In other words, the diagrams are derived from interpretation of narrow peaks that remain in cross-correlograms after subtraction of shift-control correlograms.)

Note that some logical connections (M2,M3 or N1,N2 etc.) are present during all three stimulus conditions. Other connections are present only in some of the stimulus conditions ((M1,N4), or may even switch direction (M1,N1). Note that a long cascade (N1,M1,M2,M3,N4) appears ONLY for the "both" stimulus conditions. Repeated transitions among these different states of the neuronal assembly take place on a time scale of 100 milliseconds, depending on the time structure of the stimulus presentation sequence.

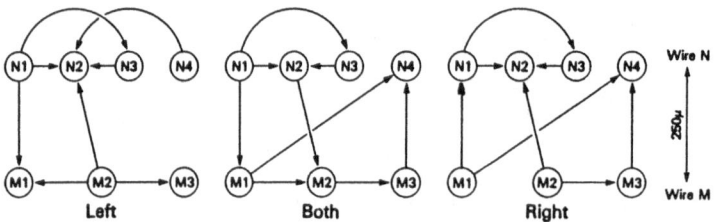

Fig. 5: Effective direct connections among these seven neurons depend on stimulus conditions. Shared input and direct stimulus effects are omitted. See text.

The data shown in Fig 5 emphasize an important unsolved theoretical problem: We need to measure and categorize the types and changes of logical connectivity in a more numerical way than the anecdotal description provided by logic diagrams. Presumably these are state changes and order-disorder transitions in the neural assembly; an appropriate quantitative tool would allow recognition of similar states and transitions in different assemblies. Exploration of some simple ideas from graph theory have not been successful [16].

We have noted another important assembly state phenomenon in the experiments using "melody" stimuli. Consider three tone sequences ABC; these can be presented in six permutations of order. If we examine the effective connectivity of an assembly during (repeated) presentation of, for example, tone C, the results vary depending on whether we are examining the state during C as first, second or third tone. In particular, if C is the third tone of our melody, it was preceded either by AB or by BA. These two circumstances produce different effective connectivities. Thus, the present state of a neural assembly depends not only on the current stimulus situation, but also on the recent (1 second or so) stimulus history. Thus we have evidence for a physiological mechanism that can account for the invoevement of auditory cortex in melody-discrimination tasks. Note again that we are here dealing with an emergent property of neural assemblies. There is no evidence that their single neuron components individually exhibit melody discrimination.

5. Global Analysis of Neural Assemblies

All the work we have described so far has been based on the pair-wise analysis of firing patterns of simultaneously recorded neurons. When larger numbers of neurons are observed, the analysis nevertheless proceeds by pairs. As we have pointed out above, this means a combinatorial explosion in the data processing. This is a serious problem not only in terms of computation time, but because it requires the experimenter to mentally combine far too much material in order to comprehend phenomena at the assembly level. A more global approach that will treat the whole assembly of neurons as a single entity is needed.

We have developed precisely this type of calculation over the last few years; it is based on the idea of gravitational clustering originally described by WRIGHT [17]. The basic idea is to transform the original N parallel point processes in time into an N-body problem in an N-space. At first glance this might not seem like a reduction of complexity. However, this transformation takes the problem from a form where the experimenter can not see patterns (in time) to a form where the experimenter can see patterns (in space). In other words, the transformation makes it easier for our own brains to visualize and interpret the activity of a neural assembly.

The basic representation [18], [19] associates each of N observed neurons with a particle in a fictitious N-space. The ith particle is located at x_i, and carries a time-varying charge q_i. This charge is a (low pass) filtered and normalized transformation of the train of action potentials from the corresponding neuron; the charge is incremented by a fixed amount at the moment of each action potential, and subsequently decays with an appropriate time-constant. Forces between the charged particles will cause relative movement. Those particles which correspond to neurons which tend to fire in close time relationship will simultaneously have high values of (decaying) charge, and under the appropriate rules will tend to aggregate. Different clusters of particles and the dynamics of the aggregation process will signify different functional groups of interacting neurons.

Calculation of propulsive fields, forces and displacements are straightforward. At the location of a particular particle, we obtain the propulsive field caused by the charges of all the other particles by making a vector summation. Force on that particle is given by the product of the propulsive field and the local particle's charge. Formally, the total force on the ith particle is:

$$\mathbf{F}_i = q_i \sum_{j \neq i} q_j \, \mathbf{r}_{ij} \tag{1}$$

where $\mathbf{r}_{ij}$ is the unit vector pointing from particle i to particle j. Note that the force magnitude is independent of interparticle distance. Particles are initially placed at equal distances in the N-space, i.e. on the corners of a hypercube. Subsequent positions are calculated with a high viscosity model, so that velocity is proportional to force.

We have shown [18], [19] that the N-space representation is enormously sensitive. With typical neural interaction strengths, useful aggregations occur after some 50-100 spikes from each neuron. With appropriate projection to a two-dimensional plane, we have visualized the spatial arrangement of aggregated particles for various simulated neural assembly circuits. The aggregations form something like a Venn diagram of the logical connectivity of the neural assembly.

The original representation described above defined an attractive force between particles when their corresponding neurons fired in approximate temporal coincidence. This would be the situation for excitatory interactions among neurons. Only a slight modification is required to deal with inhibitory interactions: reverse the sign of the force. Now, since we normalize to zero-mean charge, aggregation will occur for combinations of spike and silence, as appropriate for inhibition. In a way similar to other cross-correlation related calculations, the sensitivity to the inhibitory situation is far less than for the excitatory situation [8].

An inadequacy of the original representation is that aggregation occurs for particles whose corresponding neurons fire in approximate temporal coincidence without regard for order. Therefore it is impossible to identify which neurons are pre-synaptic (earlier), and which are post-synaptic (later). Again, a modification of the original rules yields the desired characteristics. Let us define TWO forms of charge for each particle, as shown in Fig. 6. The first form (as before) is a decaying exponential which is incremented at the moment of each action potential; the second charge form is reversed in time, i.e. the "decay" of the exponential is towards earlier time. We now maintain separately the two charge histories for each particle.

Now consider equation (1), which defines the force on a particle. Here the (single) charge is used in two ways: First we have the charges on all the "other" particles which are added vectorially to obtain the local field, and second we have the charge on the local particle which is acted upon by that field. Let us separate these uses of charge: let the charges on the distant particles be called "effector charges" q_e, while the charge on the local particle is called an "acceptor charge" q_a. The force equation then becomes:

$$\mathbf{F}_i = q_{ai} \sum_{j \neq i} q_{ej} \, \mathbf{r}_{ij} \tag{2}$$

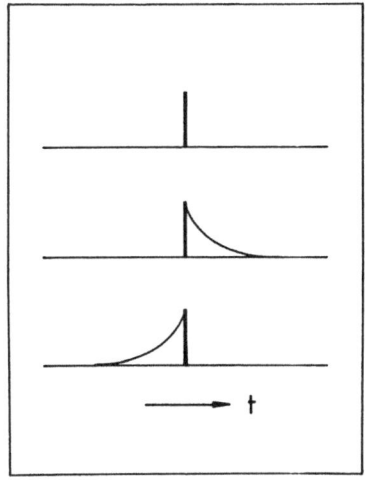

Fig. 6: Two charge forms used separately for each particle. See text. From [19].

Fig. 7: Forces and displacements for two particles under the three different charge rules. Temporal overlap of charges is indicated by stippling; such overlap results in force and displacement. See text. From [19].

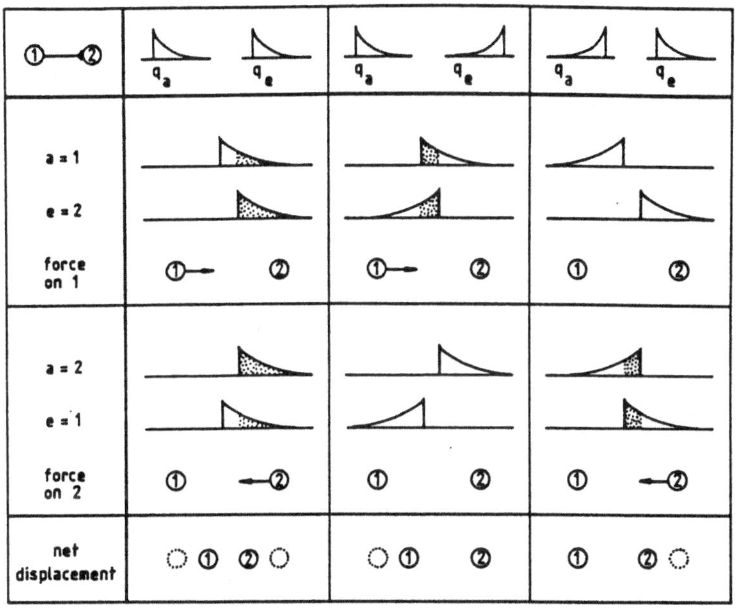

We are now free to assign "effector" and "acceptor" charges from the two charge forms defined in Fig. 6. Three possible assignments are shown in the top row of Fig. 7 for the simple case of neuron 1 synaptically driving neuron 2; the signature of such a circuit is a spike from neuron 1 shortly before a spike from neuron 2. The second row shows determination of the force on particle 1; the third row shows determination of the force on particle 2. Note that the force is determined by the overlap of the two charges, and that only in the first column (original representation) are the forces on particle 1 and 2 identical and oppositely directed. Thus we have now a system which does not obey Newton's Third Law; since the representation is an entirely imaginary construct, this should cause little anxiety. The bottom row of Fig. 7 shows the resulting final displacements of both particles. Note that for the second charge assignment, the particle representing the presynaptic neuron has moved towards a stationary particle representing the post-synaptic neuron. In the third charge assignment (third column) the reverse occurs.

The implications for trajectories and final aggregations of particles representing the various possible three-neuron connectivities are shown in Fig. 8. The original charge assignment (q and q identical) does not allow differentiation of most of the circuits. The two charge assignments, however, remove this degeneracy. It is generally necessary to calculate for both of the two charge assignments, and to take into account both the trajectories and final aggregations in order to reach robust conclusions about the organization of the observed neural assembly.

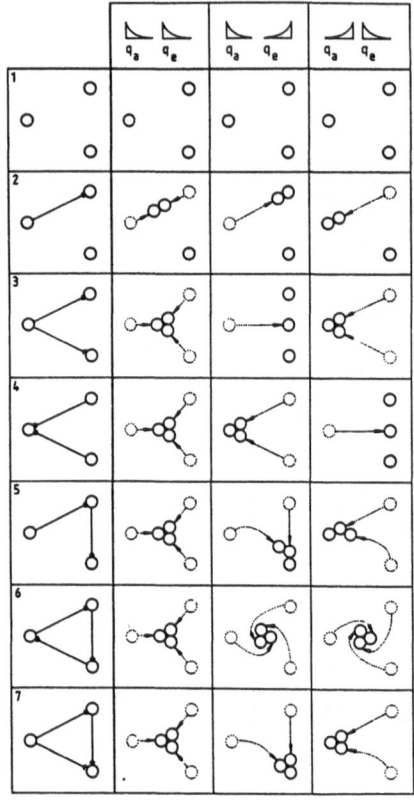

Fig. 8: Trajectories and final positions for simple three-neuron circuits under the three sets of charge rules (top row). The first column shows the circuits. See text. From [19].

Additional theoretical work is still needed on the fundamental behavior of the "gravity" representation. Even if there are no particular favored timing relations between the various spike trains which would result in aggregations, the particles will not remain fixed at their initial (equidistant) positions in the N-space. Rather, there will be displacements reminiscent of diffusion or Brownian motion. These displacements are demonstrated in Fig. 9. Calculations were made for 10 realizations of a simulation with 10 independent neurons. The resulting population of interparticle distances is shown in the upper set of curves, which represent the 10th to 90th percentiles of distance. The lower set of curves is a similar treatment of the population of distances that particles have moved from their starting locations. Note that although the mean interparticle distance stays con-

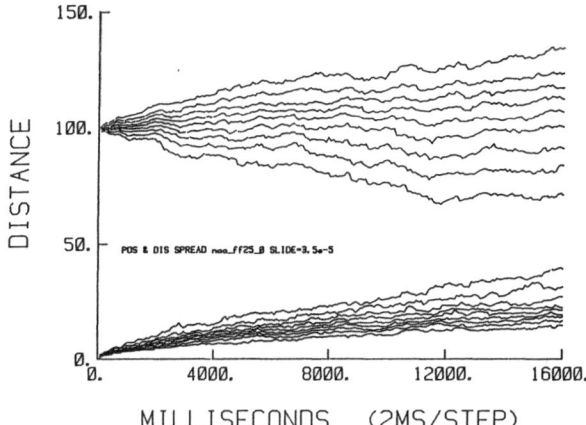

DISTANCE

150.

100.

50.

0.

POS & DIS SPREAD naa_ff25_B SLIDE=3.5e-5

0.   4000.   8000.   12000.   16000.

MILLISECONDS   (2MS/STEP)

Fig. 9: Even for the case of independent neurons, there is displacement of the particles from their equidistant starting positions. Upper curves are percentiles of a population of interparticle distances, lower curves are percentiles of distance moved from starting position. Behavior is reminiscent of a diffusion process.

stant, the width of its distribution increases approximately linearly with time. Similarly, behavior is observed for the mean distance of particles from their initial locations. This type of linear dependence on time is characteristic of diffusion processes, so that it should be possible to make explicit calculations of these distributions in terms of parameters used in the "gravity" representation. In essence, the distributions in Fig. 9 define the limits of sensitivity and "signal to noise" ratio of the "gravity" representation, since any real aggregation process has to be unambiguously visible on this random background movement.

The above brief description summarizes the "gravity" transformation as a useful global tool for analyzing an entire neural assembly as a single entity. The calculation is sufficiently demanding that a mini rather than a micro computer is required to make computing time approach quasi-real time. Such speed is essential in order that intelligent changes can be made in the stimulus or other conditions during the experiment. We are presently applying this representation to actual physiological recordings rather than to the simulated spike trains which allowed calibration of sensitivity and selectivity.

6. Conclusion

Many processes in the nervous system are carried out by assemblies of neurons acting together, rather than by individual neurons acting alone. The most direct experimental approach to the study of assembly properties of neurons involves separable recordings from many neurons simultaneously. We have reviewed here some of the technical and theoretical problems associated with such multi-neuron experiments. Our experimental data on auditory cortex indicates that the organization of neural assemblies changes rapidly in time (on several different time scales) and that there are emergent properties of assemblies that can not be predicted from the properties of individual neurons. Although there is much to be done at several levels, it seems that the present developments, if used in appropriately designed behavioral experiments, should allow significant progress in understanding the role of neuronal assemblies in the processing of information by the brain.

Acknowledgments

This work was mainly supported by the National Institutes of Health, Grant NS-05606 and by the System Development Foundation, Grant SDF-00013.

References

1. E.M. Glaser and D.S. Ruchkin: Principles of Neurobiological Signal Analysis, Academic Press, New York, 1976.
2. V.B. Mountcastle: J. Neurophysiol. 20, 408-434, 1957.
3. D.H. Hubel and T.N. Wiesel: J. Physiol. 160, 106-154, 1962
4. R.S. Creed,D.E. Denny-Brown, J.C. Eccles, E.G.T. Liddell, and C.S. Sherrington: Reflex Activity of the Spinal Cord, Oxford Univ. Press, London, 1932.
5. D.O. Hebb: The Organization of Behavior, Wiley, New York, 1949.
6. G.L. Gerstein, M.J. Bloom, I.E. Espinosa, S. Evanczuk and M. Turner: IEEE Transactions on Systems, Man, and Cybernetics, SMC-13, 668-676, 1983.
7. E.M. Schmidt: J. Neuroscience Methods, 12, 95-111, 1984.
8. A.M.H.J. Aertsen and G.L. Gerstein: Brain Research, in press, 1985
9. W.M. Jenkins and M.M. Merzenich: J. Neurophysiol. 52, 819-847, 1984.
10. J.C. Middlebrooks and J.D. Pettigrew: J. Neuroscience 1, 107-120, 1981.
11. J.C. Middlebrooks and E.I. Knudsen: J. Neuroscience 4, 2621-2634, 1984.
12. W.D. Neff, I.T. Diamond, and J.H. Casseday: "Behavioral Studies of Auditory Discrimination: Central Nervous system", in: Handbook of Sensory Physiology Vol. V, part 2, 307-396, Springer Verlag, Berlin, 1975.
13. T. McKenna, D.M. Diamond, J. Peerson and N.M. Weinberger: Soc. for Neuroscience Abstracts 10, 244, 1984.
14. M.J. Bloom and G.L. Gerstein: Soc. for Neuroscience Abstracts: 10, 245, 1984.
15. I.E. Espinosa and G.L. Gerstein: Soc. for Neuroscience Abstracts: 10, 245, 1984.
16. S. Evanczuk: PhD. Dissertation, Dept. of Physiology, University of Pennsylvania, 1985.
17. W.E. Wright: Pattern Recognition 9, 151-166, 1977.
18. G.L. Gerstein, D.H. Perkel and J.E. Dayhoff: J. Neuroscience 5, 881-889, 1985.
19. G.L. Gerstein and A.M.H.J. Aertsen: J. Neurophysiology, in press, 1985.

Investigation of a Small Volume of Neocortex with Multiple Microelectrodes: Evidence for Principles of Self-Organization

J. Krüger

Neurologische Universitätsklinik, Hansastraße 9,
D-7800 Freiburg, Fed. Rep. of Germany

Introduction and Methods

Cells in all organs can interact with their neighbours, but in the brain a cell can communicate if the need arises not with its neighbours but with some selected cells farther away. The communication network thus created is by many orders of magnitude more complex than the one possible with compact cells contacting each other. Thus the trick of the brain is essentially in the spatial dimension at the intercellular (and not at the intracellular) level. Therefore, in order to investigate essential aspects of brain function, intercellular interactions must be studied while the brain works.

This was done with 30 closely spaced fine microelectrodes [1,2,3] (fig. 1) inserted into the primary visual cortex of lightly anaesthetized, paralyzed vervet monkeys (cercopithecus aethiops). The electrodes formed a 5 x 6 array with an interelectrode distance of 160 micrometers. The array covered an area of about 0.5 mm² which corresponded to the surface of a cortical module, or hypercolumn [4]. Lateral interactions were expected to be limited to about one module width.

Spikes from single neurones or sometimes from more than one neurone were recorded simultaneously during about three hours in a given cortical layer, while various visual stimuli were applied. The latter were used to characterize the neurones in terms of stimulus preferences. Four or five layers were explored subsequently in this way. The present report covers results from essentially three monkeys, but recordings from layer VI are from two animals only.

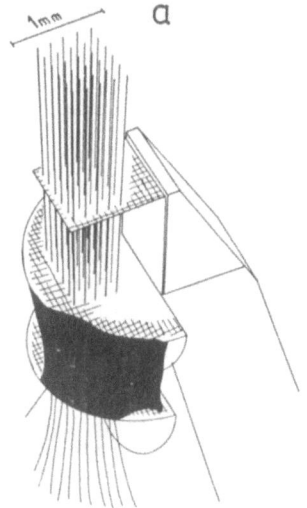

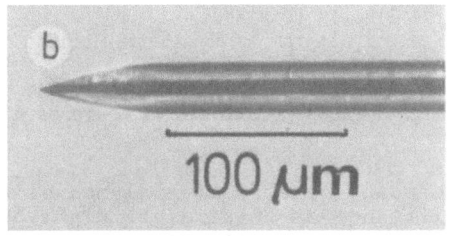

Fig. 1 a) Multiple microelectrode. The tips lie in a plane perpendicular to the electrode shafts.
b) Enlarged view of one microelectrode. Platin-Iridium-wire 5 micrometers thick is covered with quartz-glass with an outer diameter of 30 micrometers. The tips are manually ground.

Experiments

One of the intentions of these experiments was to conclude from temporal relationships in the spike trains the functional connections in the neuronal network, and to relate these connections to response properties of the neurones.

In principle, a rather direct connection between two neurones would show up in a spike train cross-correlogram as a peak centered at zero (common input), a peak laterally displaced by 1 msec (synaptic connection; fig. 2a), or some further structures in the millisecond domain [5]. After having seen some hundreds of cross-correlograms large, clear structures seemed rare. However, I had also calculated correlograms with compressed abscissae, where the spike pairs were lumped in delta-time bins 11 msec wide. There were many more well-defined structures seen on this compressed scale. Mostly, they were peaks about 60 msec wide. Subtraction of a control correlogram ("shift predictor", [6]) permitted to eliminate the synchronizing effect of common stimulation. Figure 2b shows an example. At left is a normal spike train cross-correlogram with 1 msec bins showing no clear structure in the central bins. The same correlogram is compressed, and covers a correspondingly wider range (right), and a clear peak can be seen. The control correlogram below shows no structure, proving that the peak is not due to common stimulation.

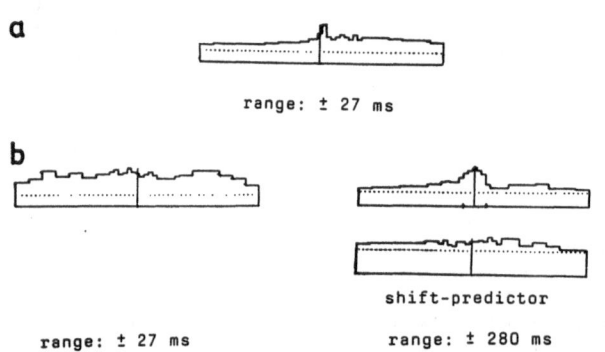

a

range: ± 27 ms

b

shift-predictor

range: ± 27 ms   range: ± 280 ms

Fig. 2 a) Example of a cross-correlogram between two cortical spike trains. Bin width 1 ms. In the spike trains there was an excessive number of cases where a spike from the one neurone was followed one millisecond later by a spike in the other neurone. In the correlogram, a corresponding peak in the bin 1 ms displaced from the centre line is obtained.
b) Example of a correlogram as in a) but without clear structures (left). Same correlogram but on a compressed abscissae, each bin being 11 ms wide (top right). A clear structure ("wide peak") can be seen. The control histogram (shift predictor; bottom right) shows no structure. Neurones different from those in 2a.

The wide peaks were first analyzed, although the interpretation of their significance at first sight might be difficult. This, however, was not so. For each electrode taken as a reference electrode, the peak areas (divided by the geometric mean of the total numbers of spikes) were represented by rectangles in a matrix formed by all other electrodes. Two of the 30 matrices are shown in fig. 3a for layer IVb. Figure 3b shows the distribution of eye dominance in the recording area, as determined by the comparison of response magnitudes when the neurones were stimulated by the two eyes separately. Comparison of 3a and b reveals that a reference electrode ◉ located in the right eye stripe mainly correlates with other right-eye domi-

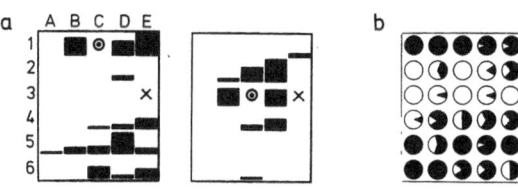

Fig. 3 a) Correlations of the activities at two reference electrodes ⊙ with those at the remaining electrodes. The arrangement corresponds to the electrode array. The sizes of the rectangles are proportional to the areas of the wide correlogram peaks as shown in fig. 2b. b) Distribution of ocular dominance in the recording array. This diagram was obtained by comparing response magnitudes to stimuli delivered to the left and right eyes separately. A full black (white) circle corresponds to an electrode where responses only via the right (left) eye could be elicited, other symbols depict the ratio of the responses from the two eyes.

nated neurones, and left-eye neurones correlate with other left-eye neurones. This relationship is most pronounced in or near layer IVc where most fibers from the eyes (via lateral geniculate nucleus) terminate.

From this finding, it is concluded that the wide peaks reflect correlations taking place in lateral networks of the retina, where they actually have been observed [7,8].

The same result is shown in fig. 4 in a different way: the area of wide peaks is depicted as a function of electrode distance and eye dominance difference (for further details see figure caption) for layer IVc (left) and layer VI (right) cumulated for two animals different from the one of fig. 3. It may be noted that in layer IVc there is a reversal of the relationship seen as the electrode distance in-

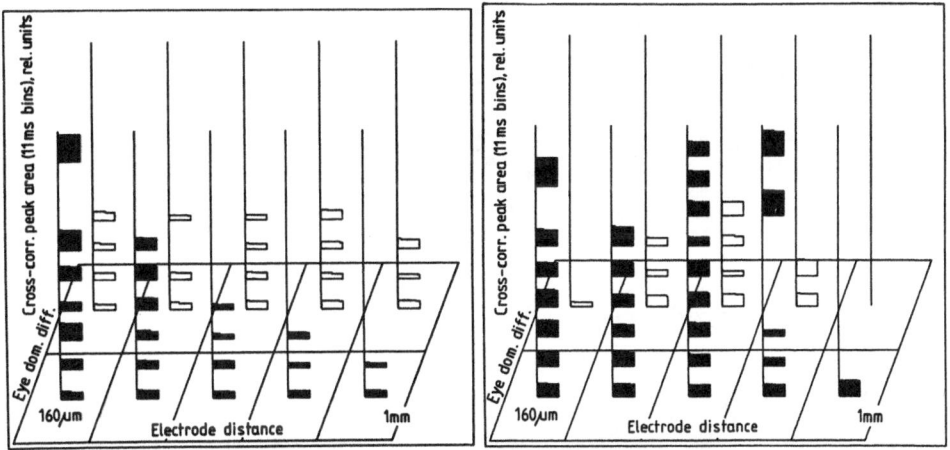

Fig. 4 Dependency of wide correlation peaks on cortical tangential distance and on the difference of eye dominance between two neurones. The axes are subdivided into bins, and the sizes of the rectangles are proportional to the relative number of cases in a bin (relative to the absolute number of cases along each coordinate). The first bin on the distance axis contains only the cases with the minimal distance 160 micrometers, the remaining axis is subdivided into equidistant bins, the largest distance being 1030 microns. The correlation axis is in relative units, only cases with nonzero positive peak areas are included. Left frame: layer IVc, right frame: layer VI.

creases, namely, neurones dominated by <u>different</u> eyes show <u>more</u> correlations at the largest distances. In <u>layer VI</u> a predominance of correlation for low eye dominance difference can be seen,but there is no
such reversal. Moreover, taking all eye dominance values together,
there is no clear decrease of correlation with distance up to about
800 micrometers, pointing to a spatial reorganization of retinal input.

In the remainder of this article, only layer VI of two monkeys is
considered; further recorded data still await evaluation.

Analyzing the distribution of asymmetries of the wide correlation
peaks shows a further point of interest: In fig. 5 all electrode pairs,
between which a correlation with a laterally displaced peak is observed, are depicted in a double matrix arrangement. The triangle in
each small matrix represents one electrode selected as reference
electrode, stars indicate that spikes at a running electrode come
more often <u>later</u> than those in the reference electrode, and vice
versa for <u>dots</u>. The special feature of this picture is that stars and
dots are not randomly distributed,but some reference electrodes are
associated mainly with only one symbol. This means that spikes from
some neurones (e.g. second from bottom, left) are <u>followed</u> by some
extra spikes in <u>most</u> other neurones. Spikes from other neurones such
as the one whose matrix is in the left column, third from top are
<u>preceded</u> by extra spikes in many other neurones. Therefore, all neu-
rones can be arranged on a kind of velocity scale based on the idea
that a spontaneous neuronal excitation somewhere in the retina simultaneously influences all nearby retinal neurones, but this influence

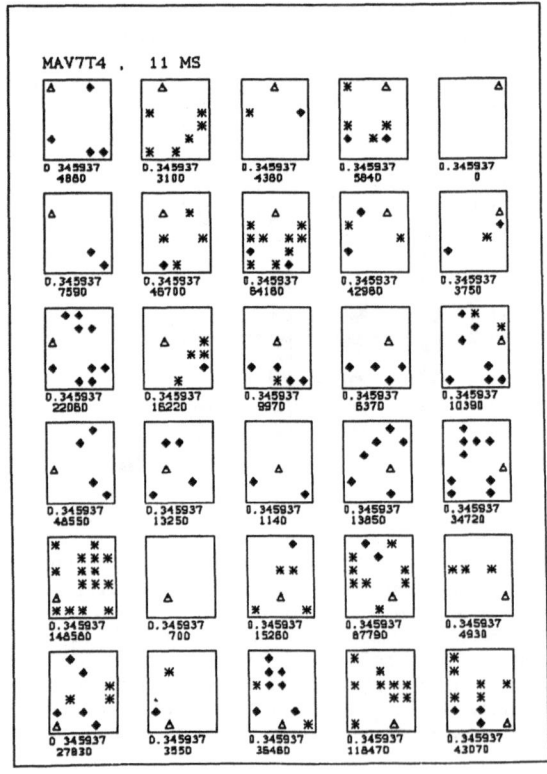

Fig. 5 Distribution of lateral
displacements of wide correlation
peaks. Each small matrix is for
one electrode selected as a
reference electrode (Δ). An excess
of spikes occurring earlier/later
than a spike on the reference
electrode is marked by ◆/✳

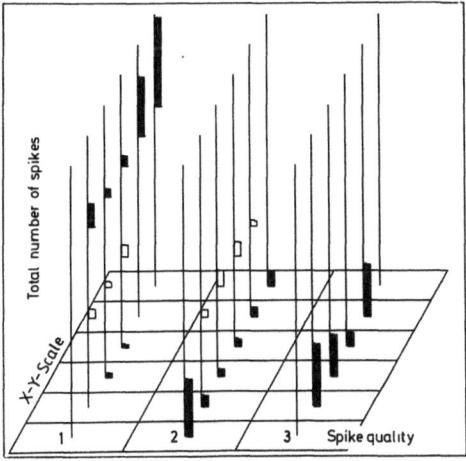

Fig. 6 Relationship between "spike quality", the X-Y ranking and the total number of spikes for layer VI neurones. The principle of the figure corresponds to that of fig. 4. The "spike quality" axis is segregated into three bins, the left-most containing the largest, best isolated spikes. The vertical axis is the square root of the total number of spikes, otherwise the distribution of actual values on a few bins would have been too uneven. Values range between about 2000 and 200 000 spikes per three hours of recording time. For clarity, the contents of the bins along the vertical axis has been differentiated by hatching.

reaches some cortical neurones on faster pathways, and others on slower ones. This corresponds to the well-established distinction between the retinocortical Y- and X-channels [9,10,11]. The former are known to be distinguished from the latter by faster conduction, higher ongoing activity, and larger neurones. Figure 6 shows the relationship between the velocity (or X-Y-)scale derived from the cross-correlograms, the total number of spikes emitted during the recording time, and the "spike quality". The latter parameter, noted during the experiment by inspection of the oscilloscope traces, gives a judgment of whether there was one well-isolated spike, one possibly contaminated spike or clearly more than one spike. Spike quality essentially depended on spike size. Cases where two large spikes could not be discriminated formed a negligible minority. Spike size in turn is known to be related to neurone size [12,13]. In fig. 6 it can be seen that indeed large spikes (spike quality "1") with high spike rates rank highest on the velocity (or X-Y-)scale. Thus, the observations are well anchored in the network of knowledge about cortical input organization.

There is controversy as to whether the X-Y-distinction is lost as processing goes on in the visual cortex. The present results (fig.7)

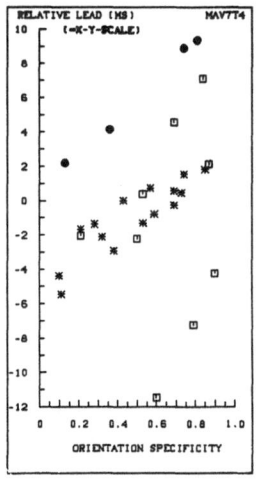

Fig. 7 Dependency of X-Y ranking on orientation tuning sharpness for layer VI neurones from one monkey. Neurones with more than 60 000 (●) and less than 7000 (◻) spikes per 3 hours, and intermediate range (✳).

show that clear traces of segregation are still visible in layer VI:
For each neurone from one monkey the angular tuning sharpnesses for
responses to oriented moving light bars (abscissa) and the position
on the X-Y-scale (ordinate) are plotted. Y-like neurones high up on
the ordinate, in addition discriminated by spike rates of more than
60 000 per total recording time (●) show a linear relationship in
this graph. A parallel distribution is found about 10 msec lower for
more X-like neurons with spike rates between 60 000 and 7000 (✳),
suggesting that some orientation-related processing is performed
independently in the two subpopulations. Neurones with still lower
spike rates (□) probably do not form a separate group but rather,
their ranking on the ordinate might be uncertain,because with less
spikes available the correlogram structures upon which ranking is
based become less clearly discernible. Results from another monkey
show the same relationship.

After having seen that wide peaks in correlograms could be interpreted consistently,I turned again to correlogram structures a few
milliseconds wide,because intracortical interactions, if not mediated
by too many interneurones, were expected to manifest themselves in
such structures. Again, only layer VI is considered here.

The most interesting relationship found is the one involving
orientation specificity, i.e. the ability of cortical neurones to
respond preferentially to particular angles of inclination of lumi-
nous bars presented in the receptive field. In fig. 8 it can be seen
that at short interelectrode distances the neurones correlate most
strongly when the difference of preferred angles is small, whereas
beyond distances of about 200 micrometers there is stronger correlation when the angle difference exceeds 45 degrees. (In the previously treated domain of wide peaks,the corresponding relationship
is only a stronger correlation for small angle differences without
reversal.)

In the domain of narrow correlation peaks there is an excess of
correlation for neurones dominated by the same eye, and this excess
decreases with distance without reversal.

Receptive field position scatter [14] is another neuronal parameter whose function is certainly important but so far not understood
and which might be systematically related to cross-correlation struc-

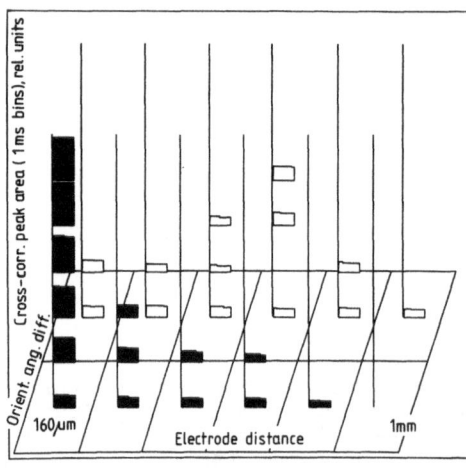

Fig. 8 Dependency of narrow correlation
peaks on cortical tangential distance
and on the difference between the angles
of preferred orientation of two neurones.
Otherwise the explanation given in the
legend of fig. 4 applies.

tures. Receptive field midpoints could be deduced with great accuracy
from responses to oriented bars. In layer VI, the receptive field
positions could be compared to those in layer IVc where an extremely
regular progression with distance was found. Some well-defined small
receptive fields in layer VI were displaced by considerable visual
angles,corresponding to a lateral displacement in the cortex of up
to 2 mm, as deduced from the measured cortical magnification factor.

Although considerable effort, involving further variables, was
spent with receptive field scatter, so far there was no more relationship to correlations than an absence of wide-peak correlations when
both fields were strongly displaced, except when such fields coincided.
This again points to a retinal origin of the wide correlation peaks.

Some further relationships so far not observed are worth mention-
ing: Firstly, there were many cases of asymmetries of narrow peaks
of the kind interpretable in terms of synaptic connection from one
neurone to another. However, these asymmetries were not found to be
related in systematic ways to any other variable so far examined.
Secondly, there seems to be no correlation of any type related systematically to the selectivity of movement direction, or to colour
selectivity. Thirdly, the position of a neurone vis-à-vis the angle
of preferred orientation of another neurone was not found to be linked
to any correlation parameter.

Discussion. Relations to Theory
In the past most researchers tried to study interactions between
neurones in the brain with two electrodes, but due to the weakness of
the interactions, the limited recording time, and the uncertainty of
electrode tip positions,so far no general picture of cortical interactions emerged. (For example, two neurones separated by 300 micrometers typically have cross-correlograms with areas of narrow peaks
containing only about 60 spike pairs per 3 hours, i.e. a single event
from which a connection between these neurones may be deduced occurs
only once every 3 minutes.)

With multiple microelectrodes, the recording stability is high,
so that long spike trains can be examined, and the number of neurone
pairs is about proportional to the square of the number of electrodes,
so that many cross-correlograms are available. However, thousands of
such curves had to be inspected and analyzed by eye,since at least
for the beginning a computerized feature extraction was premature.

The results are in agreement with a theory of the self-organization of the visual cortex [15,16,17]. This theory is concerned with
the question of how during ontogenesis the fibres coming from the
lateral geniculate nucleus find their proper place in the cortex,so
that a topographical map and ocular dominance stripes are generated,
and a regular spatial arrangement of angles of preferred orientation
is created [4].

Clearly, the synaptic links required for such a network cannot be
specified individually,but there must be global laws governing the
ordering process. The theory was originally formulated to operate on
the basis of chemical gradients serving as indicators for which
endings should become connected to which ones. However, mathematically,the theory is equivalent (v.d. Malsburg, personal communication)
when it is assumed that Hebb's rule [18] governs the strengthening of
synaptic connections. (This rule simply states that a synaptic contact
tightens if the pre- and the postsynaptic neurones are simultaneously
active.)

Actually there are two versions of the theory, one for topography and eye dominance, and another for orientation angle-ordering. The former [16] requires that there is correlation between retinal neurones for short lateral distances. Independently, cortical neurones receiving these fibres (the lateral geniculate nucleus is omitted here for simplicity) should also correlate in about the same way, and in addition there may be anticorrelation at greater lateral distances but this is not necessary. Both retinae project to the same cortical area, and, of course, the short-distance correlation is only present in the fibres coming from the same eye. Self-organization proceeds by reinforcing synapses between active incoming fibres and cortical neurones which are simultaneously activated by adjacent pathways, while some normalization rules have to be obeyed. When the organizing process has reached its final state, the distribution of intracortical lateral correlation is an overall decrease with lateral distance but at short distances the correlations are highest for neurones dominated by the same eye, whereas at greater distances neurones depending on different eyes are more strongly correlated.

This theoretical prediction corresponds precisely to the relationship observed in layer IVc between electrode distance, eye dominance difference, and the wide correlation peaks shown in the left diagram of fig. 4.

The theory involving orientations [15,17] operates in a very similar way: There is a "cortical sheet" in which short-range excitatory lateral connections (and possibly longer-range inhibitory connections) first organize themselves into stable stripe-like periodic connectivity patterns. In a second step "retinal" fibres, again showing short-range correlations, connect themselves by competition based on Hebb's rule to the cortex, with visual input alone [15], or in addition with a minor genetic component [17] serving as a guide. This theory predicts that after completion of the self-organizing process,there should be correlation of neurones with similar angles of preferred orientation at short lateral distances, whereas at greater distances dissimilar angles should lead to the larger correlations. In addition there is an overall decrease of all correlations with distance.

The relationship between correlation strength, distance, and orientation angle difference depicted in fig. 8 corresponds precisely to the theoretical prediction.

It is remarkable that the relationship is now observed for the narrow (± 4 ms) correlation peaks, and it is found in layer VI where strong orientation selectivity exists [19], whereas the correlational evidence for ocularity self-ordering is seen in layer IVc,where the afferent visual fibres terminate and where the strictest ocularity segregation is observed [19]. The latter process operates on more loosely timed correlations. The difference in temporal synchronization accuracy,together with the layer difference,is likely to be the cause of the independency of the spatial distributions of ocular dominance and orientation preference,but temporal sequencing during ontogenesis may be an additional factor.

The theory considers fibres from a "retina" connecting to a "cortical sheet". I conclude from the experimental findings that for the ocularity and topography ordering the "retina" is indeed the retina, and the cortical layer IVc is the "cortical sheet": the correlations observed in cat retinal ganglion cells can show broad peaks up to 50 ms wide [7,8], and certainly these are still broadened by the passage through the lateral geniculate body. Widening of synchronization accuracy also occurs when there is even a slight tendency of bursting. For Hebb's rule to operate, layer IVc must contain mechanisms (complex circuitry or long-lasting excitatory post-synaptic potentials) which ensure that retinal signals with loose relative timing are still accepted as synchronous.

For the orientation ordering, the "retina" is a cortical layer from which layer VI receives most of its input, and the "cortical sheet" is layer VI. Relatively direct intracortical synaptic interactions (the ones supposed to organize the stripes) correspond to the narrow correlation peaks, which show the dependency on orientation angles as required by the theory.

A further feature of the theories considered here is that all connections are assumed to operate bidirectionally. On the average this was found to be the case, since no variable was found so far to be involved in dependencies on asymmetries of narrow correlation peaks. If this is a general principle, then it is not astonishing that there is no systematic relationship between the position of a neurone in the cortical plane, the orientation angle of another neurone, and the cross-correlation between these neurones. The reason is that such a relationship is asymmetric with respect to interchanging the two neurones, and systematically asymmetric cross-correlations would then be likely.

It is very intriguing that we observe systematic relationships between spike train correlations, lateral distance, and further variables only for those variables which are known to show spatial order in tangential directions. These variables are retinal position, contra/ipsilateral eye, and orientation selectivity. No spatial order is known for receptive field position scatter, or for the specificity to the direction of movement at least in the monkey (in the cat there is a positive report [20]), and essentially no systematic relationships to cross-correlations were found. Colour selectivity is reported to show a spatial order [21], but in the infragranular layers the ordering is less pronounced, and moreover there is a multiple definition of neuronal properties to be associated with the "colour blobs", namely colour selectivity and lack of orientation specificity, so that I was unable to define clearly whether the observed properties of a given neurone corresponded to those reported for neurones within these blobs.

As a conclusion, the hypothesis may be forwarded that spatial order of a variable in tangential directions implies that the cross-correlations depend in a specific way on this variable and on distance.

When this investigation was started, I was first reluctant to trust the weak correlations observed, but the great regularity of the results even permits to select a given neurone, and to <u>predict</u> its physiological properties from the strengths of the connections to all other neurones, as deduced from cross-correlations. Thus, with the aid of the theory of cortical self-organization, a major goal of brain research has been reached, namely to explain physiological properties of neurones by observed neuronal connections.

References

1. J. Krüger, M. Bach: Simultaneous recording with 30 microelectrodes in monkey visual cortex. Exp. Brain Res. 41, 191-194 (1981)
2. J. Krüger: A 12-fold microelectrode for recording from vertically aligned cortical neurones. J. Neurosci. Meth. 6, 347-350 (1982)

3. J. Krüger: Simultaneous individual recordings from many cerebral neurons: techniques and results. Rev. Physiol. Biochem. Pharmacol. 98, 177-233 (1983)
4. D.H. Hubel, T.N. Wiesel: Sequence regularity and geometry of orientation columns in the monkey striate cortex. J.Comp. Neurol. 158, 267-294 (1974)
5. G.P. Moore, J.P. Segundo, D.H. Perkel, H. Levitan: Statistical signs of synaptic interaction in neurons. Biophys. J. 10, 876-900 (1970)
6. D.H. Perkel, G.L. Gerstein, G.P. Moore: Neuronal spike trains and stochastic point processes. II. Simultaneous spike trains. Biophys. J. 7, 419-440 (1967)
7. D.N. Mastronarde: Correlated firing of cat retinal ganglion cells. I. Spontaneously active inputs to X- and Y-cells. J.Neurophysiol. 49, 303-324 (1983)
8. D.N. Mastronarde: Interactions between ganglion cells in cat retina. J. Neurophysiol. 49, 350-365 (1983)
9. B. Dreher, Y. Fukuda, R.W. Rodieck: Identification, classification and anatomical segregation of cells with X-like and Y-like properties in the lateral geniculate nucleus of old-world primates. J. Physiol. (Lond.) 258, 433-452 (1976)
10. G.H. Henry: Receptive field classes of cells in the striate cortex of the cat. Brain Res. 133, 1-28 (1977)
11. P. Lennie: Parallel visual pathways: A review. Vision Res. 20, 561-594 (1980)
12. F.S. Grover, J.S. Buchwald: Correlations of cell size with amplitude of background fast activity in specific brain nuclei. J. Neurophysiol. 33, 160-171 (1970)
13. Ü. Tan, C. Marangoz, F. Şenyuva: Antidromic response latency distribution of cat pyramidal tract cells: Three groups with respective extracellular spike properties. Exp. Neurol. 65, 573-586 (1979)
14. D.H. Hubel, T.N. Wiesel: Uniformity of monkey striate cortex: A parallel relationship between field size, scatter, and magnification factor. J. Comp. Neurol. 158, 295-306 (1974)
15. C. v.d. Malsburg: Self-organization of orientation sensitive cells in the striate cortex. Kybernetik 14, 85-100 (1973)
16. C. v.d. Malsburg: Development of ocularity domains and growth behaviour of axon terminals. Biol. Cybern. 32, 49-62 (1979)
17. C. v.d. Malsburg, J.D. Cowan: Outline of a theory for the ontogenesis of iso-orientation domains in visual cortex. Biol. Cybern. 45, 49-56 (1982)
18. D.O. Hebb: Organization of behaviour (Wiley, New York 1949)
19. D.H. Hubel, T.N. Wiesel: Receptive fields and functional architecture of monkey striate cortex. J. Physiol. (Lond.) 195, 215-243 (1968)
20. B.R. Payne, N. Berman, E.H. Murphy: Organization of direction preferences in cat visual cortex. Brain Res. 211, 445-450 (1981)
21. M.S. Livingstone, D.H. Hubel: Anatomy and physiology of a color system in the primate visual cortex. J. Neurosci. 4, 309-356 (1984)

New Light on the Mind-Brain Problem:
How Mental Events Could Influence Neural Events

J.C. Eccles

Max-Planck-Institut für Biophysikalische Chemie,
D-3400 Göttingen, Fed. Rep. of Germany

A. Introduction

It has long been recognized, that, if non-material mental events, such as the intention to carry out an action, are to have an effective action on neural events in the brain, it has to be at the most subtle and plastic level of these events. Attention has to be focussed on the biological units of the brain, the neurones or nerve cells, and on the manner of their communication at specialized sites of close contact, the synapses. An introduction to conventional synaptic theory leads on to an account of the manner of operation of the ultimate synaptic units. These units are the synaptic boutons that, when excited by an all-or-nothing nerve impulse, deliver the total contents of a single synaptic vesicle, not regularly, but probabilistically. This quantal emission of the synaptic transmitter molecules (about 5,ooo to 1o,ooo) is the ultimate functional unit of the transmission process from one neurone to another. This refined physiological analysis leads on to an account of the ultrastructure of the synapse, which gives clues as to the manner of its unitary probabilistic operation. The essential feature is that the effective structure of each synapse is a paracrystalline presynaptic vesicular grid, which acts probabilistically in quantal release.

When considering mental events in relation to a possible influence on these subtle neural events, it is essential to avoid mental events with levels of complexity that result in a confusion of neural happenings which are beyond analysis. There will be an account of several recent studies in which mental events result in such simple neural events that correlation with probabilistic quantal operation of synapses is possible. Conventional neurophysiology enables one to transcend unitary synaptic actions by operations of the known neural pathways, so that a mental intention to move can become fulfilled by the desired movement.

In the final stage of this enquiry it has to be considered how a non-material mental event, such as an intention to move, can influence the subtle probabilistic operations of synaptic boutons. On the biological side, attention will be focussed on the paracrystalline presynaptic vesicular grids as the targets for non-material mental events. On the physical side, attention will be focussed on the probabilistic fields of quantum mechanics which carry neither mass nor energy, but which nevertheless can exert effective action at microsites. The new light on the mind-brain problem comes from the hypothesis that the non-material mental events, the World 2 of Popper, relate to the neural events of the brain (the World 1 of matter and energy) by actions in conformity with the physics of quantum theory. This hypothesis opens up an immense field of scientific investigations both in quantum physics and in neuroscience. The materialist objections to dualist-inter-

actionism are as dead as the materialist philosophy based on nine-
teenth century physics.

B. The integrative action of Ia impulses on a motoneurone

A simple introduction to the synaptic concept of brain action is
provided by an account of the mode of action of impulses in Ia nerve fibres on a motoneurone as shown diagrammatically in Fig. 1A.
These large nerve fibers come from the annulospiral endings in a
muscle and directly excite the motoneurones of that muscle as indicated by the intracellular recordings in Fig. 1B to J. It is the
neuronal system responsible for the simple knee jerk. As the stimulus applied to the bundle of Ia fibres was progressively increased
to excite more and more of the nerve fibres (see the sharp upper
traces of nerve spikes) converging on that motoneurone, there was
a corresponding increase in the brief depolarizations recorded intracellularly (note ms time scale). The intracellular microelectrode shown diagrammatically in Fig. 1A recorded a resting potential
across the membrane of the motoneurone, about - 7o mV (internal
negativity) and synaptic stimulations caused a brief diminution of
this resting potential, the excitatory postsynaptic potential
(EPSP), which increased up to 7 mV in Fig. 1G to J when all the Ia

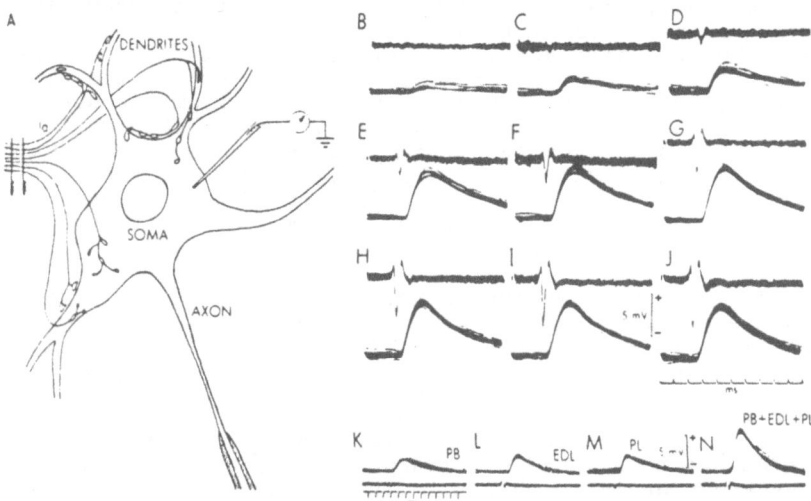

Fig. 1: Monosynaptic excitation of motoneurons by the group Ia
afferent pathway. A. A drawingof a motoneuron showing the central
dendritic regions, the soma, the initial segment (IS) of the axonal origin and the beginning of the axonal medullation. On the
dendrites and soma are shown the excitatory synaptic endings of
seven group Ia afferent fibers that have an applied stimulating
electrode (actually in the peripheral muscle nerve). The intracellular microelectrode recording is shown diagrammatically.
B-J. The upper traces give the size of the afferent volley as it
enters the spinal cord, and the lower the simultaneously recorded
EPSPs. All records are formed by the superimposition of about 25
faint traces. K-M. The EPSPs recorded in another motoneuron
(peroneus longus) in response to maximum group Ia volleys in the
nerves to three muscles - peroneus brevis, extensor digitorum
longus and peroneus longus. N. All three muscles combined.(Eccles
et al., 1957.)

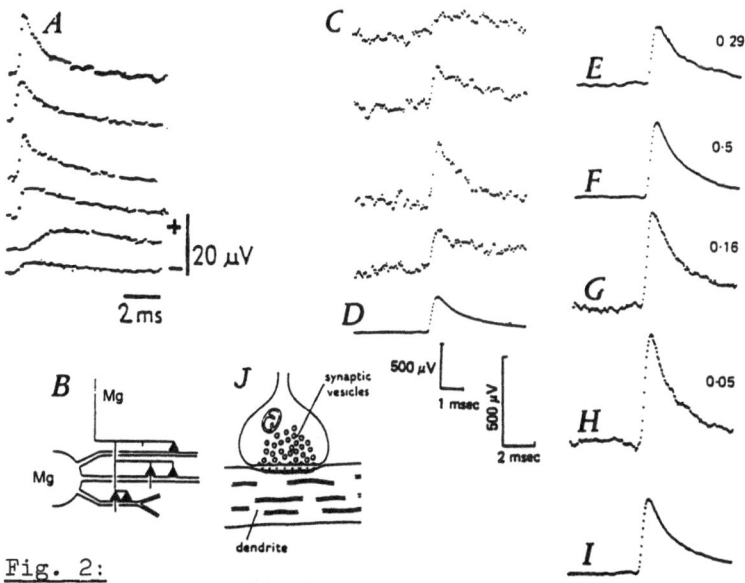

Fig. 2:

A. Averaged recordings of EPSPs produced by impulses in the same Ia fiber terminating on six different motoneurones (Mendell and Henneman, 1971).
B. Summary diagram of the location of Ia synapses from a single medial gastrocnemius Ia fibre on to a medial gastrocnemius motoneurone at five sites on three different dendrites as indicated (Brown, 1981).
C. Four individual EPSPs selected from a population of 8oo responses.
D. is the average of all the 8oo responses
E. is component 1 of the EPSP derived from fluctuation analysis.
F,G,H are components 2,3 and 4 of this same fluctuation analysis. The probabilities of the occurrence of these components are indicated to the right of each. I is the reconstructed EPSP obtained by adding, the weighted sum of E,F,G,H; 0.29 E + 0.5 F + 0.16 G + 0.05 H. (Jack et al., 1981a).
J. Drawing of a synapse on a dendrite to show the bouton with vesicles and the synaptic cleft.

fibres were stimulated. In Fig. 1G to J there was an apparent discrepancy, the nerve fibre volley increasing with no corresponding increase in the EPSP, but this was due to the excitation of another class of nerve fibres (Ib) that do not contribute to the monosynaptic EPSP. Fig. 1K to N show that the Ia EPSPs sum linearly when small, N = K+L+M.

When the bundle of Ia fibres shown in Fig. 1A was cut down to one, the EPSP's were very small, but can be amplified by successive addition in a computer. In that way it was shown that a single Ia fibre was distributed very widely to motoneurones of its muscle of origin. For example in Fig. 2A the same group Ia fibre produced EPSPs in six different motoneurones (Mendell and Henneman, 1971). By a double horseradish peroxidase (HRP) technique it has been possible to identify all the synapses made by a single Ia fibre on a motoneurone (Burke et al., 1979; Brown, 1981). For example the diagram of Fig. 2B represents the locations of synaptic endings by a single Ia afferent from medial gastrocnemius muscle on a medial gastrocnemius motoneurone. The 5 synaptic endings are on 3 different dendrites, two being rather close to the soma, the others more distant. Wide ranges of distribution are found, but there is a tendency for clustering. This will account for the range in time course of EPSP's produced by a single Ia fibre in Fig. 2A. On some motoneurones there was clustering of its boutons close to the soma, on others more dispersal, and on the second lowermost trace there would be a synaptic clustering far out on dendrites.

C.   The Quantal emission of a bouton

A still more refined level of enquiry concerns the EPSP produced by a single bouton, as indicated in Fig. 2J, which is a component of the several bouton endings of a single Ia fibre (Fig. 2B). By a technique of fluctuation analysis (Redman, 1980; Hirst et al., 1981; Jack et al., 1981a, 1981b) it has been possible to distinguish between the EPSP's generated by each bouton on a motoneurone when activated by a single Ia impulse. For example Fig. 2C illustrates the wide range of fluctuating EPSP's produced by a single Ia impulse and in Fig. 2D there is summation of 800 such responses to give a typical unitary EPSP, such as those of Fig. 2A. By the fluctuation analysis this EPSP is shown to be composed of elements, each generated by a single bouton. It is very rare that an activated bouton liberates more than one vesicle which is the quantal package of the synaptic transmitter (cf. Fig. 2J). In Fig. 2E-H there are shown 4 EPSP's derived by the fluctuation analysis of Fig. 2D, each arising from a single bouton. The probability of quantal emission from a bouton of a single synaptic vesicle ranges from 0,5 to 0,05, Fig. 2E to H showing the time courses of the EPSP's generated by the synaptic emission from each bouton. In sequence for E-H, sizes of the quantal EPSPs are 302, 406, 505 and 607 uV and when summated with allowance for probability, there is an accurate reconstruction fo the EPSP produced by a single fibre, Fig. 2I, which is identical with Fig. 2D when allowance is made for the different voltage scales. It can be assumed that the four derived EPSPs of Fig. 2E-H are each produced by a single bouton at various distances from the soma, H closest, E most remote, as in Fig. 2B.

From this remarkable analysis (Jack et al., 1981 a) there are derived two conclusions on the presynaptic functioning of a single Ia fibre on a motoneurone.

1. There is a wide gradation of intermittency,for example 0,5 to 0,05 in Fig. 2E to H. Some boutons may even approach a probability of 1, but above 1 is not observed.

2. Usually a Ia fibre gives 3 to 5 boutons to a motoneurone (cf. Fig. 2B), but the observed range is 1 to 10.

In a subsequent analysis (Redman and Walmsley, 1983a) by a double horseradish peroxidase technique it was possible to identify all the synaptic boutons which the Ia fibre gave to the motoneurone under investigation. Unfortunately the bouton number was usually much larger than the number delivered by the fluctuation analysis. It could be that some boutons are so closely related on a dendrite that the analysis fails to distinguish between them. However Redman and Walmsley (1983 b) prefer the hypothesis that many boutons have a very low or zero probability of emission.

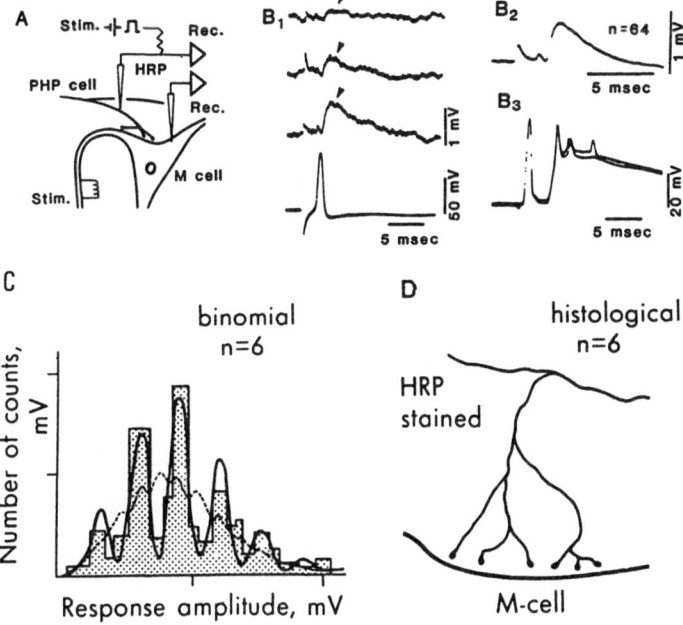

Fig.3: Evidence for quantal fluctuations of unitary IPSPs. (A).
Experimental arrangement used for simultaneous intracellular recordings (Rec.) from the M cell and a presynaptic inhibitory interneuron (PHP cell), both neurons being identified by their
characteristic responses to antidromic stimulation (Stim.) of the
M axon in the spinal cord. The presynaptic electrode was also used
for intracellular stimulation (Stim.) and subsequent staining with
HRP. (B1 to B3). Properties of depolarizing IPSPs recorded in a
Cl⁻ injected M cell throughout the same experiment. (B1) Variable
amplitude of unitary IPSPs (arrows, upper three traces) following
single presynaptic impulses directly evoked at a frequency of 1 per
sec. Only one presynaptic spike is shown (lower trace). (B2) Compu-
ter-averaged unitary IPSP (N=64). (B3) The maximum amplitude IPSP
following antidromic activation of the recurrent collateral network
was large enough to fire the M cell (Korn et al., 1981). C and D.
Correlation of mathematical and histological results provided in
the same experiment. (C) Following successive stimulations of a
physiologically identified interneuron (rate: 1 per sec.), the re-
sultant amplitude histogram of fluctuating unitary IPSPs (shaded)
was analysed with a computer program that, taking noise into consideration, gave the best possible fits based upon the theoretical
Poisson (dashed line) and binomial (continuous curve) equations.
Obviously, the latter provided a better approximation. The six
peaks correspond to the binomial term n, which defines the number
of quanta. (D) Drawing of the camera lucida reconstruction of the
terminal arborization of the investigated HRP-filled presynaptic
cell, indicating that the number of boutons established on the M-
cell (histological n) was equivalent to the binomial one. This type
of result led to the conclusion that each terminal knob is an all
or none releasing unit. (Faber and Korn, 1982.)

Korn and associates (Korn et al., 1982; Korn and Faber, 1985) have studied a very different synapse, the inhibitory synapses on the Mauthner cell in the fish spinal cord (Fig. 3A). It was possible to carry out a fluctuation analysis of the inhibitory postsynaptic potentials (EPSPs) produced by a single presynaptic inhibitory fibre (Fig. 3B). They employed a different technique from that used by Redman and associates (Fig. 2), a binomial analysis, which showed a composition of 6 quanta (Fig. 3C) in the response amplitude, and this number is in agreement with the histologically determined number of boutons (Fig. 3D). A critical evaluation of these two techniques will be attempted after there has been an account of synaptic structure.

D. The structure of a chemically transmitting synapse in the central nervous system of vertebrates

A synapse is formed usually by an expansion of a fine nerve fibre to form a bouton that makes a close contact with the surface of a dendrite or soma of a neurone. Fig. 4A shows the commonest type of synapse, where the bouton makes contact across a narrow space, the synaptic cleft (~2oo Å), with the expanded end of a dendritic spine (Gray, 1982). Noteworthy are the synaptic vesicles and the dense triangular projections about 1000 Å apart from the presynaptic membrane into the bouton. A tangential section just above the synaptic cleft shows that the dense projections are arranged in a triangular pattern to form a plate (Fig. 4B, Gray, 1983).

Fig. 5A is an electronmicrograph transversely across the synaptic cleft as in Fig. 4A showing the dense projections and the synaptic vesicles, while in the inset is a tangential section, as in Fig. 4B, to show the triangular arrangement of the dense projections (dp) with the synaptic vesicles (sv) forming a hexagonal array (Akert et al., 1969). In Fig. 5B is a diagrammatic reconstruction of a tangential section to show the triangular arrangement of the dense projections that appear to be ordering the hexagonal array of the synaptic vesicles (Akert et al., 1969). This structure forms a paracrystalline plate, the presynaptic vesicular grid, whereby the bouton confronts the synaptic cleft as indicated in Fig. 4A and 5A.

It is important to state at the outset that, in accordance with Hubbard (1970), the synaptic vesicles observed in all boutons in the central nervous system of vertebrates are the morphological correlates of the quantal emission of transmitter that occurs in all chemically transmitting synapses made by boutons on nerve cells. The synaptic vesicles are recognized as quantal packages of the preformed transmitter molecules (about 5,000 to 1o,000) which are ready for release as a quantal package into the synaptic cleft in a unitary operation (Fig. 5B).

Subsequent study by the freeze etching technique (Akert et al., 1972; 1975; Akert, 1973) confirmed and clarified the original findings of Figs. 4 and 5 so that an idealized bouton could be drawn in perspective (Fig. 6), with to the left the triangular array of the dense projections (az) ordering the hexagonal array of synaptic vesicles (sv) confronting the synaptic cleft in a paracrystalline structure. Other vesicles lie further back, being in reserve. On the right side, stripping of the grid shows the vesicle attachment sites (vas) to the presynaptic membrane, while in the centre window are seen across the synaptic cleft the particle aggregations (pa) that probably relate to the transmitter receptors on the postsynaptic membrane. On each side of the nerve terminal

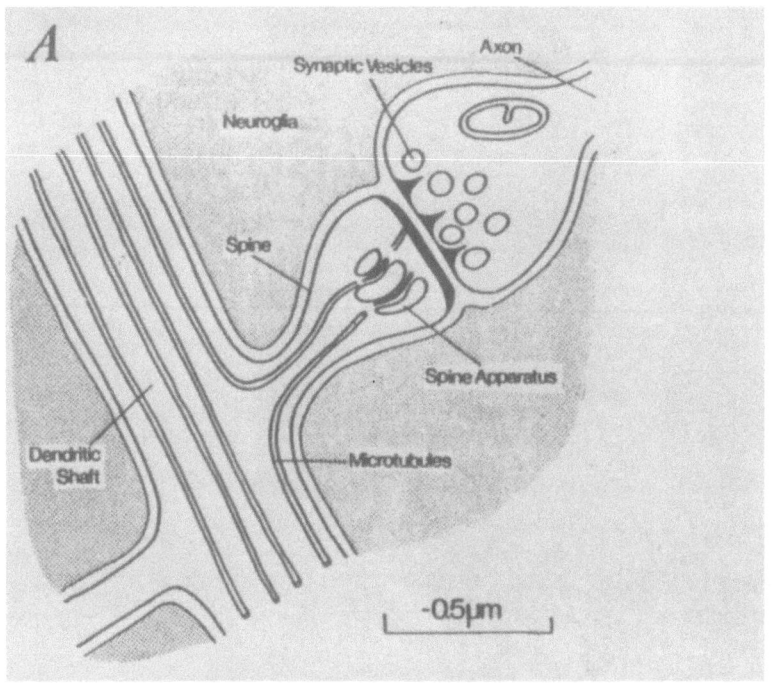

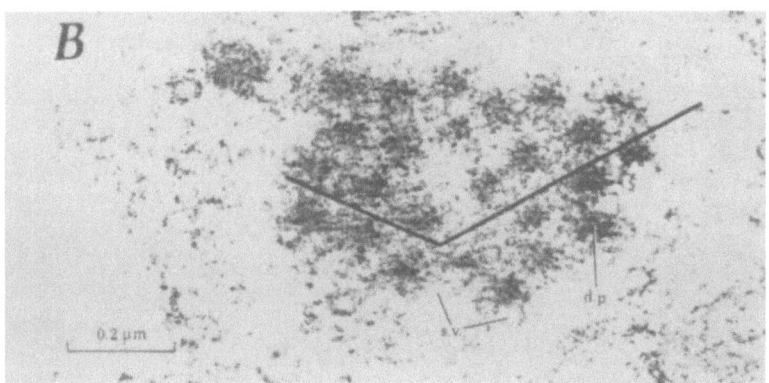

Fig. 4: A. Drawing of a synapse on a dendritic spine. The bouton contains synaptic vesicles and dense projections on the presynaptic membrane (Gray, 1982). B. Presynaptic dense projections d.p. seen en face on the presynaptic membrane, sv synaptic vesicles (Gray, 1983).

there are shown in rectangular perspective the presynaptic vesicular grid and the presynaptic membrane after it has been stripped off.

Figs. 4 and 5 and 6 are synapses from the mammalian brain. Essentially similar synapses are observed in a very different neuronal system, the inhibitory synapses on the Mauthner cell in the fish spinal cord (Triller and Korn, 1982). Fig. 7A shows the dense projections confronting the synaptic cleft with spaces for synaptic vesicles indicated by the crossed arrows, while Fig. 7B,C shows the pattern of dense projections in a tangential section

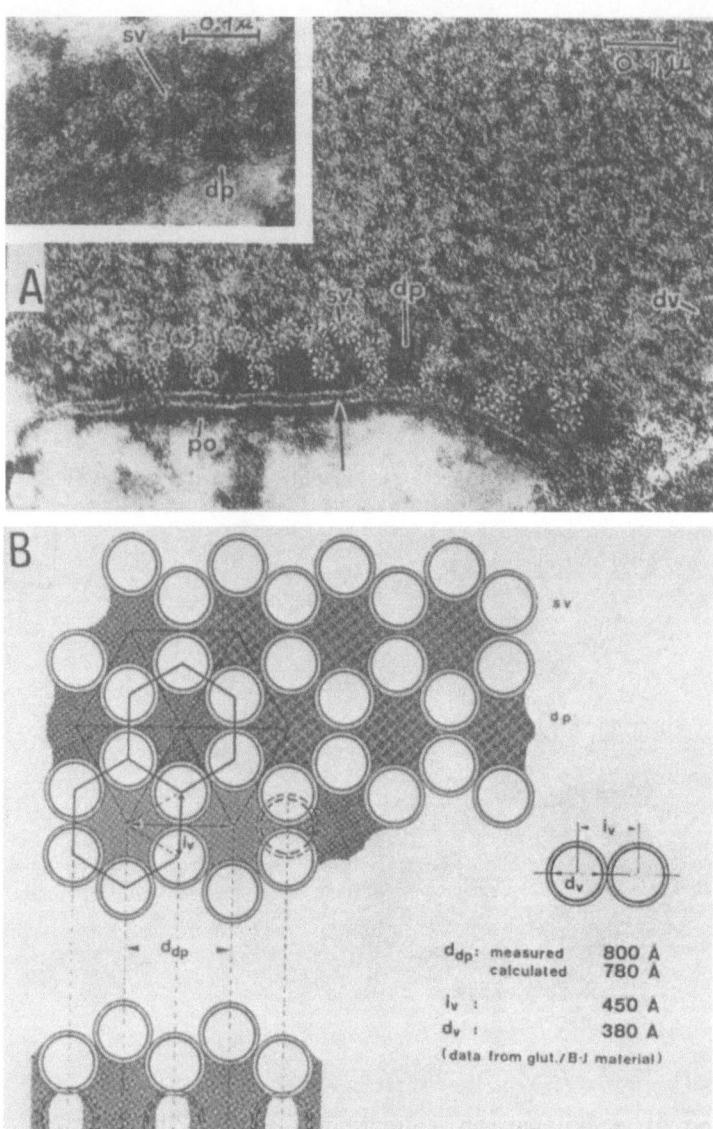

Fig.5: Transverse section across the synaptic cleft.
A. Presynaptic dense projections of Gray and the formation of a
hexagonal vesicular grid. The spiny appearance of dense projections is due to "fuzz coat" surrounding the adjacent synaptic
vesicles. dv=dense-cored vesicle; po=postsynaptic density; arrow=
synaptic cleft with intracleft lines. Inset: Tangential section
through vesicular grid. Dark areas represent presynaptic dense
projections (dp) in hexagonal arrangement with interconnecting
filaments. Clear profiles between dense projections represent
synaptic vesicles (sv), forming a rosette-like pattern. Note that
each hole of the grid accommodates one single vesicle (cf. Fig.5B).
Primary magnification 4o,ooox.

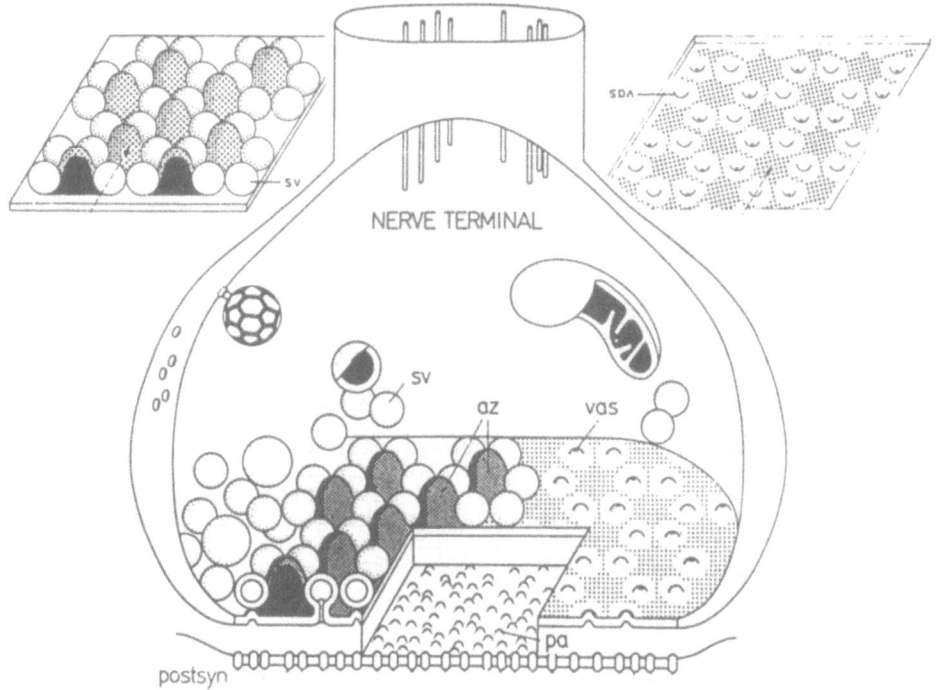

Fig. 6: Schema of the mammalian central synapse. The active zone
(az) is more complex and allows far more vesicle attachment sites
(vas) per square unit of surface than the motor end plate. The postsynaptic aggregation of intramembraneous particles is restricted to
the area facing the active zone. sv=Synaptic vesicles, pa=particle
aggregations on postsynaptic membrane (postsyn.). Further description in text. (Akert et al., 1975.)

with a predominantly triangular pattern of the dense projections
that structure the presynaptic vesicular grid much as in Figs. 4B
and 5.

    Figs. 4-7 illustrate an important feature, namely that as a rule
a synaptic bouton has only one presynaptic vesicular grid which occupies only a fraction of the total area of synaptic contact. It
may be a single oval plate (Figs. 4B, 5B, 6, 7B), but it could be
a ring (Fig. 7C) or a horseshoe. Triller and Korn (1982) report
that at least 95 % of the inhibitory boutons on Mauthner cells
have only a single presynaptic vesicular grid. There is evidence
that disorders of the synapses may be accompanied by cyclic changes

B.   The presynaptic vesicular grid. Reconstruction of geometrical
relationships between presynaptic dense projections (dp) and synaptic vesicles (sv). Upper diagram represents tangential section
of the grid, lower diagram represents cross-section. d=diameter;
i=interval. The measured and calculated dimensions of grid and
vesicles are within close range. The relationship between vesicles
and dense projections as viewed in the cross-sectional reconstruction below can be detected in BI stained cross-sections of synapses
under optimal conditions (see A). (Akert et al., 1969.)

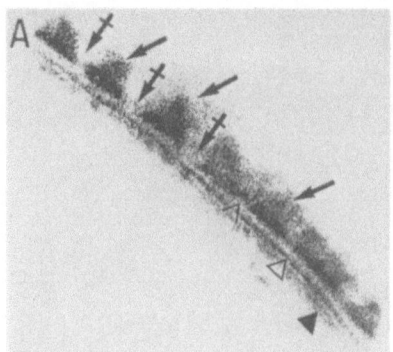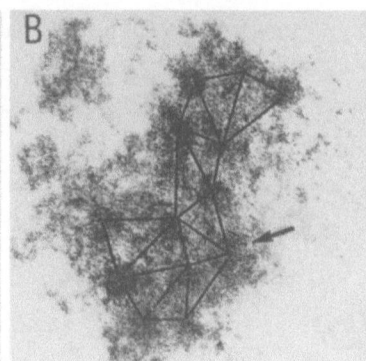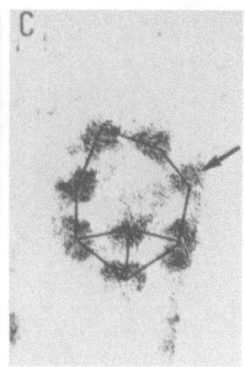

Fig. 7: Presynaptic dense projections of a synapse on a Mauthner cell.
A. High power magnification of a presynaptic differentiation. Note that the roughly triangular PDPs (arrows) are separated by a free space for vesicles (crossed arrows) and that their bases can be linked by a dense material (V-shaped arrowheads) (x176,000).
B,C. Tangential section of presynaptic grids with macular (B) and annular (C), respectively. (x150,000). (Triller and Korn, 1982).

in the presynaptic vesicular grid – from a small oval to a large oval, to a ring, to a horseshoe, to fragmentation, to regeneration of an oval (Nieto-Sampedro et al., 1982; Carlin and Siekevitz, 1983).

There are still great problems involved in the origin of synaptic vesicles, their charging by the specific transmitter molecules, the movement through the bouton for incorporation in the presynaptic vesicular grid, which we may call the firing zone (Fig. 8A) and the paracrystalline properties of this grid,whereby of the large number of its constituent synaptic vesicles no more than one is emitted by a nerve impulse (Fig. 8B) and that only with a probability usually far below one.

Since there is usually only one presynaptic vesicular grid to a single bouton, it is possible to estimate for a bouton the number of synaptic vesicles incorporated into the grid,and hence immediately available for release by a presynaptic nerve impulse. Unfortunately there are in the literature few references to this important number, though counts can be made of synaptic attachment sites in published illustrations of freeze etching findings. Akert (1973) gives the number of dense projections for small grids at about 12, with numbers up to 150 for large grids. So the synaptic vesicle attachment sites would range from about 25 to a rarely observed maximum of up to 300. However, counting in illustrations gives no more than 30 to 50. Triller and Korn (1982) give estimates ranging from 44 to 83 for synaptic vesicles close to the presynaptic membrane of inhibitory boutons on Mauthner cells.

After this introduction to the fine structure of a bouton we have to consider how it comes about that, from this relatively large population of synaptic vesicles on the firing line of the presynaptic vesicular grid, rarely, if ever, more than one is discharged by a presynaptic impulse, and usually the probability is 0,5 or less (Jack et al., 1981a; Korn et al., 1982; Korn et Faber, 1985; Redman et Walmsley, 1983b). Triller and Korn (1982) made two suggestions. In one it is assumed that only a small fraction of

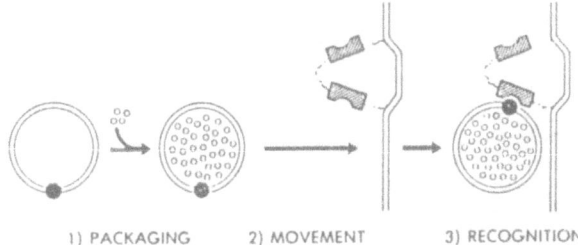

1) PACKAGING   2) MOVEMENT   3) RECOGNITION

B

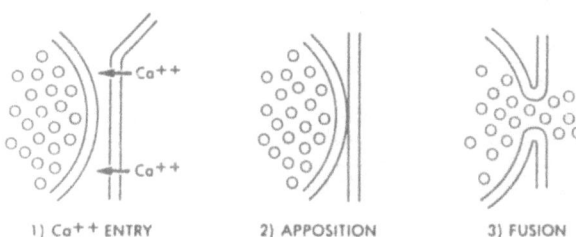

1) Ca^{++} ENTRY   2) APPOSITION   3) FUSION

Fig. 8: Stages of synaptic vesicle development, movement and exocytosis.
A. The three steps involved in filling a vesicle with transmitter and bringing it to attachment to a presynaptic dense projection of triangular shape. B. Stages of exocytosis with release of transmitter into the synaptic cleft.depicting the essential role of Ca^{2++} input from the synaptic cleft (Kelly et al., 1979).

synaptic vesicles in the grid are in a state of high probability for release. The other is that the emission of one vesicle crowds the path through the dense projections for all other vesicles, the presynaptic grid being a paracrystalline structure. They admit that this is a rather simplistic mechanical model. I think it has to be proposed that the presynaptic vesicular grid (cf. Figs. 5B, 6) is a subtle dynamic structure designed to limit the emission of synaptic vesicles by a presynaptic impulse. By its unitary operation it sets the probability of vesicular emission from its whole structure and thus from the bouton.

It has to be remembered that there are probably no more than 20,000 synaptic vesicles in a bouton, which is adequate for only a few minutes of normal operation.

According to the analysis of Redman and associates (Jack et al., 1981A; Redman and Walmsley, 1983b) the probability of quantal release may be almost one for some boutons and down to zero for others, with all ranges in between. According to the analysis of Korn and associates (Korn et al., 1982; Korn and Faber, 1985) the mean probability for the different synapses ranged from 0,17 to 0,62. Various factors modify the probability, the most important being the increase due to Ca^{2+} ion (Fig. 8B), which acts after combination with the protein calmodulin, 4 Ca^{2+} ions to one molecule (de Lorenzo, 1981). Other factors are the enhancing effect by a prior stimulus (Hirst et al., 1981; Jack et al., 1981b) probably due to the residual Ca^{2+} in the bouton, and by 4-aminopyridine (Jack et al., 1981b), and the depressant effect of high-frequency stimulation, halving at about 33 Hz (Korn and Faber, 1985)

Despite the general agreement that the probability of release of a vesicle (quantum of transmitter) from a bouton for a single presynaptic impulse is less than unity, there are minor conflicts between the findings of Redman and associates and of Korn and associates. These derive from the assumptions used in the deconvolution techniques from the recorded EPSPs. Redman's group (Jack et al., 1981a; Redman and Walmsley, 1983b) assumes that the EPSP's produced by quantal emission from the boutons of a Ia fibre on a motoneurone are of a standard size, the deconvolution on this basis giving the probability of emission for each bouton. This analysis yields a bouton number usually much less than that actually determined by horseradish peroxidase analysis, so the assumption has to be made that some boutons have a zero probability of emission. On the contrary Korn's group (Korn et al., 1982; Korn and Faber, 1985) used binomial theory in their analysis of the synaptic responses of a Mauthner cell to a single inhibitory impulse, and found a very good quantitative relationship between the analytical bouton number and the histological number (cf. Fig. 3C,D); hence they assume the correctness of this deconvolution technique.according to which all the bouton endings of a single axon on a Mauthner cell have the same probability of emission of a single vesicle, the mean probability value being 0,38 for the 18 axons studied.

In conclusion I would suggest that neither deconvolution technique is fully acceptable. The analysis by Redman and associates leads to a highly improbable solution of bouton number, with many boutons at zero quantal probability. The solution also requires that a quantal emission activates all the transmitter sites on the postsynaptic membrane. This may appear to be in conflict with the hypothesis that in long-term potentiation there is an increase in the activated postsynaptic receptor sites; but this increase could be due either to the uncovering of existing transmitter sites, as diagrammed by Lynch and Baudry (1984), or to the creation of additional receptor sites. The solution of Korn and associates is superficially the most attractive.because the binomial analysis gives results in good accord with the number counted in horseradish peroxidase preparations (Fig. 3C,D). However,the assumption of identical probability for all the boutons of an axon seems unlikely. Crucial testing of the two deconvolution techniques must await fortunate circumstances, where for example there are no more than 3 or 4 bouton sites of a Ia fibre widely dispersed on the dendrites of a motoneurone, so that unitary discrimination could be more critically tested than was possible by Redman and Walmsley (1983b). Meanwhile it is fortunate that both deconvolution techniques deliver essentially the same information, namely that in response to a single presynaptic impulse a single synaptic bouton is a quantal emitter of vesicles with a probability usually well below one and never above one.

This refined story of the quantal operation of synaptic boutons indicates that there should be a reexamination of the responses of neurones at the highest level of the central nervous system for which there is evidence of responses evoked by a mental intention or mental attention.

E. Introduction to the mind-brain problem

There are many materialist theories of the mind, as summarized in the four entries of Fig. 9. Radical materialism eliminates itself. The three other materialist theories recognize the existence of mind or mental events but give it no independent status. According to the above three materialist theories of the mind, mental states are an attribute of matter or the physical world, either

DIAGRAMMATIC REPRESENTATION OF BRAIN-MIND THEORIES

World 1 = All of material or physical world including brains

World 2 = All subjective or mental experiences

World 1_P is all the material world that is without mental states

World 1_M is that minute fraction of the material world with associated
        mental states

Radical Materialism: World 1 = World 1_P; World 1_M = 0; World 2 = 0.

Panpsychism: All is World 1-2, World 1 or 2 do not exist alone.

Epiphenomenalism: World 1 = World 1_P + World 1_M

                           World $1_M \rightarrow$ World 2

Identity theory: World 1 = World 1_P + World 1_M

                           World 1_M = World 2 (the identity)

Dualist - Interactionism: World 1 = World 1_P + World 1_M

                           World $1_M \Longleftrightarrow$ World 2; this interaction occurs

                           in the liaison brain, LB= World 1_M.

                           Thus World 1=World 1_P+ World 1_{LB}, and

                           World $1_{LB} \Longleftrightarrow$ World 2

Fig. 9: Diagrammatic representation of brain-mind theories that incorporates the World 1 and World 2. The essential features of the materialist theories of the mind are summarized for Panpsychism, Epiphenomenalism and the Identity Theory. This latter theory has a variety of names according to the whims of the creators of the minor varieties of what are essentially parallelist theories. The subdivision of World 1 into World 1_p and World 1_M helps in clarification of their specific features. World 1_M is assumed to be restricted to special states of the brain in Epiphenomenalism, the Identity Theory, and Dualist-Interactionism. The essential and unique feature of Dualist-Interactionism is shown by the reciprocal arrows between World 1_M and World 2 in the second line.

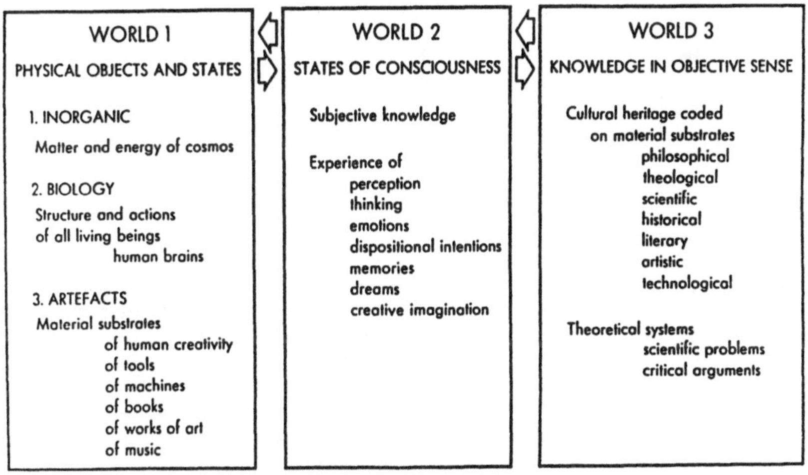

WORLD 1	WORLD 2	WORLD 3
PHYSICAL OBJECTS AND STATES	STATES OF CONSCIOUSNESS	KNOWLEDGE IN OBJECTIVE SENSE
1. INORGANIC Matter and energy of cosmos	Subjective knowledge	Cultural heritage coded on material substrates philosophical theological
2. BIOLOGY Structure and actions of all living beings human brains	Experience of perception thinking emotions dispositional intentions memories	scientific historical literary artistic technological
3. ARTEFACTS Material substrates of human creativity of tools of machines of books of works of art of music	dreams creative imagination	Theoretical systems scientific problems critical arguments

Fig. 1o: Tabular representation of the contents of the three worlds in accordance with the philosophy of Karl Popper. These three worlds are non-overlapping, but are intimately related as indicated by the large open arrows at the top. They contain everything in existence and in experience. World 1 is material, World 2 and 3 are immaterial.

of all matter as in panpsychism, or of matter in the special state
in which it exists in the highly organized nervous systems of animals and man. One variety of this, epiphenomenalism need not be
further considered, having been replaced in recent decades by the
identity theory that was first fully developed by Feigl (1967).
Popper (1977) states that "all four assert that the physical world
(World 1) is self-contained or closed. ... This physicalist principle of the closedness of the physical World 1 ... is of decisive
importance ... as the characteristic principle of physicalism or
materialism". Popper then goes on to give a critical account of all
materialistic theories of the mind. Fig. 1o gives diagrammatically
Popper's 3-World system.

It has been difficult to discover statements by philosophers
that relate to the precise neural events which are assumed to be
identical with mental events. The clearest expression was given by
Feigl (1967). On page 79 he states:

"The identity thesis which I wish to clarify and to defend as-
serts that the states of direct experience which conscious human
beings "live through", and those which we confidently ascribe to
some of the higher animals, are identical with certain (presumably configurational) aspects of the neural processes in those
organisms ... processes in the central nervous system, perhaps
especially in the cerebral cortex ... The neurophysiological
concepts refer to complicated highly ramified patterns of neuron discharges."

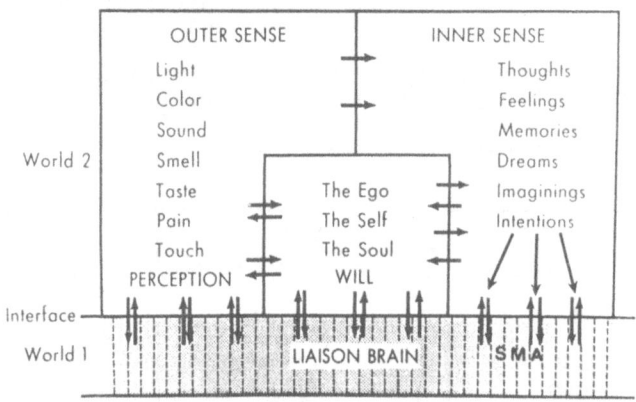

Fig. 11: Information flow diagram for mind-brain interaction in
human brain. The three components of World 2: Outer Sense; Inner
Sense; and the Ego, Self or Soul are diagrammed with their communications shown by arrows. Also shown are the lines of communication across the interface between World 1 and World 2, that is
from the liaison brain to and from these World 2 components. The
liaison brain has the columnar arrangement indicated by the vertical broken lines. It must be imagined that the area of the liaison brain is enormous, with open or active modules numbering over
a million, not just the two score here depicted. The supplementary
motor area, SMA, is shown specially related to intentions of World
2, with the three arrows giving some suggestion of the potential
specificity of action of the intention on the modules of the SMA,
as discussed in the text. World 2 is shown above World 1, but this
is a diagrammatic device without spatial significance. If World 2
is to be given any spatial location, it will be placed where it
acts.

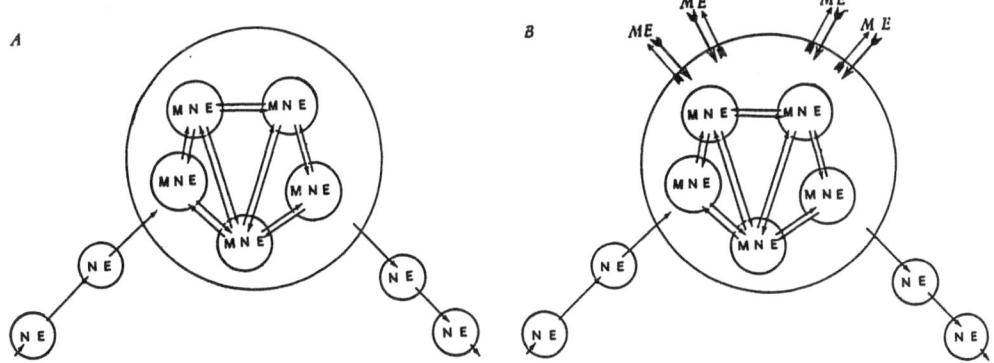

Fig. 12: Diagrams of mind-brain theories. A. The identity theory.
B. Dualist-interactionism. Assemblages of neurones are shown by
circles. NE represents the conventional neurones which respond
only to neural events. MNE are neurons that are associated with
both mental and neural events and are grouped in a larger circle
representing the higher nervous system. In B,ME arrows represent
mental influences acting on the neural population that is associated both with mental and neural events. All other arrows in A
and B represent the ordinary lines of neural communication which
are shown in reciprocal action.

We can raise the question whether there could be experimental
testing of predictions from the Dualist-interactionist hypothesis
(Fig. 11) on the one hand and the Identity hypothesis on the other.
A simple diagram (Fig. 12A) embodies the essential features of the
Identity hypothesis. In accord with Feigl there was agreement that
mental-neural identity occurs only for neurones or neurone systems
at a high level of the brain, especially in the cerebral cortex.
These neurones can be called mental-neural event (MNE) neurones,
whereas other neurones in the brain, and in particular neurones
on the input and output pathways, would be no more than simple
neural event (NE) neurones, as in the diagram (Fig. 12). It would
be predicted from the identity hypothesis that MNE neurones would
be distinctive,because in special circumstances their firing would
be in unison (the identity) with mental events. But of course this
firing would be in response to inputs from other neurones, MNE or
NE, and is in no way determined by or modified by the mental events.
This is the closedness of the physical world referred to above by
Popper (1977).

It is remarkable that neurones or neurone systems in the cerebral cortex have been discovered experimentally that would seem to
be MNE neurones, being distinctively related to intentional or
attentional mental states (Roland et al., 1980; Roland, 1981). In
the diagram (Fig. 12A) the identity theorists would have to postulate that the firing of the MNE neurones would be entirely neurally
generated and explicable as responses either to NE inputs or to
other MNE neurones of the higher brain.

What then is the position of the dualist-interactionist? It
would be the crucial difference that the MNE neurones would have,
in addition to the MNE and NE inputs, an input from mental events
(ME) per se as is shown by the additional arrows in the diagram
Fig. 12B. The firing of MNE neurones would exhibit a response that
is different from what it would be in the absence of the mental
events of intention or attention. There is evidence that this in-

deed does occur both when intention activates neurones of the
supplementary motor area (SMA) (Brinkman and Porter, 1979; Roland
et al., 1980) and when attention activates neurones, probably of
the mid-prefrontal area, that project to the appropriate somato-
topic neurones of the somatosensory area (Roland, 1981).

F. Action of Intention on the Supplementary motor area (SMA)

   Fig. 13 shows the position of the supplementary motor area (SMA)
of the left cerebral hemisphere in the medial part of the frontal
cortex just anterior to the motor area of the hind limb and extending deep on the medial side. By a radioXenon technique Roland et
al. (1980) recorded the regional cerebral blood flow (rCBF) over a
cerebral hemisphere, there being an assembled pattern from 254
Geiger counters for recording the detailed spatial pattern of radio-
emission following a brief injection of radio-Xenon into the internal carotid artery. It is now established that any regional increase
in rCBF is a reliable signal of an increased neuronal activity in
that area. The subject was trained to make a complex pattern of
finger-thumb movements for the full duration (45 sec) of the Geiger
counting. In Fig. 14A there was a strong activation of the contralateral motor and sensory areas for the thumb and fingers, as would
be expected, but there was just as strong an activation of the SMA
and that was bilateral. The primacy of the SMA is revealed in Fig.

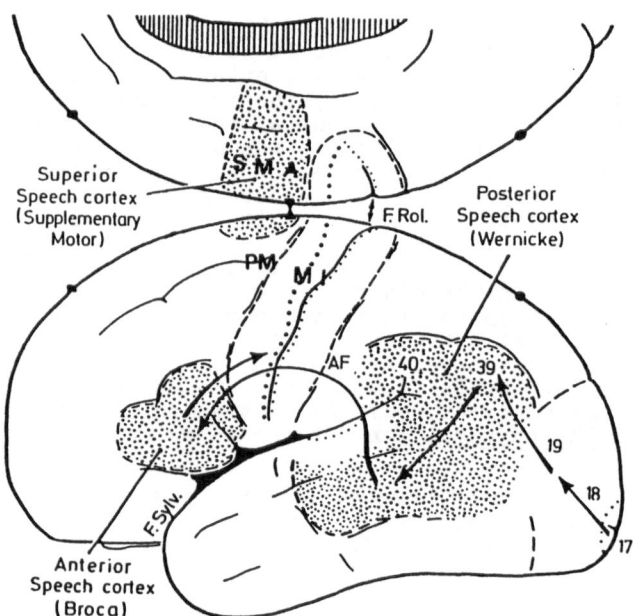

Fig. 13: The left hemisphere from the lateral side with frontal
lobe to the left. The medial side of the hemisphere is shown as
if reflected upwards. F.Rol. is the fissure of Rolando or the central fissure. F.Sylv. is the fissure of Sylvius. The primary motor
cortex MI is shown in the precentral cortex just anterior to the
central sulcus and extending deeply into it. Anterior to MI is
shown the premotor cortex, PM, with the supplementary motor area,
SMA, largely on the medial side of the hemisphere (modified from
Penfield and Roberts, 1959).

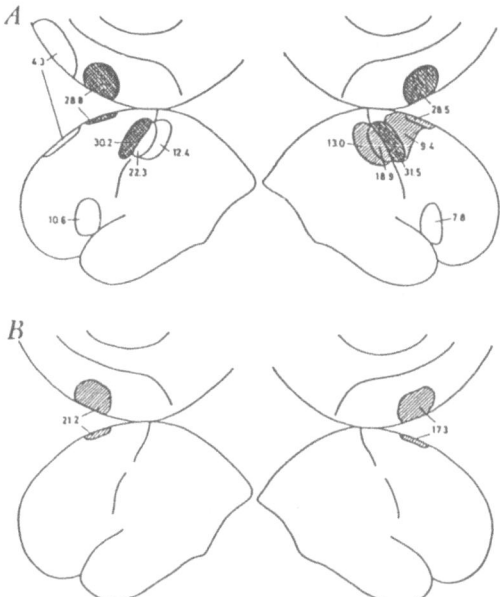

Fig. 14: A. Mean increase of the rCBF in percent during the motor-sequence test performed with the contralateral hand, corrected for diffuse increase of the blood flow. Cross-hatched areas have an increase of rCBF significant at the 0,0005 level. Hatched areas have an increase of rCBF significant at the 0,005 level, for other areas shown the rCBF increase is significant at the level 0,05. Left: Left hemisphere, five subjects. Right: Right hemisphere, 1o subjects. B. Mean increase of rCBF in per cent during internal programming of the motor-sequence test, values corrected for diffuse increase of the blood flow. Left: Left hemisphere, three subjects; Right: Right hemisphere, five subjects (Roland et al. 1980).

14B when during the radio-Xenon test the subject was making no movement but merely carrying out the learned motor task mentally. A highly significant 2o% increase in neuronal activation was restricted to the SMA on both sides, and was nowhere else. The subject was at complete rest with eyes and ears closed. This rCBF increase is an index of an increase in neuronal activity of the SMA under the influence of a mental intention by the subject.

By means of an implanted microelectrode it has been possible to study the responses of SMA neurones of a monkey while it was carrying out a voluntary movement (Fig. 14A,C; Brinkman and Porter, 1979). There was an increase in the discharge rate of many neurones that eventually would cause the willed movement as signalled by the electromyogram of Fig. 14B (cf. Eccles, 1982a; 1982b). Ethical considerations preclude the carrying out of such an experiment on a human subject. However the recording of electric and magnetic fields over the human scalp during repetitive voluntary movements (Deecke and Kornhuber, 1978) also point to the SMA as the site of action of the mental intention.

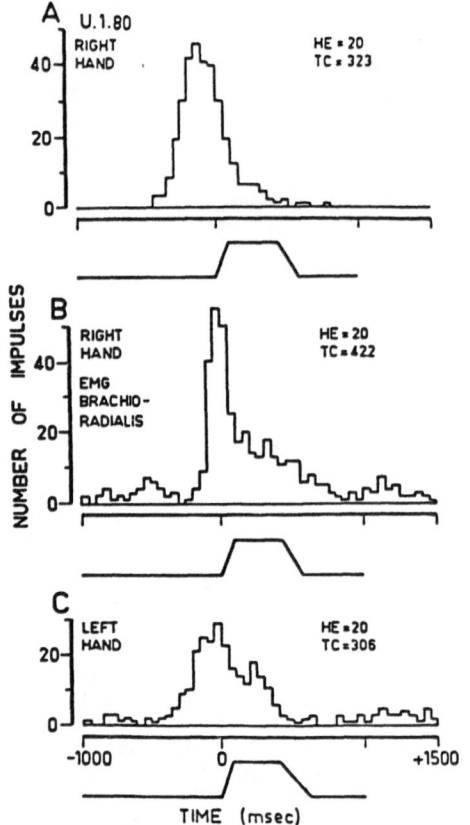

◄Fig. 15:
Illustration of the discharges patterns
of a neurone associated with flexion of
the elbow during the lever pull for
both the right (A) and left hand (C).
B is the periresponse time histogram
demonstrating the EMG activity of a
representative elbow flexor, m.
brachioradialis, in the right arm
during the same 20 pulls as those in A,
and shows that the neurone increased
its discharge well before EMG activity
increased. This was the case for the
majority of neurones in which the
discharge pattern could be compared
with EMG changes (Brinkman and Porter,
1979).

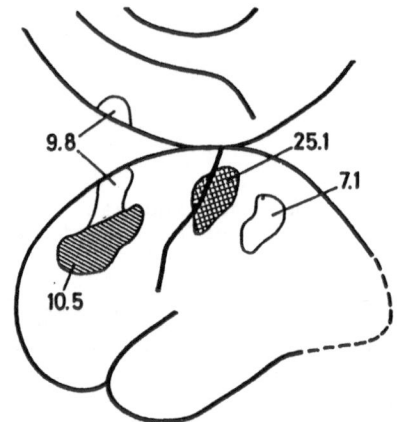

Fig. 16: Mean increase of rCBF in percent during pure selective
somatosensory attention; that is, somatosensory detection without
peripheral stimulation. The size and location of each focus shown
is the geometrical average of the individual focus. Each individual
focus has been transferred to a brain map of standard dimensions
with a proportional stereotaxic system. The cross-hatched areas
have an increase of rCBF significant at the o,ooo5 level (Student's
test, one-sided significance level). For the other areas shown the
rCBF increase is significant at the o,o5 level. Eight subjects
(Roland, 1981).

G. The Action of Attention on the prefrontal lobe

Fig. 16 illustrates a remarkable finding of Roland (1981) that,
when the human subject was attending to a finger on which a just
detectable touch stimulus was to be applied, there was an increase
in the rCBF over the finger touch area of the postcentral gyrus
of the cerebral cortex as well as in the mid-prefrontal area.
These increases must result from the mental attention because no
touch was applied during the recording. Thus Fig. 16 is a clear
demonstration that the mental act of attention can activate appropriate regions of the cerebral cortex. A similar finding occurs
with attention to the lips in expectation of a touch, but of course
the activated somatosensory area is now for the lips.

A related finding is that, when the subject is attending to
simple counting or other arithmetical mental activities in complete
silence, there was an increased rCBF in a medial strip of the fron-

tal cortex anterior to the SMA (N.A. Lassen, personal communication).

H. The Mind-Brain problem

The identity theorist is committed to the doctrine that mental events per se cannot contribute to the generation of neural events (Figs. 14B; 15A,C), which is the doctrine of the closedness of World I. The experience of an intention to move must therefore be attributable to the activity of an assemblage of MNE neurones in the SMA (cf. Fig. 12A). One may ask what causes these neurones to fire impulses and so bring about a voluntary movement? To this question the identity theorist has no answer except to say that there must be other as yet unidentified neuronal centres containing MNE neurones that fire before the SMA, which is merely an evasion of the problem. Another possible answer is that the SMA activity is due to chance, but can one accept chance as the invariable cause of our voluntary movements?

By contrast,the experimental findings are in accord with predictions from dualist-interactionism. As indicated in Fig. 11 by the arrows,each mental event of intention (Fig. 12B, ME) initiates (shown by arrows) the firing of a set of neurones of the SMA that, through the various known pathways indicated diagrammatically in Fig. 17, cause the "correct" motor cortical neurones to discharge impulses down the pyramidal tract to cause the desired voluntary movement (Eccles, 1982a; 1982b).

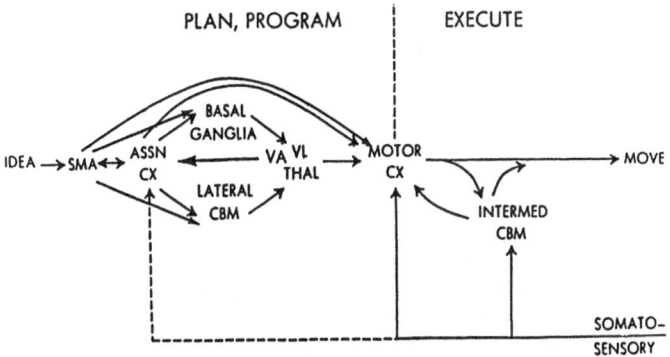

Fig. 17:Diagram showing the pathways concerned in the execution and control of voluntary movement; ASSN CX, association cortex; lateral CBM, cerebellar hemisphere; intermed. CBM, pars intermedia of cerebellum. The arrows represent neuronal pathways composed of hundreds of thousands of nerve fibres.

The specific cortical activities produced by attention (Fig.16) in the absence of all neural inputs or overt activity of the cerebral cortex are comparable to the effects of internal programming of intention (Fig. 14B), and likewise present an apparently inexplicable explanatory problem for the identity theorist. The dualist-interactionist is able to account for the action of the mental event of attention on the neurones of the prefrontal lobe as indicated in Fig. 18. It is assumed, but not proven, that this is the site of the interaction,and that the neurones of the touch area of the cerebral cortex were secondarily activated (Roland, 1981). A similar diagram could be made for the action of internal counting on the medial prefrontal cortex.

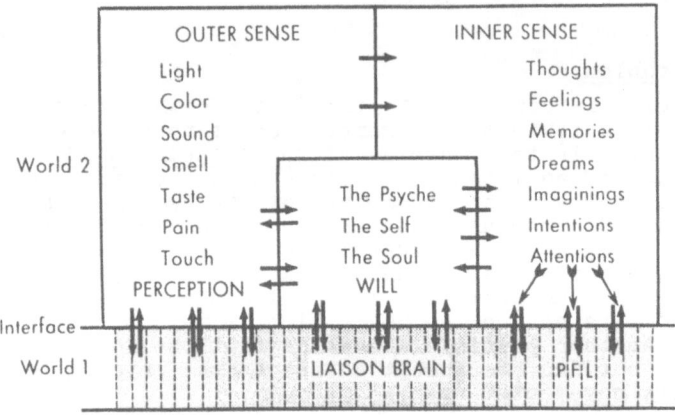

Fig. 18: Information flow diagram for brain-mind interaction in human brain. The three components of World 2: outer sense; inner sense; and the psyche, self or soul; are diagrammed with their communications shown by arrows. Also shown are the lines of communications across the interface between World 1 and World 2, that is from the liaison brain to and from these World 2 components. The liaison brain has the columnar arrangement indicated by the vertical broken lines. It must be imagined that the area of the liaison brain is enormous, with open or active modules numbering over a million, not just the two score here depicted. As shown by the arrows, attentions are specially directed to the modules of the prefrontal lobe (PFL).

However, the materialist critics argue that insuperable difficulties are encountered by the hypothesis that immaterial mental events can act in any way on material structures such as neurones as is diagrammed in Figs. 11, 12B. Such a presumed action is alleged to be incompatible with the conservation laws of physics, in particular of the first law of thermodynamics. This objection would certainly be sustained by 19th Century physicists, and by neuroscientists and philosophers who are still idealogically in the physics of the 19th Century, not recognizing the revolution wrought by quantum physicists in the 2oth Century. Unfortunately it is rare for a quantum physicist to dare an intrusion into the brain-mind problem. But in a recent book the distinguished quantum physicist Margenau (1984) makes a fundamental contribution. It is a remarkable transformation from 19th Century physics to be told (page 22) ... "that some fields, such as the probability field of quantum mechanics carry neither energy nor matter ". He goes on to state (page 96):
"In very complicated physical systems such as the brain, the neurons and sense organs, whose constituents are small enough to be governed by probabilistic quantum laws, the physical organ is always poised for a multitude of possible changes, each with a definite probability; if one change takes place that requires energy, or more or less energy than another, the intricate organism furnishes it automatically. Hence, even if the mind has anything to do with the change, that is if there is a mind-body interaction, the mind would not be called upon to furnish energy". In summary Margenau states that (page 97): "The mind may be regarded as a field in the accepted physical sense of the term. But it is a nonmaterial field, its closest analogue is perhaps a probability

field. It cannot be compared with the simpler nonmaterial fields that require the presence of matter (i.e. gravity) ... Nor does it necessarily have a definite position in space. And so far as present evidence goes it is not an energy field in any physical sense, nor is it required to contain energy in order to account for all known phenomena in which mind interacts with brain". These considerations should eventually convince the materialists or physicalists who are still imbued with 19th Century physics.

In formulating more precisely the dualist hypothesis of mind-brain interaction, the initial statement is that the whole World of mental events (World 2) has an existence as autonomous as the World of matter-energy (World 1) (Fig. 1o). The present interactionist hypothesis does not relate to these ontological problems, but merely to the mode of action of mental events on neural events, that is to the nature of the downward arrows across the frontier in Figs. 11, 12, 18. Following Margenau (1984) the hypothesis is that mind-brain interaction is analogous to a probability field of quantum mechanics which has neither mass nor energy, yet can cause effective action at microsites.

More specifically, it is proposed, that intentions (Fig. 11) or attentions (Fig. 18) of the non-material mind can cause neural events by a process analogous to the probability field of quantum mechanics. Furthermore, these induced events are in the first instance alterations in the probabilities of quantal emission by the presynaptic vesicular grids acting on the target neurones (cf. Figs. 4, 5, 6, 7).

It should be pointed out that there are two distinct uses of the word quantum, that in quantum physics and that in neuroscience for the unit of synaptic transmission, a synaptic vesicle (cf. Fig.17).

It is not postulated that the firing of neurones is the target of the mental events, but merely that this firing is modified by alterations of the probabilities of quantal emission of those synapses that are engaged in actively exciting them. This is an important limitation in the target for mental events such as intention and attention.

On the hypothesis it would be predicted that the effectiveness of mental events would be reduced to zero when the presynaptic background was reduced to zero. Loss of consciousness would occur, and be irreversible unless there would be revival to a considerable degree of the impulse discharges in the cerebral cortex. An example is "vigil coma" that supervenes when brain injury to the mid-brain turns off the reticular activating system (Hassler,1978; Eccles, 1980, p. 16o). In fact, the principal role of the reticular activating system may be to provide a background of excitatory impulses into the cerebral cortex, with an immense array of probabilistic quantal emissions that are targets for the quantal probabilistic fields of mental influences.

Unconsciousness also occurs at the opposite extreme, the intense neuronal activity of an epileptic seizure (Eccles, 1980, p.157). It is to be expected that the postulated subtle probability fields would be ineffective on the cerebral cortex in such extreme states However, they could still be effective on areas of the cerebral cortex not subjected to the seizure. The unconsciousness of sleep, strangely enough, is still controversial with respect to the levels and patterns of cortical activity that are responsible for the various types of sleep (cf. Eccles, 1980, pp. 153,154).

A general criticism of the hypothesis could be that a change in probability of quantal emission of one synaptic bouton would be orders of magnitude too small for an effective action on the patterns of neuronal activity in large areas of the brain. For example in Fig. 15A and C the intention of the monkey to pull the lever resulted in a great increase in the frequency of firing of that SMA neurone and also of many similar neurones in other recordings. However, there are many thousands of similar boutons on such a pyramidal cell of the SMA. The hypothesis is that the probability field of the mental intention is widely distributed not only to the synapses on that cell, but also to the synapses of a multitude of other cells with similar actions. Such a widespread excitatory influence is exhibited by the increased rCBF of the SMA in Fig. 14A and B. Likewise in the mental attention of Fig. 16 there must be an enormous increase in neuronal discharges, which may be attributed to a widespread distribution of the probability fields of mental influences in concentrated attention (the downward arrows of Fig.18) with consequent changes in the quantal probability of emissions from an immense number of boutons.

So we can assume that, in a global manner, the mental events achieve interaction with the neural events of spatio-temporal patterns of activity (Eccles, 1982b) of the awake cerebral cortex. Even in one cortical module with its 4,000 or so neurones, there must be an on-going intense dynamic activity of unimaginable complexity. Although we know the outlines of the neuronal structure of a module (Szentagothai, 1978, 1983), there has as yet been only a very limited study of the physiology. All that we can surmise is that mental events acting as a field in the manner postulated by Margenau could effect changes in the spatio-temporal activity of a module by changing the probability of emission in many thousands of active synapses. There need be no violation of conservation laws.

A further objection has been made by Bunge (1980) that "every information flux rides on some physical process", i.e. information from nonmaterial mental events could not be transmitted to the brain. But as stated above this objection cannot apply to the probability field of quantum mechanics, and according to Margenau the mind may be regarded as some such field.

One can ask how does the monkey set up the immense synaptic barrage that results in the neuronal firings in Fig. 14A and C, and that through the well-known complex pathways schematically shown in Fig. 17 result in the desired motor action. The only answer is that this performance is at the end of a long line of training sessions. Motor learning is essential for all skilled actions devolving from the cerebral cortex (Eccles, 1985). Memory of some kind is required for all conscious experiences and actions.

The suggestion that the mind operates as a quantal probability field brings into consideration earlier conjectures of Eddington (1935, 1939) based upon the uncertainty principle of Heisenberg, that were further developed by Eccles (1970) in the light of the discoveries that the synaptic unit of action was a vesicle of transmitter of about 4oo Å in diameter with a mass of about 3×10^{-17}g. If the uncertainty principle is applicable to an object of this size, then it was calculated that the uncertainty in the position of such a vesicle would be of the order of 30 Å in 1 millisecond. The intricate process of quantal emission of transmitter from a synaptic vesicle involves action across the presynaptic membrane (Fig. 8B), which is about 50 Å across, and the times of action are in the order of a millisecond. It was assumed that a synaptic vesicle was a freely moving unit. This situation has now been transformed by the discovery that the synaptic vesicle is strongly controlled by being incorporated in a presynaptic vesicular grid (Figs. 4,5,6,7), which is a paracrystalline structure in which about 40 vesicles are held in close relation to the presynaptic membrane. We have seen in Section D that this grid somehow acts in a unified manner to set the quantal emission of the whole structure at a probability level that is never more than one per impulse (Jack et al., 1981a; Redman and Walmsley, 1983b; Korn et al., 1982; Korn and Faber, 1985), and usually it is 0,2 to 0,7. Can we envisage the presynaptic vesicular grid as the synaptic structure "tuned" into the quantal probability field of a mental influence?

It may be objected that the quantal emission of synapses has been rigorously investigated only at lower levels of the central nervous system, the motoneurones of the cat spinal cord and the fish Mauthner cells. However, investigations on the mammalian hippocampus, which is a primitive cerebral cortex, show quantal responses to single impulses resembling those of Figs. 2C and 3B (McNaughton et al., 1981). Moreover,electronmicroscopy demonstrates that the synapses of the mammalian cerebral cortex also have presynaptic vesicular grids (Gray, 1983; Akert et al., 1969, 1975).

It is suggested that we develop some new insight somewhat related to Bohr's Complementarity Principle, whereby the probability waves of quantum mechanics (analogous to mental events) relate to the quantal probabilities of quantal emission from the presynaptic vesicular grid (the neural events). This could be a challenge to creative thinking on the flow of information across the most important of all frontiers, that between the mental events of World 2 and the neural events of World 1, as depicted in Figs. 11 and 18.

It can be concluded that the quantum physics of the 20th Century provides possible clues as to the manner in which mental events such as intentions and attentions increase the activity of cortical neurones (neural events). There is revival of the free-will problem with the important corollary of moral responsibility. Thus,there seems to be at last some new light on the mind-brain problem that had been impenetrably dark since Descartes' mistaken hypothesis that the soul influenced the body by its action on the pineal body!

I. Summary

Attention is concentrated on the mode of operation and the structure of the unitary transmitting elements of the brain, the synaptic boutons. Intracellular records of the responses of a neurone to an impulse in a single presynaptic fibre have been analysed to reveal the response evoked by a single bouton. This has been shown to act in a quantal manner, corresponding to the emission of a single vesicle of transmitter. However, the response is probabilistic, the emission occurring with a probability of usually 0,5 or less and never above 1,0. The microstructure of the bouton reveals that those synaptic vesicles awaiting exocytosis are embedded in a paracrystalline structure, the presynaptic vesicular grid, which in an unknown manner acts globally to ensure that the probability of emission is below one for the ensemble of 40 or more vesicles embedded in it. There is only one such grid for a bouton in the central nervous system.

There is an account of the brain-mind theories with special reference to the manner in which the concepts of quantum mechanics have transformed the problem. According to Margenau the nearest physical analogue to non-material mental events is the probability field of quantum mechanics, which has neither mass nor energy, yet it can effectively act at microsites without violating the conservation laws. The action on special sites of the cerebral cortex by mental events of intention and attention is described and explained on analogy with the probability field of quantum mechanics.

The hypothesis of mind-brain interaction is that mental events act by a quantal probability field to alter the probability of emission of vesicles from presynaptic vesicular grids. There must be an immense operation in parallel on the thousands of presynaptic vesicular grids confronting a neurone, with in addition many neurones similarly activated. Then by conventional neuronal circuitry mental events of attention achieve the desired brain response, leading to the desired motor movements. There is a similar explanation of the action of concentrated mental attention activating special areas of the cerebral cortex. Thus, the mind-brain interaction postulated in dualism has been shown to be in accord with quantum physics. It does not violate natural laws as has been maintained by its critics.

REFERENCES

Akert, K. 1973 Dynamic aspects of synaptic structure. Brain Res. 49, 511-518

Akert, K., Moor, H., Pfenninger, K. & Sandri, C. 1969 Contributions of new impregnation methods and freeze etching to the problems of synaptic fine structure. In Progress in Brain Research (ed. K. Akert and P.G. Waser). vol. 31, pp.223-240. Amsterdam: Elsevier Publ.Co.

Akert, K., Peper, K. & Sandri, C. 1975 Structural organization of motor end plate and central synapses. In Cholinergic Mechanisms (ed.E.G.Waser). pp. 43-57. New York: Raven Press.

Akert, K., Pfenninger, K., Sandri, C. & Moor, H. 1972 Freeze etching and cytochemistry of vesicles and membrane complexes in synapses of the central nervous system. In Structure and Function of Synapses (Ed. G.P. Pappas & Purpura, D.F.) pp. 67-86. New York: Raven Press.

Brinkman, C. & Porter, R. 1979 Supplementary motor area in the monkey: activity of neurons during performance of a learned motor task. J. Neurophysiol. 42, 681-7o9.

Brown, A.G. 1981 Organization in the spinal cord: The anatomy and physiology of identified neurones. pp. 238. Berlin, Heidelberg, New York: Springer Verlag.

Bunge, M. 1980 The mind-body problem. Oxford: Pergamon Press.

Burke, R.E., Walmsley, B. & Hodgson, J.A. 1979 HRP anatomy of group Ia afferent contacts on alpha motoneurones. Brain Res. 160, 347-352.

Carlin, R.K. & Siekevitz, P. 1983 Plasticity in the central nervous system: do synapses divide? Proc. Natl.Acad.Sci USA, 8o, 3517-3521

Deecke, L. & Kornhuber, H.H. 1978 An electrical sign of participation of the mesial 'supplementary' motor cortex in human voluntary finger movement. Brain Res. 159, 473-476.

DeLorenzo, R.J. 1981 The calmodulin hypothesis of neurotransmission. Cell.Calcium 2, 365-385.

Eccles, J.C. 197o Facing Reality. New York, Heidelberg, Berlin: Springer Verlag.

Eccles, J.C. 198o The Human Psyche. Berlin, Heidelberg, New York Springer International.

Eccles, J.C. 1982a The initiation of voluntary movements by the supplementary motor area. Arch.Psychiatr.Nervenkr. 231, 423-441.

Eccles, J.C. 1982b How the self acts on the brain. Psychoneuro-endocrinology 7, 271-283.

Eccles, J.C. 1985 Learning in the motor system. In Oculomotor and Skeletalmotor system (ed. J.Noth). In Press

Eccles, J.C., Eccles, R.M. & Lundberg, A. 1957 Synaptic actions on motoneurones in relation to the two components of the group I muscle afferent volley. J. Physiol. 136, 527-546.

Eddington, A.S. 1935 New Pathways in Science. London: Cambridge University Press.

Eddington, A.S. 1939 The Philosophy of Physical Science. London: Cambridge University Press.

Feigl, H. 1967 The "mental" and the "physical". Minneapolis, Minn.: University of Minnesota Press.

Gray, E.G. 1982 Rehabilitating the dendritic spine. Trends in Neurosciences 5, 5-6

Gray, E.G. 1983 Neurotransmitter release mechanisms and micro-tubulus. Proc. R. Soc. Lond. B 218, 253-258.

Hassler, R. 1978 Interaction of reticular activating system for vigilance and the truncothalamic and pallidal systems for directing awareness and attention under striatal control. In Cerebral correlates of conscious experience. (ed. P.A. Buser & Rougeul-Buser, A.). pp. 110-129. Amsterdam: Elsevier North Holland.

Hirst, G.D.S., Redman, S.J. & Wong, K. 1981 Post-tetanic poten-tiation and facilitation of synaptic potentials evoked in cat spinal motoneurones. J. Physiol. 321, 97-109.

Hubbard, J.I. 1970 Mechanism of transmitter release. Progress in Biophysics and Molecular Biology 21, 33-124.

Jack, J.J.B., Redman, S.J. & Wong, K. 1981a The components of synaptic potentials evoked in cat spinal motoneurones by im-pulses in single group Ia afferents. J. Physiol. 321, 65-96.

Jack, J.J.B., Redman, S.J. & Wong, K. 1981b Modifications to synaptic transmission at group Ia synapses on cat spinal moto-neurones by 4-aminopyridine. J. Physiol. 321, 111-126.

Kelly, R.B., Deutsch, J.W., Carlson, S.S. & Wagner, J.A. 1979 Biochemistry of neurotransmitter release. Ann. Rev. Neurosci. 2, 399-446.

Korn, H., Mallet, A., Triller, A. & Faber, D.S. 1982 Transmission at a central inhibitory synapse. II. Quantal description of release, with a physical correlate for binomial n. J. Neuro-physiol. 48, 679-707.

Korn, H. & Faber, D.S. 1985 Regulation and significance of prob-abilistic release mechanisms at central synapses. In New In-sights into synaptic function. (ed. G.M. Edelman, W.E. Gall & W.M. Cowan). New York: Neurosciences Research Foundation Inc.: J. Wiley & Sons Inc.

Lynch, G. & Baudry M. 1984 The biochemical intermediates in memory formation: a nes and specific hypothesis. Science 224, 1057-1063.

McNaughton, B.L., Barnes, C.A. & Andersen, P. 1981 Synaptic efficiency and EPSP summation in granule cells of rat fascia dentata studied in Vitro. J. Neurophysiol. <u>46</u>, 952-966.

Margenau, H. 1984 The Miracle of Existence. Woodbridge (Conn.): Ox Bow Press.

Mendell, L.M. & Hennemann, E. 1971 Terminals of single Ia fibers: Location, density and distribution within a pool of 3oo homogeneous motoneurons. J. Neurophysiol. <u>34</u>, 171-187.

Niet-Sampedro, M., Hoff, S.F. & Cotman, C.W. 1982 Perforated postsynaptic densities; Probable intermediates in synapse turnover. Proc.Natl.Acad.Sci. USA <u>79</u>, 5718-5722.

Penfield, W. & Roberts, L. 1959 Speech and Brain Mechanisms. Princeton: Princeton University Press.

Pfenninger, K., Sandri, C., Akert, K. & Engster, C.H. 1969 Contribution to the problem of structural organization of the presynaptic area. Brain Res. <u>12</u>, 1o-18.

Popper, K.R. 1977, p. 51 of Popper and Eccles 1977 "The Self and its Brain". Berlin, Heidelberg, New York, London: Springer International.

Redman, S.J. 198o Mechanisms of transmitter release at Ia afferent terminations. Adv. Physiol. Sci. <u>1</u>, 93-1oo. In Regulatory functions of the CNS. Principles of Motion and organization. (ed. J. Szentagotahi, M. Palkovits, J. Hamori). Oxford: Pergam Pergamon Press.

Redman, S. & Walmsley, B. 1983a The time course of synaptic potentials evoked in cat spinal motoneurones at identified group Ia synapses. J. Physiol. <u>343</u>, 117-133.

Redman, S. & Walmsley, B. 1983b Amplitude fluctuations in synaptic potentials evoked in cat spinal motoneurones at identified group Ia synapses. J. Physiol. <u>343</u>, 135-145.

Roland, P.E. 1981 Somatotopical tuning of postcentral gyrus during focal attention in man. A regional cerebral blood flow study. J. Neurophysiol. <u>46</u>, 744-754.

Roland, P.E., Larsen, B., Lassen, N.A. & Skinhøj, E. 198o Supplementary motor area and other cortical areas in organization of voluntary movements in man. J. Neurophysiol. <u>43</u>, 118-136.

Szentágothai, J. 1978 The neuron network of the cerebral cortex. A functional interpretation. Proc. R. Soc. London B <u>2o1</u>, 219-248.

Szentágothai, J. 1983 The modular architectonic principle of neural centers. Rev. Physiol. Biochem. Pharmacol. <u>98</u>, 11-61.

Triller, A. & Korn, H. 1982 Transmission at a central inhibitory synapse. III. Ultrastructure of physiologically identified and stained terminals. J. Neurophysiol. <u>48</u>, 7o8-736.

Spin Glasses as Model Systems for Neural Networks

W. Kinzel

Institut für Festkörperforschung der Kernforschungsanlage Jülich,
D-5170 Jülich, Fed. Rep. of Germany

Spin glasses are disordered magnetic materials with unusual magnetic properties.
The complicated structure of the phase space, especially the existence of infinitely many metastable states, leads to long time relaxational processes. Important experiments and models are briefly outlined. Analytical and numerical results for a
mean-field-model of spin glasses are discussed. This model has the interesting
property to learn and process information, hence it has recently been suggested that
it may be a simple model for neural networks.

1. Introduction

In the last two decades there is a still increasing activity to understand the physics of disordered materials. While most of the phenomena of pure crystals are now
well understood, this is not true for many materials of our daily life, like glass,
polymers, and metallic alloys. The absence of translational symmetry in such systems makes it impossible to apply standard experimental and theoretical tools of
solid state physics [1].

One example of a class of disordered materials are spin glasses. Spin glasses
are magnetic systems with unusual magnetic properties [2]. Their name, which was
created in 1968, indicates that the cooperative behaviour of the magnetic moments
(= spins) is similar to the one of atoms and molecules in structurally disordered
systems (= glasses). Hence one hopes that the research on spin glasses will contribute to the general physics of disordered materials.

In fact, spin glass properties are rather universal. They are found in metallic
crystals like $Au_{1-x}Fe_x$, in nonmetallic crystals like $Eu_xSr_{1-x}S$ and amorphous metals
like Gd_xAl_{1-x} for a wide range of the concentration x of the magnetic component.
This universality already indicates that only a few ingredients are responsible for
the spin glass properties. In fact, it turns out that *disorder* and *competition* of
the magnetic interactions are common to all spin glasses. This allows to describe
the real systems by a very simple model introduced by EDWARDS and ANDERSON (EA) in
1975 [3]. Although only in the special unrealistic case of infinite range of the
magnetic interactions (= mean-field theory) analytical results are available, much
is known about this model from computer simulations. Hence spin glass research is
characterized by intensive exchange of results and ideas between many kinds of different experiments, computer simulations and analytic work. But although at the
moment this great activity results in about one publication per day on spin glasses,
main and fundamental problems are still not well understood and still are a challenge to solid state physics.

The complex phase space of spin glasses may be used to store and process information. In fact, simple models can learn many patterns in their network of interactions. The memory is nonlocal and very insensitive to defects. By an asynchronous
dynamics a noisy input pattern can be recognized very fast. The emergent computational properties of such models are very different from the ones of present day
computers. It was recently suggested by HOPFIELD [4] that such systems may be simple models for neural networks. The state of a neuron is modelled by the spin vari-

able and the synapses by the magnetic interactions. The adjustment of the neurons to the potential from the connected neighbours corresponds to the zero-temperature relaxation of the magnetic energy.

Of course, this simplification is a physicist's point of view, and neurobiologists will have to estimate its use for the understanding of the brain. Nevertheless, the investigation of the computational properties of such models with the tools of condensed matter theory has its own fascination,and surely will be continued in the future, also with regard to other complex problems occurring, for instance, in combinatorial optimization [5].

This paper briefly outlines some main results and problems of spin glasses. In the following section, experimental facts and the results of computer simulations of the EA model are discussed. Sect. 3 presents recent analytical results of the infinite range EA model. The computational properties of this model are outlined in Sect. 4. They are demonstrated by recent numerical and analytical results of the author.

2. Experiments and Models

In usual magnetic systems like ferromagnets or antiferromagnets, one observes a sharp phase-transition from a disordered state at high temperatures to an ordered at low temperatures. In the ordered phase, the magnetic moments are - up to thermal fluctuations - aligned periodically in the material. At the transition temperature a measure of the order, the order parameter, either jumps to zero (first-order transition) or goes continuously to zero (critical transition). The order parameter may be the thermal average of the magnetization in ferromagnets or of the sublattice magnetization in antiferromagnets. In the continuous case, one observes power law singularities in the thermodynamic observables like order parameter, specific heat, magnetic susceptibility, or correlation length. Note that an ordered phase is a consequence of the cooperative behaviour of the *infinite* system, in a finite system there is no order possible due to the thermal fluctuations.

In spin glasses, the situation is different. There a freezing temperature T_f exists below which the magnetic moments freeze into *randomly distributed* directions, as sketched in Fig. 1. For simplicity, we consider only anisotropic moments which can either point up or down. This will be characterized by the spin variable $S_i = +1$ (up) or $S_i = -1$ (down). This freezing is a cooperative effect. The random directions are due to the network of magnetic interactions of both signs, i.e. there are positive couplings which want to align the spins parallel,and negative ones which prefer an antiparallel alignment. But in the disordered network there is *frustration* [6], i.e. the spins obviously cannot align to *all* of their interactions. This frustration leads to the spin glass properties.

The important difference of spin glasses to usual magnets is not the random alignment but the degeneracy of aligned spin states. In usual (anisotropic) magnets, there are only two possible states (up or down in ferromagnets). In spin glasses it turns out that there are *infinitely* many possible states. These states may be

 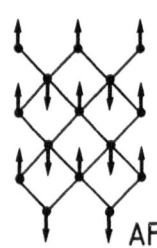 

F   AF   SG

Fig. 1. Alignment of the magnetic moments (= spins) in ferromagnets, antiferromagnets and spin glasses at low temperatures

metastable, that means, although each single spin points into its local field created by its neighbours, the whole system has a higher energy than the equilbrium state. Therefore such a metastable state decays very slowly to one of the many states of thermal equilibrium.

These features of spin glasses manifest themselves, for instance, in the following experimental facts [2]:
(a) There is a sharp cusp in the magnetic susceptibility as a function of temperature T at a freezing temperature T_f .
(b) The specific heat is smooth at T_f .

(c) Neutron scattering does not show any periodic order.
(d) Below T_f the field-cooled susceptibility is different from the zero-field cooled susceptibility, i.e. the magnetization in an external magnetic field depends on the history of the sample.
(e) T_f depends on the time-scale of the experiments which can vary from 10^{-11} sec for neutron scattering up to days for magnetization measurements. However, this variation is not large.
(f) There exists a remanent magnetization which decays slowly to zero.

Especially the last observation is important for the computational properties of disordered and frustrated spin systems, hence it is sketched in Fig. 2. In thermal equilibrium,the global magnetization M is zero due to the random alignment. If an external magnetic field is switched on, M jumps to a value depending on temperature T and field strength H . If then the field H is switched off, then M does not jump to its equilibrium value M=0 but the spin configuration is trapped in a state with nonzero magnetization. At T=0 this state is stable but for $0<T<T_f$ M(t) decays slowly to zero due to thermal activation. Hence M is irreversible and depends on time, t, temperature, T, the previously applied field, H, and the history of the sample (field or zero-field cooled).

Fig. 2. Magnetization M of spin glasses as a function of time t if an external magnetic field is switched on and off, as indicated

As already mentioned, spin glass properties are rather universal; they follow from disorder and competition of the magnetic interaction. Hence spin glasses are described by the simple EA model 3 with energy

$$H = - \sum_{ij} J_{ij} S_i S_j - h \sum_i S_i \quad , \tag{1}$$

where the spin variables $S_1 = \pm 1$ may be placed on a regular lattice whose sites are labelled by i and j , h is an external field and J_{ij} are randomly distributed magnetic couplings. Usually a Gaussian distribution for the J_{ij} is taken

$$P(J) \propto \exp(-J^2/2\Delta J^2) \quad . \tag{2}$$

Hence the energy scale is given by the width ΔJ of P(J) . Note that the sign of the couplings is randomly distributed leading to frustration. In addition to the Hamiltonian (1) and (2), a simple relaxational dynamics is introduced,as used in usual computer (Monte Carlo) simulations [7].

Obviously, this model is not a good microscopic model of, say, AuFe or EuSrS. Nevertheless, it has been found that it describes the experimental data in surprisingly many details. If the couplings J_{ij} are short ranged, for instance, nearest neighbour bonds on a square lattice, then there is still no reliable theory available to describe the properties of the model. But computer simulations have shown that all of the findings listed above are reproduced by this model qualitatively in many details [2,7,8,9].

Numerical calculations have shown that in two space dimensions the EA model (1) has no phase-transition [10]. Hence the freezing process is a dynamic phenomenon of local correlated regions. However, the size of those regions and their typical relaxation time τ diverge very fast with decreasing temperature; one finds, for instance [9,11]

$$\ln \tau \sim T^{-\nu z} \quad , \tag{3}$$

with an exponent $\nu z \approx 2$. Hence for temperatures T of the order of ΔJ the relaxation time is of the order of any observation time and the systems seem to be frozen into random spin orientations with typical spin glass behaviour. Obviously it is hard to distinguish such a process from a true phase-transition at a nonzero critical temperature T_c. Therefore the problem of the existence of a phase-transition in our three-dimensional world is still not resolved, neither in real spin glasses nor in the EA model [12].

Finding the state $\{S_i\}$ with lowest energy H of (1) is a complex problem of combinatorial optimization [5]. Recently it has been shown that such problems belong to the class of NP-complete problems [13], and therefore probably no algorithm exists which finds a solution in a time bounded by a polynomial of the number N of spins. However, experience with computer simulations shows that a good estimate of the solution can be obtained by the Monte Carlo method by slowly cooling down the system from temperatures T larger than ΔJ to $T=0$. This method can also be applied to other complex optimization problems [5].

3. Mean-Field Theory

Many problems of statistical mechanics can be solved in the limit of infinite range interactions, only. Although in this limit (which is also called mean-field theory, MFT) important fluctuations disappear, such a MFT gives a useful qualitative picture of the cooperative behaviour of the system.

The MFT of the EA model (1) has been introduced [14] and studied extensively [2]. In this case each spin S_i interacts with every other spin S_j with the coupling J_{ij} which is again randomly distributed by (2). To get a meaningful energy the width ΔJ has to go to zero like $\Delta J = J/\sqrt{N}$ in the thermodynamic limit $N \to \infty$ (N is the number of spins).

In the last 5 years it has been found that this model has a rather complex behaviour which is still not quite understood. There has been great progress in developing analytical tools and calculating static and dynamic properties [2]. Again there is a remarkable qualitative agreement with experiments.

It turns out that the MFT gives a sharp phase-transition in thermal equilibrium at the temperature $T_c=J$. For $T < T_c$ the phase space has a complex structure as sketched in Fig. 3. In a MFT the free energy F (which is the energy H at $T=0$) is a function of continuous variables m_i. The local minima of F ($\{m_i\}$) are the physical states of the system. In ferromagnets (for $h=0$), there are only two minima at $m_i=M$ and at $m_i=-M$ which correspond to the two magnetic states with a spontaneous magnetization $+M$ or $-M$. The order parameter is just given by $M(T)$. In spin glasses, there are infinitely many local minima $m_i^\alpha = <S_i>_\alpha$, where the last bracket $<...>_\alpha$ denotes the thermal average of the spin S_i over the available phase space α. There are many states which belong to thermal equilibrium ($\hat{=}$ global minimum), and even more states which are metastable and result in long-time relaxational processes of spin glasses.

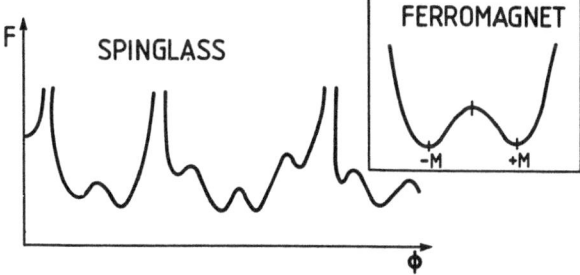

Fig. 3. Free energy as a function ϕ of one coordinate in the N-dimensional space of configurations $\{m_i\}$. The local minima $\{m_i\}^\alpha$ are the physical states. While there are only two possible stable states in ferromagnets, there are many stable and metastable states in spin glasses

Such a complex space of possible configurations is described by an order parameter function $q(x)$ which can be obtained analytically and gives the thermodynamic quantities [15]. It has the following physical meaning: Consider two states $\{<S_i>_\alpha\}$ and $\{<S_i>_\beta\}$, i.e. two global minima in Fig. 3. Then a distance

$$d_{\alpha\beta} = \frac{1}{2} (1 - q_{\alpha\beta}) \tag{4}$$

between these states α and β can be defined by the overlap $q_{\alpha\beta}$ given by

$$q_{\alpha\beta} = \frac{1}{N} \sum_i <S_i>_\alpha <S_i>_\beta \quad . \tag{5}$$

Since the spin glass has many states with different overlaps $q_{\alpha\beta}$ one has to consider the distribution $P(q)$ of overlaps q . It turns out that one has

$$P(q) = dx/dq \quad . \tag{6}$$

Hence in thermal equilibrium a spin glass can be described by the distribution of distances between all low-lying valleys in the configurational space.

With respect to such a metric, (4) and (5), the spin glass states have an interesting topology which is called ultrametric geometry [16]. Namely, given three arbitrary states α, β and γ they form an isosceles triangle, i.e. one always has $d_{\alpha\beta} = d_{\beta\gamma} \geq d_{\alpha\gamma}$ (after relabelling the indices).

This structure may be illustrated in Fig. 4. There the points at the bottom stand for the spin glass states (the valleys in Fig. 3). The distance $d_{\alpha\beta}$ be-

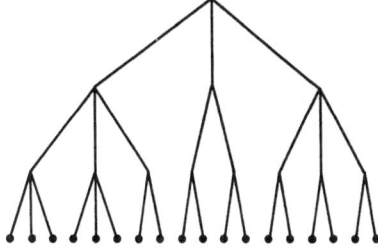

Fig. 4. Illustration of the ultrametric topology of spin glass states which are represented by the bottom sites. The level of common ancestors defines a distance between states

tween points α and β is given by the level of common ancestors. Obviously such a distance has the ultrametric property and leads to interesting clustering effects. not only in configurational space but also in real space [17]. Of course, Fig. 4 is very simplified since in spin glasses the distance is continuous and the number of bifurcations is infinite.

It is quite interesting that this ultrametric structure is also observed in other complex systems like in travelling salesman problems [18] or in computational structures [19]. The implication of this topology on the dynamics of spin glasses has to be worked out, yet.

4. Learning and Pattern Recognition

HOPFIELD [4] suggested a simple model of neural networks which is similar to the EA model of spin glasses: Each neuron is just characterized by a "on-off" variable $S_i = \pm 1$, where $S_i = +1$ (-1) means that a neuron at site i is firing (not firing). Each neuron i receives a potential E_i from its connected neighbour neurons j which is just given by the superposition

$$E_i = \sum_j J_{ij} S_j + h_i + \text{const} \quad . \tag{7}$$

Here J_{ij} characterizes the potential which is created by the corresponding synapses connecting neuron i to neuron j. h_i is some external potential, for instance, from the sensory system. Since there exist excitatory and inhibitory synapses which can adjust their strength, J_{ij} varies in strength and sign.

Each neuron i adapts its state according to the potential E_i. It fires only if E_i is larger than a threshold value E_0. If the constant in (7) is chosen such that $E_0 = 0$, this means that the state of a neuron i is given by

$$S_i E_i > 0 \quad , \tag{8}$$

or that the function

$$H = - \sum_i S_i E_i = - \sum_{ij} J_{ij} S_i S_j - \sum_i h_i S_i \tag{9}$$

has a local minimum. In the case of $J_{ij} = J_{ji}$ one recovers the spin glass model (1) with locally varying fields h_i. Hence the states of the neural network correspond to the metastable states of the spin glass (valleys in Fig. 4).

Here we want to discuss the computational properties of this model. First, let us assume that the system is completely unorganized, for instance the connections $J_{ij} = J_{ji}$ are randomly and independently distributed as in (2). Even then, the model has some kind of short-time memory. Namely, if a strong external pattern $\underline{h} = (h_1, h_2, \ldots, h_N)$ is applied, the system is in the state $\underline{S}^0 = (\text{sgn } h_1, \text{sgn } h_2, \ldots, \text{sgn } h_N)$ by the mechanism of (8). If the external signal is switched off, the system runs into a state $\underline{S}^1$ which has a large overlap to the initial information $\underline{S}^0$, namely one has $\underline{S}^0 \cdot \underline{S}^1 = O(N)$. To see this, note that $\underline{S}^0$ is not correlated to the random connections J_{ij}. Therefore $\underline{S}^0$ is equivalent to an initial signal $\underline{h} \propto (1,1,\ldots,1)$, which corresponds to the external field $\underline{h}$ in the spin glass model (1). Hence the overlap $\underline{S}^0 \cdot \underline{S}^1$ is just the remanent magnetization discussed in Fig. 2 [20]. Therefore, even the unorganized network remembers part of the initially applied information. In Fig. 3, this means that an initially applied information ($\hat{=}$ configuration or state $\underline{S}^0$) relaxes into a neighbour valley bottom $\underline{S}^1$ which still has a large overlap to $\underline{S}^0$.

To store information longer, the system has to adapt its connections J_{ij} to the applied patterns. Let us assume we want to store m-patterns $\underline{S}^1 = (S_1^1, S_2^1, \ldots,$

112

S_N^1) , $\underline{S}^2$, $\underline{S}^3$, ..., S^m simultaneously with the same weight. This can be done by the adaption process [4]

$$J_{ij} = \sum_{\nu=1}^{m} S_i^\nu S_j^\nu \quad .$$ (10)

For an illustration we have taken the letter A as pattern $\underline{S}^1$, as shown in Fig. 5. The number of variables S_i was N=20×20 and, in addition to A , 29 other random-ly generated patterns were stored in the network of J_{ij} by (10). Now, if a noisy A is presented to the system as initial information and then switched off, the sys-tem runs into the learned pattern which has a large overlap to the initial pattern, in very few steps (spin flips per site) as is shown in Fig. 5. Hence the system can quickly recognize noisy patterns.

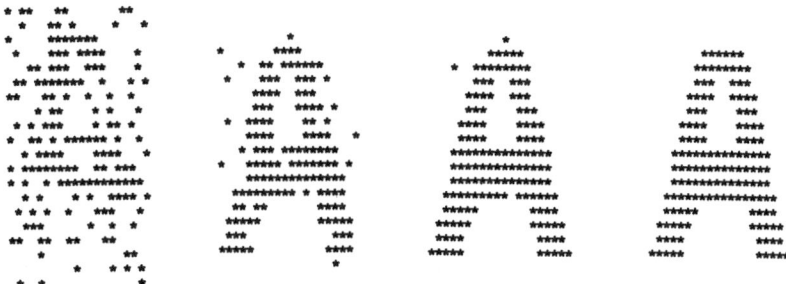

Fig. 5. Pattern recognition of the Hopfield model. The variables S_i are placed on a 20×20 lattice. S_i=+1 is shown by a star, S_1=-1 is not shown. The network has learned 30 patterns. If an initial noisy pattern is applied, the system relaxes in a few adaptions per site to its pure learned pattern (from left to right)

For random patterns, one can derive an approximate formula to see how many pat-terns one can store by (10) and how well the system can recognize patterns as a function of noise. Namely, if too many patterns are stored the system learns pat-terns which have some errors with respect to the m-patterns S^1, S^2, ..., S^m which it should learn. The fraction of errors may be quantified by the Hamming distance $d_{\alpha\beta}$ defined by (4) and (5) between the initial pattern $\underline{S}^\alpha$ and its relaxed state

S^β . One obtains d approximately from the implicit equation [20] (in the limit of large N values)

$$d = \frac{1}{2}\left[1 - \phi\left(\frac{1-2d}{\sqrt{m/z}}\right)\right]$$ (11)

where $\phi(x) = 2/\sqrt{2\pi} \int_0^x e^{-(t^2/2)} dt$ is the probability integral and z is the number of connections from a site i . In the case of infinite range connections, one has z=N and one can store infinitely many patterns for N → ∞ with some error d . For instance, if d = 0.5 % one can store 0.15 N patterns. Each pattern can be re-cognized with almost 50 % noise. However, (11) shows that for a finite number z of connections per site, the amount of patterns to be stored is rather limited. Nu-merical simulations [20] show that (11) is a good approximation.

One can see already analytically that the mechanism of (10) does not work for correlated patterns, say, for several letters, for example. This is demonstrated in Fig. 6 where we have chosen patterns which are similar, like R and P or E and F . However, in this case, we have found a mechanism [20] which was motivated by the fact that in the first few years of life many synapses are dying off, although the amount of learned information is large and permanent [21].

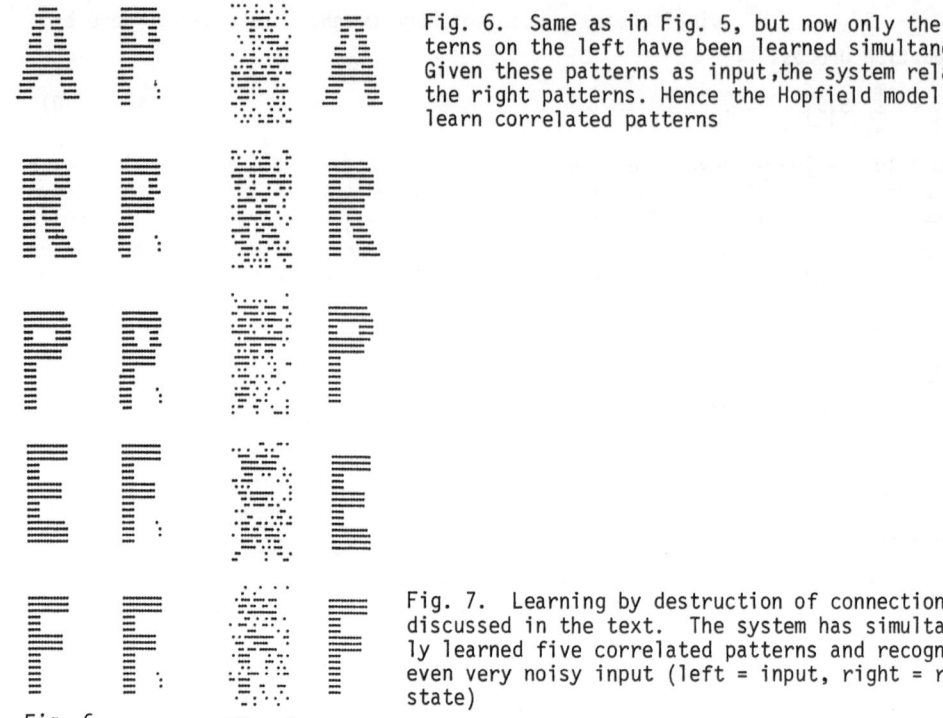

Fig. 6. Same as in Fig. 5, but now only the 5 patterns on the left have been learned simultaneously. Given these patterns as input, the system relaxes to the right patterns. Hence the Hopfield model cannot learn correlated patterns

Fig. 7. Learning by destruction of connections as discussed in the text. The system has simultaneously learned five correlated patterns and recognizes even very noisy input (left = input, right = relaxed state)

Fig. 6   Fig. 7

Thus we start from a completely unorganized network with infinite ranged and Gaussian distributed bonds J_{ij} as in the spin glass model, Eqs. (1) and (2). Then all bonds are destroyed which are frustrated in one of the m patterns to be learned, i.e. J_{ij} is taken to be zero if $J_{ij} S_i^\alpha S_j^\alpha < 0$ for any pattern $\underline{S}^\alpha$. For instance, in Fig. 7, five patterns A,R,P,E, and F were learned simultaneously in a 20×20-network by this mechanism. The system recovers even very noisy patterns, again in a few steps per site of asynchronous adaption as in (8). However, the number of patterns which can be learned is very small, it increases only like $\ell n\ N$.

Let us just list some of the qualitative properties of the Hopfield model:
   (a) Learning and pattern recognition is a spontaneous cooperative property of the whole system without the use of a complex processing system. Even a random network has a simple short-time memory.
   (b) The whole network of connections stores all of the learned information; storage is nonlocal.
   (c) The system is very insensitive to defects. Even a part of the system works as the whole system, only the maximum number of learned pattern is reduced.
   (d) Information is processed, i.e. the system relaxes to a learned pattern which has the largest overlap to the initial signal. Therefore, information is retrieved by content and not by address as in computers. Hence the system can recognize noisy patterns very fast.
   (e) The model works with a parallel asynchronous dynamics.
   (f) Information is learned more or less precisely, depending on how strongly a pattern contributes to the connections. If the bonds J_{ij} are bounded then the system can even "forget" information when new patterns $_{ij}$ are learned.
   (g) As spin glasses already show, the properties of the model are rather insensitive to model details. For instance, if the J_{ij} can take only the values ±1 , the system still can learn and recover information.

Up to now we have discussed a deterministic dynamics only, as given by (8) and (9). However, even with a stochastic dynamics, the model has still the valley structure of phase space which is responsible for its computational behaviour [22].

Hence, the Hopfield model has several remarkable computational properties which are very different from the way of storage and retrieval of information in present day computers but which may be compared with properties of biological computation.

References

1. R. Zallen: The Physics of Amorphous Solids (J. Wiley, N.Y. 1983)
2. For a review see: K.H. Fischer: phys.stat.sol. (b) 116, 357 (1983); phys. stat.sol. (b) 130, xxx (1985); a short outline of recent results may be found in W. Kinzel: Lecture Notes in Physics 206, 113 (1984)
3. S.F. Edwards and P.W. Anderson: J. Phys. F 5, 968 (1975)
4. J.J. Hopfield: Proc. Natl. Acad. Science USA 79, 2554 (1982)
5. S. Kirkpatrick, C.D. Gelatt, and M.P. Vecchi: Science 220, 671 (1983)
6. G. Toulouse: Comm. Phys. 2, 155 (1977)
7. K. Binder and K. Schröder: Phys. Rev. B 14, 2142 (1976)
8. W. Kinzel: Phys. Rev. B 19, 4595 (1979)
9. W. Kinzel and K. Binder: Phys. Rev. Lett. 50, 1509 (1983); Phys. Rev. B 29, 1300 (1984)
10. I. Morgenstern and K. Binder: Phys. Rev. Lett. 43, 1615 (1979); Phys. Rev. B 22, 288 (1980)
11. K. Binder and A.P. Young: Phys. Rev. B 29, 2864 (1984)
12. Recent contributions to this question are: Ref. 11; N. Bontemps et al: Phys. Rev. B 30, 6514 (1984); A.T. Ogielski and I. Morgenstern: Phys. Rev. Lett. 54, 9238 (1985); R. Bhatt and A.P. Young: Phys. Rev. Lett. 54, 924 (1985)
13. C.P. Bachas: J. Phys. A 17, L709 (1984)
14. D. Sherrington and S. Kirkpatrick: Phys. Rev. Lett. 35, 1792 (1975)
15. G. Parisi: Phys. Rev. Lett. 43, 1754 (1979); 50, 1946 (1983)
16. M. Mezard, G. Parisi, N. Sourlas, G. Toulouse, and M. Virasoro: Phys. Rev. Lett. 52, 1156 (1984)
17. M. Mezard and M.A. Virasoro: to be published
18. S. Kirkpatrick and G. Toulouse: to be published
19. B.A. Huberman and M. Kerszberg: J. Phys. A 18, L331 (1985)
20. W. Kinzel: to be published
21. I thank M. Kerszberg for bringing this fact to my attention
22. D.J. Amit, H. Gutfreund, and H. Sompolinsky: to be published; P. Peretto: Biol. Cybern. 50, 51 (1984)

Strange Attractors in the Dynamics of Brain Activity

A. Babloyantz

Faculté des Sciences, Université Libre de Bruxelles,
Campus Plaine, Bd. du Triomphe, B-1050 Brussels, Belgium

1. Introduction

Mammalian brain is certainly one of the most complex systems encountered in nature. It is made of billions of cells endowed with individual electrical activity and interconnected in a highly intricate network. The cell-cell interaction is characterized by nonlinear dynamics in which time delays must also be considered [1].

The average electrical activity of a portion of the brain may be recorded in course of time and is called electroencephalogram (EEG). The EEG reflects the sum of elemental self-sustained neuronal activities of a relatively long period (of the order of 0.5 to 40 Hz). Recordings from the human brain show that to various stages of brain activity there correspond characteristic electrical wave forms. Several protocols are available which prescribe the medical implications of particular forms of the EEG [2].

Whether these average activities obey to well-defined, reproducible phase relationships or, rather, they appear as incoherent irreproducible events, is obviously a major question pertaining to the very nature of the brain function. Whence the need to characterize, as sharply as possible, the nature of the EEG viewed as the manifestation of an underlying dynamical system.

Recent progress in the theory of nonlinear dynamical systems have provided new methods for the study of complex systems [3,4]. These methods are specially valuable for the analysis of experimental data obtained from a single variable (one-dimensional) time series. Their chief merit is to discriminate between random or determiministic nature of the dynamical system. For instance, they allow one to determine the minimum number of variables necessary for description of the dynamical system. Furthermore, they give criteria for the existence of attractors, which characterize deterministic dynamics, as well as information about such quantitative properties as dimensionality.

Such an approach has been applied with success to hydrodynamical and-chemical processes and to complex natural systems such as climatic variability [5] or the activity of the humain brain [6,7]. Several examples of this kind of study are presented in this volume (see contributions by C.D. Jefferies and S. Grossmann).

In the present paper we extend the study to the various stages of the brain activity. For example during the awake-sleep cycle, the nature of EEG changes in a well-defined cyclic fashion. A resting and alert brain shows activity of an average frequency of about 10 cycles per second and amplitudes of the order of 10 microvolts called α waves (see Fig. 1a). During a normal night's sleep α waves give way to other repeated cycles of activity each marked by several stages.

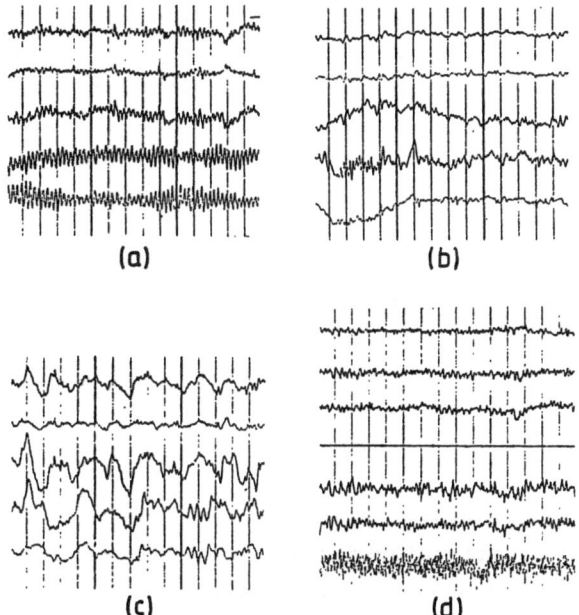

(a)

(b)

(c)

(d)

Fig.1 EEG recordings from human brain. (a) awake state. (b) sleep stage two. (c) sleep stage four. (d) REM state.

In stage one, the individual drifts in and out of sleep. In stage two (see Fig. 1b), the sleeper is disturbed by the slightest noise. In stage three a loud noise would be needed to rouse the sleeper. Finally, the deep sleep of stage four sets in. Afterwards the cycle is reversed back through stages three and two. After this stage the sleeper enters the phase of rapid eye movement sleep (REM) in which he dreams. This episode is followed by stage two and a new cycle begins. The sleep cycles continue through the night; however the periods of REM get longer and those of deep sleep shorter.

As sleep sets in the fast $\propto$ waves gradually give way to slower and higher amplitude δ waves. At deep sleep stage four the average frequency of the so called δ waves is of the order of 3-5 cycles per second and amplitudes are of several hundred microvolts (see Fig. 1c). During REM sleep intense bursts of high frequency activity appears (see Fig. 1d).

2. Phase Space Description

Let $X_0(t)$ be the set of values of the variable recorded during EEG. One shows that from such a unique time series and without reference to any particular model, a set of variables describing the dynamics of the system could be defined. These variables are obtained by shifting the original time series by a fixed lag τ ($\tau = m\Delta t$, where m is an integer and τ is the interval between successive samplings). These variables span a phase space which allows the drawing of the phase portrait of the system or more precisely its projection to a low dimensional subspace of the full phase space.

In the phase space, the instantaneous state of the system is characterized by a point, a sequence of such states followed in time defines the phase space trajectory. If the dynamics of the system is reducible to a set of deterministic laws, the system reaches in time a state of permanent regime. This fact is reflected by the convergence of families of phase trajectories towards a subset of the phase space. This invariant subset is called an attractor.

Presently we must answer to the following questions :

(i) It is possible to identify an attractor for a given EEG? In other words, can the salient features of neuronal activities be viewed as the manifestation of a deterministic dynamics (possibly very complex one) or, rather, do they contain an irreducible stochastic element?

(ii) Provided that an attractor exists, what is its dimensionality d? The latter provides us with valuable information on the system's dynamics. For instance, d = 1 describes self-sustained periodic oscillations; if d = 2 we are in the presence of quasi-periodic oscillations of two incommensurate frequencies; and if d is noninteger and larger than 2 (referred to as a fractal), the system exhibits a chaotic oscillation featuring a great sensitivity to initial conditions, as well as an intrinsic unpredictability.

(iii) What is the minimal dimensionality n, of the phase space within which the above attractor is embedded? This defines the minimum number of variables that must be considered in the description of the underlying dynamics. Note that d is necessarily smaller than n.

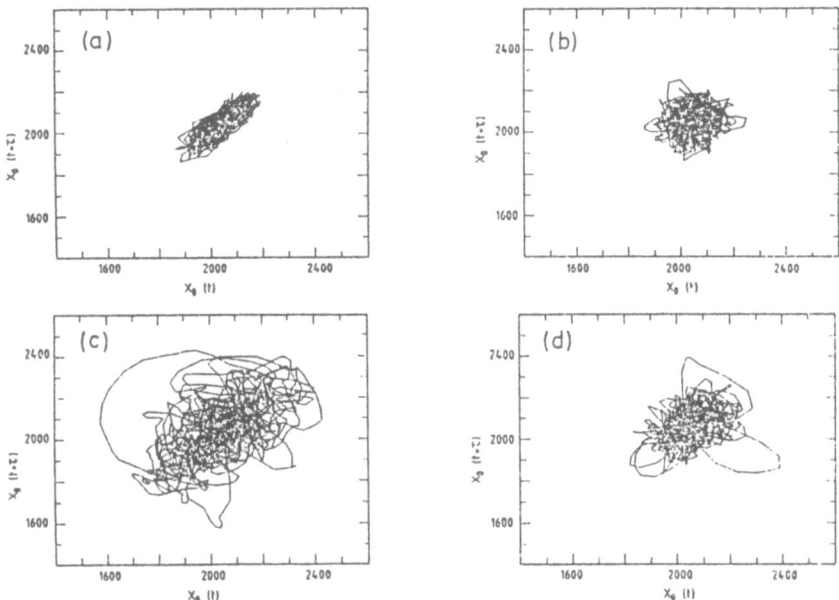

Fig.2 Two-dimensional phase portraits derived from the EEG of (a)
        an awake subject (b) sleep stage two (c) sleep stage four
        (d) REM sleep. The time series $X_0(t)$ is made of N=4000
        equidistant points. Central EEG derivation C4-A1 according
        to Jasper system. Recorded with PDP 11-44, 100 Hz for 40
        sec. The value of the shift from 2a to 2d is τ = 10 Δt.

Fig. 2a-2d depict the phase portraits corresponding respectively to
the awake state, sleep stage two, sleep stage four and REM activity of
the human brain. A total of six sleep episodes taken from different
individuals were analysed. The complexity of these graphs reflects the
complexity of the brain activity. However the most striking feature
is the evolution of this complexity as the sleep cycle unfolds.
The phase portrait of the awake subject is densely filled and occupies
a small portion of the phase space (Fig 2a). The representative point
undergoes deviations from some mean position in practically all directions. At the sleep stage two, already a tendency towards a privileged direction is seen and a larger portion of the phase space is visi-
ted (Fig 2b). This tendency is amplified in the sleep stage four and
one sees preferential pathways suggesting the existence of reproducible relationships between instantaneous values of the pertinent variables (Fig 2c). This phase portrait is the largest and exhibits a
maximum "coherence" which diminishes again when REM sleep sets in (Fig
2d). Our next task is to characterize this difference more sharply,
using the recent advances in the theory of dynamical systems. In

119

particular we must find out if the phase portraits of Figs. 2a-2d are the manifestation of deterministic dynamics or rather are the result of an irreducible stochastic element?

3. The EEG Attractor

We introduce a vector notation : $\vec{X}_i(t)$ stands for a point of phase space whose coordinates are $\{ X_0(t_i),\ldots\ldots,X_0(t_i+(n-1)\tau)\}$ [2]. A "reference" point $\vec{X}_i$ from these data is chosen and all its distances $|\vec{X}_i - \vec{X}_j|$ from the N-1 remaining points are computed. This allows us to count the data points that are within a prescribed distance r from the point $\vec{X}_i$ in phase space. Repeating the process for all values of i one arrives at the quantity.

$$C(r) = \frac{1}{N^2} \sum_{\substack{i,j=1 \\ i \neq j}}^{N} \theta(r- |\vec{X}_i - \vec{X}_j|) \tag{1}$$

where θ is the Heaviside function.
One shows that for small r

$$C(r) = r^d \tag{2}$$

The dimensionality d of the attractor is therefore given by the slope of log C(r) versus log r.
With the help of relation (2) a dimensionality d is computed by considering successively higher values of the embedding dimension n of the phase space. If the d versus n dependence is saturated beyond some relatively small n, the system represented by the time series should possess an attractor. The saturation value d is regarded as the dimensionality of the attractor represented by the time series. The values of n beyond which saturation is observed provides the minimum number of variables necessary to model the behavior represented by the attractor.
The above procedure has been applied to several sets of data corresponding to EEG recorded from various stages of human sleep cycle. A satisfactory saturation exists for sleep stage two. In this case for two individuals we find respectively d = 5.03 ± 0.07 and d = 4.99 ± 0.11. Saturation curves are also found for sleep stage four [6]. The analysis of EEG data of stage four of three individuals showed d = 4.05 ± 0.5, d = 4.08 ± 0.05 and d = 4.37 ± 0,1. Thus we may conclude that the dynamics underlying the EEG for sleep stage two and the deep sleep stage 4 activity of the brain possesses a deterministic attractor.

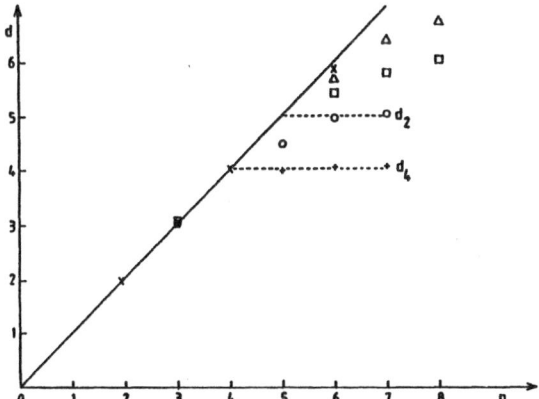

Fig. 3   Dependance of dimensionality d on the number of phase
space variables n for a white noise signal(x) the EEG
attractor of an awake subject (△), sleep-stage two (○)
sleep stage four (+) and REM sleep (□), for the same
number of data points as in Fig 2.

The saturation trend is to be contrasted with the behavior obtained
from a random process, such as a Gaussian white noise (Fig. 3, points
in crosses). In fact in this latter case d turns out to be equal to
the dimensionality of the embedding phase space, d = n. As regards
the EEG of the awake subject and REM sleep, our results do not yet
allow us to reach a clear cut conclusion. Generally speaking the
trends observed in this case are : higher attractor dimensionalities
and somewhat poor saturation of the d at least up to n = 7. This fact
may reflect two entirely different situations. Either the dynamics of
the system exhibits intrinsic stochastic behaviour or saturation
occurs for a larger number of variables. Presently we are extending
the analysis to much longer time series.
The existence of strange attractors in human sleep may be shown by
another method. If we are in the presence of chaotic motion, the
trajectories in the phase space have the tendency to diverge. One can
estimate the speed of this divergence. The average of these individual measures over a large number of data is defined as a Lyapounov
exponent. A positive Lyapounov exponent indicates unambiguously the
presence of a chaotic attractor.
Recently algorithms have been developed which allow for the evaluation
of the largest positive Lyapounov exponent from the knowledge of a
time series describing a dynamical system. Using the Fortran code
described by WOLF et al [8] we have evaluated the largest positive
Lyapounov exponents for stage two and stage four of deep sleep. For

stage two we find a positive value of λ_2 between 0.4 and 0.8. The inverse of this quantity gives the limits of predictability of the long-term behavior of the system. For stage four we find also a positive number $0.3 < \lambda_4 < 0.6$.

4. Conclusions

We have shown that the EEG of one of two characteristic stages of the brain activity, namely the sleep stage 2 and 4, may be viewed as a deterministic dynamical system involving a limited set of variables. The fact that the underlying attractor is a fractal, implies that the dynamics is a complex one. This property should be related to the very ability of the brain to generate and process information. Indeed, the instability inherent in the chaotic dynamics of the deep sleep system suggests that in the awake state sensorial inputs or internally generated excitations can be amplified through this very instability. We believe that this possiblility may be related to one of the main features of cognitive activity, namely the possibility of exploration and innovation.

REFERENCES

1. L.K. KACZMAREK and A. BABLOYANTZ, Biol. Cyber., 26, 199-208, (1977).

2. A. RECHTSHAFFEN and A. KALES, A manual standardized terminology, techniques and scoring system from sleep stages of human subject. (UCLA 1968).

3. F. TAKENS, in Lecture Notes in Mathematics, 898 (eds. Rand, D.A. and Young L.S.) (Springer, Berlin, 1981).

4. P. GRASSBERGER and I. PROCACCIA, I. Physica, 9D, 189-208 (1983).

5. C. NICOLIS and G. NICOLIS, Nature, 311, 529-532 (1984).

6. A. BABLOYANTZ and C. NICOLIS, Submitted to J. Theor. Biol.

7. A. BABLOYANTZ, J.M. SALAZAR and C. NICOLIS, Submitted to Phys. Lett..

8. A. WOLF, J.B. SWIFT, H.L. SWINNEY and J.A. VASTANO, In press Physica D (1985).

Part III

Coordination of Motion

Cooperative Phenomena in Biological Motion*

J.A.S. Kelso and J.P. Scholz

Haskins Laboratories, New Haven, CT 06511, USA

1 Introduction

The production of a "simple" utterance, such as the
syllable /ba/, involves the cooperation of a large number
of neuromuscular elements operating on different time
scales, e.g., at respiratory, laryngeal, and supralaryngeal
levels. Yet somehow, from this huge dimensionality, /ba/
emerges as a coherent and well-formed pattern. Similarly,
were one to count the neurons, muscles, and joints that cooperate to produce the "simple" act of walking, literally
thousands of degrees of freedom would be involved. Yet
again, somehow walking emerges as a fundamentally low-dimensional cyclical pattern--in the language of dynamical
systems, a periodic attractor. In physics, an infinite dimensional system, described by a complicated set of partial, nonlinear differential equations can be reduced--when
probed experimentally or analyzed theoretically--to a
low-dimensional description [1,2]. In all these cases, it
seems, information about the system is compressed--from a
microscopic basis of huge dimensionality--to a macroscopic
basis of low dimensionality.

Our particular interest is how such compression occurs
in the multidegree of freedom actions of people and
animals. How does an internally complex system "simulate"
a simpler, lower dimensional system? As we shall see, an
important feature of our efforts to understand the control
and coordination of movement is the concept of order parameter [3,4, see also 5]. Order parameters define the col-

*Research supported by ONR Contract N00014-83-C-0083,
NINCDS Grant NS-13617, and BRSG Grant RR-05596. We are
very grateful to Professor H. Haken for his continued interest and encouraging comments on this work, as well as
his coworkers in Stuttgart (particularly G. Schöner and
H. Bunz) for their helpful, informed discussions.

lective behavior of the system's many components in terms of its essential variables alone; they are few in number even in very complicated physical and chemical systems. Note how the emphasis on discovering order parameters takes us away from a focus on individual elements (regardless of the level at which these elements are described): Just as the motion of a single molecule is not relevant to the essential description of the behavior of a gas, so too, one suspects, the action of a single reflex is not relevant to the essential description of an organism's behavior.

Our focus here is on the spatiotemporal patterns formed by the ensemble activity of neurons, muscles, and joints during the performance of a coordinated act. As WEISSKOPF [6] emphasizes in a different context, such problems rest with defining relations between different aggregates of atoms or molecules, and of the modes of transition from one structure to another. The abstraction of a system's order parameters is thus of paramount importance, because it allows one to separate the essential from the non-essential, thereby enabling a complex phenomenon to become more transparent. This "macroscopic" strategy is brought to bear here on our efforts to discover the principles underlying the control and coordination of movements. In the following we first summarize briefly evidence for the existence of unitary processes in complex actions and denote some of the characteristic properties of such units. From such analysis, the phase relation among the motions of skeletomuscular components will emerge as a candidate order parameter. We then contrast various theoretical notions about pattern generation in movement and introduce some recent evidence in favor of a synergetic approach. Synergetics motivates the treatment of complicated biological motion as fundamentally a cooperative phenomenon. In support of this view, certain kinds of activities will be shown to display the features of a nonequilibrium phase transition.

2 A Unitary Process (Coordinative Structure)

For the Soviet physiologist BERNSTEIN [7], the existence of a large number of potential degrees of freedom in the motor system precluded the possibility that each was controlled individually at every point in time. Rather, he hypothesized that the central nervous system (CNS) "collects" multiple degrees of freedom into functional units that then behave, from the perspective of control, as a single degree of freedom. During a movement, the internal degrees of freedom are not controlled directly, but constrained to relate among themselves in a relatively fixed and autonomous fashion. But is it, in fact, the case that in coordinated actions, the many neuromuscular components actually function as a single degree of freedom?

Support for the hypothesis that a group of relatively independent muscles and joints forms a single functional unit would be obtained if it were shown that a challenge or perturbation to one or more members of the group was, during the course of activity, responded to by other remote (non-mechanically linked) members of the group. We have recently found that speech articulators (lips, tongue, jaw) produce functionally specific, near-immediate compensation to unexpected perturbation, on the first occurrence, at sites remote from the locus of perturbation [8]. The responses observed were specific to the actual speech act being performed: for example, when the jaw was suddenly perturbed while saying the syllable /bæb/, the lips compensated so as to produce the final /b/, but no compensation was seen in the tongue. Conversely, the same perturbation applied during the utterance /bæz/ evoked rapid and increased tongue muscle activity (so that the appropriate tongue-palate configuration for a fricative sound was achieved) but no active lip compensation.

Recent work has also varied the phase of the jaw perturbation during bilabial consonant production. Remote reactions in the upper lip were observed only when the jaw was perturbed during the closing phase of the motion, that is, when the reactions were necessary to preserve the identity of the spoken utterance. Thus the form of cooperation observed is not rigid or "hard wired": the unitary process is flexibly assembled to perform specific functions (for additional evidence in other activities, see [8]). Elsewhere we have drawn parallels between these findings and brain function in general [5]. Just as groups of cells, not single cells are the main units of selection in higher brain function [9], so too task-specific ensembles of neuromuscular elements appear to be the significant units of control and coordination of action.

Stunning evidence attesting to this self-organizational style of neural and behavioral function comes from recent microelectrode studies of somatosensory cortex in adult squirrel and owl monkeys by MERZENICH and colleagues (see [10] for review): when the middle finger of the monkey's hand was surgically removed, brain regions representing the other adjacent fingers progressively shifted (over the course of a few weeks) into the missing finger's hitherto exclusive brain region. Also, if a portion of cerebral cortex was injured, the appropriate somatosensory "map" moved to the region surrounding it--a spatial shift of nerve cell activity as it were. These data challenge a view of neural functioning that is determined by "hard-wired" or "fixed" anatomic connections established before or shortly after birth. Just as we have observed rapid "soft" forms of compensation in speech production, so it seems, the brain has a functionally fluid, self-organizing character that allows longer-term compensation for injury.

3 Characteristic Properties Of A Unitary Process

A main way to uncover the intrinsic properties of a functional unit of action is to transform the unit as a whole (e.g., by scaling on movement rate; amplitude etc.) and search for what remains invariant across transformation. The discovery of such "relational invariants" (e.g., [11]) could provide a useful step toward explicating the design logic of the motor system.

Much evidence now exists from a wide variety of movement activities that relative timing among muscles and kinematic components is preserved across scalar changes in force or rate of production. For example, when a cat's speed of locomotion increases, the duration of the "step cycle" decreases [12,13] and an increase in activity is evident in the extensor muscles during the end of the support phase of the individual limb. Notably, this increase in muscle activity (and corresponding development of propulsive force) does not alter the relative timing among functionally linked extensor muscles, although the duration of their activity may change markedly (see [12,13] for reviews).

Interestingly, there is some limited evidence that this style of organization applies also to speech production. What makes a word a word in spite of differences among speakers, dialects, intonation patterns and so on? Our view is that the key to this question lies in understanding how the coordinated movements of the vocal tract articulators structure sound for a listener. According to this view, the invariance which allows us to perceive the sounds of a language in so many different contexts exists in the functionally-defined behavior of the articulatory system. But how is such behavior to be described? It is well-known, for instance, that the same word has markedly different kinematic, electromyographic, and acoustic attributes when produced in different contexts. A solution to this dilemma may lie in the finding by TULLER, KELSO, and HARRIS [15] that the relative timing of activity in various articulatory muscles is preserved across the very substantial metrical changes in duration and amplitude of muscle activity that occur when a speaker varies his/her speaking rate and stress pattern (for evidence in other motor skills see [14]). An important extension of these earlier EMG findings is the discovery that the relative timing of articulator movements is stable across different speaking rate and stress patterns. Presently, these results apply to the cooperative relations among lips, tongue, jaw and larynx (see [16] for review).

How is the relative timing invariant to be rationalized? A popular view is that time is metered out by a central motor program (see below) which instructs the articulators when to move, how far to move and for how long. A reconceptualization and consequent reanalysis of the TULLER and KELSO [16] data, however, strongly suggests that time, per se, is not directly controlled. Using phase plane techniques to represent the motions geometrically, we have shown that critical phase angles--relating one articulator's position-velocity $(x, \dot{x})$ state to another--appear to be most crucial for orchestrating the coordination among articulators [17,18]. The beauty of this gestural phase analysis (which is autonomous and does not require an explicit representation of time) is that it provides a topological description of articulatory behavior that remains unaltered across manifold speaker characteristics. Moreover, critical phase angles are revealed by the flow of the dynamics of the system, not externally defined. Thus, they can serve as natural sources of information for guaranteeing the stability of coordination in the face of scalar (metrical) change (for more details, see [18]).

Finally, there is a strong hint that phase constancy reflects an evolutionary design principle. From the invertebrates, in which many groups employ large numbers of propulsive structures (limbs, tube feet, or cilia) for swimming and locomotion, to the vertebrates which walk, run or jump using one, two, three or four pairs of legs, the same design property is apparent, viz. all of these creatures possess processes that communicate information about the phase of activity among component structures [19,20]. Below we will develop in more detail the notion that phase is an essential parameter of complex, coordinated action. Suffice to emphasize at this point that a phase constancy indicates a functional constraint on movement, what we call a coordinative structure or unit of action (cf. [21,22,23,24]). Thus, during an activity the spatiotemporal behavior of individual components is constrained within a particular relationship. Flexibility can then be attained by adjusting control parameters over the entire unit.

4 Theories of Pattern Generation

The core idea expressed in Sections 2 and 3 above--that a system possessing a large number of potential degrees of freedom is compressed into a single functional unit of action (or coordinative structure) that requires few control decisions is unorthodox. It differs in significant ways from more conventional treatments of movement which are based on either the information processing notion of a motor program or the neurally-based notion of a central pattern generator. The motor program, by definition, is an internal representation of a movement pattern that is prestructured in advance of the movement itself. Analogous with a computer program, it constitutes a prescribed set of instructions to the skeletomuscular system. In MACKAY's [25] analysis of a dynamic activity, the locomotory step cycle, the many kinematic details are ordered a priori by a sequence of commands/instructions to the skeletomuscular apparatus whose role is to implement these instructions. The format of the program is that of a formal machine; symbol strings are employed to achieve (or explain) the order and regularity of the step cycle. As in most programming accounts, the control prescription is highly detailed and the role dynamics play in fashioning the pattern is ignored. So also is the interface between the small-scale "informational" contents of the program and the large-scale, energetic requirements of the muscle- joint system. Finally, the contents of the program are not rationalized: a principled basis for selecting desired quantities (e.g., apply flexion torque for 100 ms) is omitted.

The neural counterpart of the motor program is the central pattern generator (CPG). Here too, the order and regularity observed in the world is attributed to a device inside the CNS (a neural circuit) that when activated coordinates the different muscles to produce movement [26]. Though subject to feedback influences, the circuit is "hard-wired" and the goal is to locate the neurons that constitute the network and to define their properties and interrelations. Though an admirable enterprise, there are questions about its propriety. For example, the parameter space of a CPG, e.g., the membrane properties of its elements, synaptic connections, etc. has been variously estimated to be 46 or 55 (compare [27] to [28]; also [29]). Presumably not all of these are necessary to understand a CPG, but principles beyond those of neurophysiology are surely needed to guide the selection of relevant parameters in such a high-dimensional space. As LOEB and MARKS [30] emphasize, principles of operation constitute the knowledge for understanding a CPG and these are disembodied from the actual device (or its model). In addition, even if all the details of a putative CPG were known, the problem of relating the known microproperties to characteristic macroproperties such as the amplitude, phase and frequency of a wing beat or a step cycle would still remain.

The question then is this: where do the necessary principles come from? For some years now, we have advocated an approach in which problems of biological motion are treated in a manner continuous with cooperative phenomena in other physical, chemical, and biological systems, i.e., as synergetic or dissipative structures [5,31,32]. Common features of the latter are that--like movement--they consist of very many subsystems. Unlike the theoretical approaches discussed above, however, where the emphasis is on detailed prescriptions for control, in synergetics, when certain conditions (so-called "controls") are scaled up even in very non-specific ways, the system can develop new kinds of spatiotemporal patterns. The latter are maintained in a dynamic way by a continuous flux of energy (or matter) through the system [4]. Although there is pattern formation in the nonequilibrium phenomena treated by synergetics, e.g., the hexagonal forms produced in the Bénard convection instability, the transition from incoherent to coherent light waves in the laser, the oscillating waves of the Belousov-Zhabotinsky chemical reaction, etc., there are strictly speaking no pattern <u>generators</u>. That is, the emphasis is on the lawful basis, including the necessary and sufficient conditions, for pattern <u>formation</u> to occur. The explanation is derived from first principles: it <u>never</u> takes the form of introducing a special mechanism--like a motor program--that contains or represents the pattern before it appears.

5 Phase Transitions in Biological Motion

There are already strong hints in the motor system's literature that a highly detailed prescription from higher neural centers is not necessary to produce either a stable spatiotemporal pattern (say among the legs of a locomoting animal) or an abrupt change in ordering among the legs, as in locomotory gait changes. An early indication comes from remarkable experiments by VON HOLST [19] on the centipede <u>Lithobius</u>. By amputating leg pairs until only three such pairs were left, von Holst transformed the centipede's gait (a pattern in which adjacent legs are about one-seventh out of phase) into that of a six-legged insect. Further, when all but two pairs of legs were left, the asymmetric gaits of the quadruped were exhibited. It is hard to imagine that the nervous system of the centipede possessed stored programs or pattern generators for these gaits in anticipation of its legs being amputated by an innovative experimenter. Rather, given a novel configuration, the system appears to adopt spontaneously those modes of locomotion that are dynamically stable. Synergetics attempts to predict exactly which new (or different) modes will evolve in complex systems particularly when the system undergoes qualitative macroscopic changes [4].

More direct evidence that rather diffuse inputs ("controls") can lead to highly ordered behavior comes from Russian studies on (decerebrate) locomoting cats [33]. A steady increase in midbrain electrical stimulation was sufficient not only to induce changes in walking velocity, but also--at a critical stimulation level--abrupt gait changes as well. Interestingly, unstable regions were also noted in which the cat vacillated between trotting and galloping.

A final clue suggesting that gait transitions belong to the class of nonequilibrium phase transitions comes from

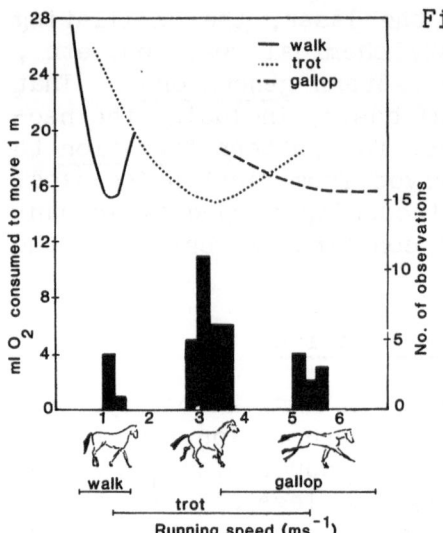

Fig. 1. Oxygen consumption and preferred speed of walk, trot and gallop of locomoting horses (See text for details). From HOYT and TAYLOR [34.]

work on the energetics of horse locomotion. It is well-known that animals use a restricted range of speeds (within a given gait) which corresponds to minimum energy expenditure. HOYT and TAYLOR [34], however, forced ponies to locomote away from these "equilibrium states" (see Fig. 1) by increasing the speed of a treadmill on which the ponies walked. As shown in Fig. 1, it becomes metabolically costly for the animal to maintain a given locomotory mode as velocity is scaled: For example, the walking mode becomes unstable, as it were, and "breaks" into a trotting mode (the next local minimum). Likewise, it is energetically expensive to maintain a trotting mode at slow velocities, a fact that appears to require switching into the walking mode (although no data on hysteresis are given). As in many other systems treated by synergetics, when a critical value is reached, the system bifurcates and a new (or different) spatiotemporal ordering emerges. Note that in Fig. 1 these locomotory mode changes are not necessarily hard-wired or deterministic. Horses can trot at speeds at which they normally gallop, but it is metabolically costly to do so.

The notion that gait shifts correspond to instabilities that arise as the system is pushed away from equilibrium would be greatly enhanced if qualitatively similar phenomena were observed in other types of activities--perhaps even of a less stereotypical "innate" kind than locomotion. The remainder of this paper will be devoted to the elaboration of a phase transition that occurs in voluntary cyclical movements of the hands [11,35]. Below we will describe the phenomenon and illustrate briefly how it has been modeled using concepts of synergetics and the mathematical tools of nonlinear oscillator theory [36]. Finally, we will show that the phenomenon contains some of the principal features of other nonequilibrium phase transitions in nature. Interestingly, this synergetic account not only handles a variety of phenomena typically described by motor programs/CPG accounts, but generates new predictions that have not come to light from either of these theories.

6 Nonequilibrium Phase Transitions In Bimanual Action

6.1 The Basic Phenomenon [5,11,35]

Consider an experiment in which a human subject is asked to cycle his/her fingers or hands at a preferred frequency using an out-of-phase, antisymmetrical motion. Under instructions to increase cycling rate, it was observed that at a critical frequency the movements shifted abruptly to an in-phase, symmetrical mode involving simultaneous activation of homologous muscle groups. When the transition frequency was expressed in units of preferred frequency, the resulting dimensionless ratio or critical value was constant for all subjects, but one (who was not naive and who purposely resisted the transition--although with certain energetic consequences, see [35]). A frictional resistance to movement lowered both preferred and transition frequencies, but did not change the critical ratio (~1.33). As an interesting aside, the ratio of transition speed to preferred speed for walk-trot and trot-gallop gait shifts, shown in Fig. 1, also gives a value ~1.32. This dimensionless number (analogous, perhaps to a Reynolds' number in hydrodynamics) may provide a rough estimate of "distance from equilibrium."

In summary, the main features of the bimanual experiments are: a) the presence of only two stable phase (or "attractor") states between the hands (see also [23,37], for further evidence); b) an abrupt transition from one attractor state to the other at a critical, intrinsically defined frequency; c) beyond the transition, only one mode (the symmetrical one) is observed; and d) when the driving frequency is reduced, the system does not return to its initially prepared state, i.e., it remains in the basin of attraction for the symmetrical mode.

6.2 Modeling [36]

In complex systems it is clearly hopeless to try to investigate the motion of each microscopic degree of freedom. Rather the challenge is to identify and then lawfully relate singular macroscopic quantities to the interactions among very many sub-components. Close to instability points, it can be shown that the the behavior of the whole system is determined by one or a few order parameters [3]. Such order parameters are not only created by the cooperation among the individual components of a complex system (e.g., by the interactions among atomic spins in a magnet), but in turn govern the behavior of those components (e.g., the magnetic field is an order parameter for a ferromagnet).

Identifying order parameters, even for physical and chemical systems, is not a trivial manner. Certain guidelines exist, however, which can be used for the selection of viable candidates. Two such selection criteria are: 1) the order parameter, by definition, changes much more slowly than the subsystems, i.e., its time constants are much longer than the time constants of the components; and 2) the order parameter's long-term behavior changes qualitatively at the critical point.

In the case of our bimanual experiments and, we suspect, many other kinds of biological motion also, relative phase, ϕ, meets these criteria quite well (cf. Section 3.0). Using relative phase as an order parameter, HAKEN, KELSO, and BUNZ [36] modeled the bimanual data by specifying a potential function, V (corresponding to the layout of attractor states defined above) and showed how that function was deformed as a control parameter (corresponding to driving frequency) was changed. The choice of V--a superposition of two cosine functions--represented the simplest form that could describe the pattern of results. The series of potential fields generated for varying values of b/a (the ratio of the cosine coefficients) is shown in Fig. 2.

It can be seen that at a critical value, ω_c, the system jumps into a local minimum, i.e., there is a transition from the anti-phase mode ($\phi = -\pi$) into the symmetric, in-phase mode, $\phi = 0$. Moreover, the system stays in that minimum even where the driving frequency is reduced below ω_c, thus exhibiting hysteresis.

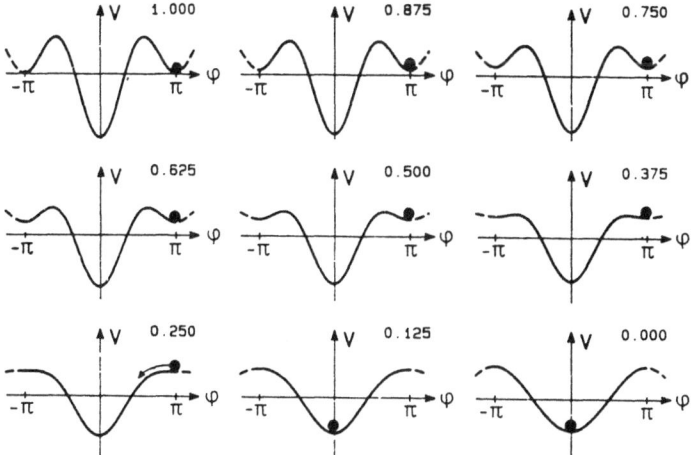

Fig. 2. The potential V/a for the varying values of b/a. The numbers refer to the ratio b/a (from [36]).

In a following analysis, HAKEN et al. [36] used nonlinear oscillator theory to show how the model equations for the potential function could be derived from equations of motion for the two hands and a nonlinear coupling between them. Since the details are published we simply illustrate briefly some recent results of a consequent computer simulation (see also [36], Figs. 6 and 7).

In Fig. 3, Lissajous portraits of the coupled oscillators are shown. The equations describing the motion are

$$\ddot{x}_1 + (\dot{x}_1^2 - 1)\dot{x}_1 + kx_1 = \alpha(\dot{x}_1 - \dot{x}_2) + \beta(\dot{x}_1 - \dot{x}_2)(x_1 - x_2)^2 + F_{noise} \tag{1}$$

$$\ddot{x}_2 + (\dot{x}_2^2 - 1)\dot{x}_2 + kx_2 = \alpha(\dot{x}_2 - \dot{x}_1) + \beta(\dot{x}_2 - \dot{x}_1)(x_2 - x_1)^2 + F_{noise} \tag{2}$$

In (1) and (2) above, the LHS corresponds to a Rayleigh-type, nonlinear oscillator (Equation 3.6 of [36]); the RHS is a Van der Pol coupling term plus some noise to simulate fluctuating forces (Equation 3.25 of [36]). The only difference between the two simulations lies in the magnitude of fluctuations. Indeed, the transition shown in Fig. 3(b) is remarkably like the behavior we observe typically (see e.g., [5]). Though we have not made a full study of the effects of initial conditions, coupling parameters and fluctuations, our impression is that--given sufficient coupling strength--fluctuations play a major role. Suffice to note at this point that the model cap-

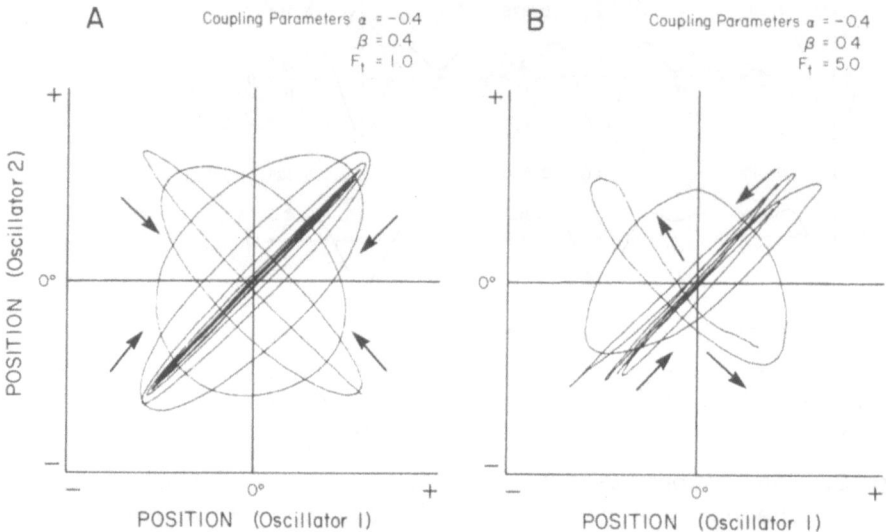

Fig. 3. Lissajous portrait of behavior of two coupled
Rayleigh oscillators (see text for details). Intrinsic frequency continuously scaled. Initial
conditions of simulations: x_1 = 25°, x_2 = -25°, $\dot{x}_1$
= $\dot{x}_2$ = 0. A and B differ only in level of noise
component (Bruce Kay performed the simulations).

tures not only observed decreases in hand movement amplitudes as ω is increased, but also the abrupt change in
qualitative behavior from antisymmetric to symmetric modes.

6.3   Theoretical Underpinnings

If the bimanual phase transition constitutes a critical
instability far from equilibrium then certain specific
predictions can be generated regarding the system's behavior near the transition. In particular, the hypothesized
order parameter (relative phase) should exhibit at least
two major properties: 1) critical slowing down as the
transition is approached, i.e., the relaxation time of the
order parameter to any perturbation should diverge at the
transition. In general, the system exhibits a symmetry
breaking instability, i.e., a constraint arises during the
transition that restricts the future configuration of the
system; and 2) enhanced fluctuations of the order parameter
in space and time near the transition. The data presented
next represent a preliminary attempt to explore the degree
to which these theoretical predictions may or may not apply
to phase transitions in hand movements.

6.4 New Experiments

Two kinds of experiment were performed. In each, subjects were seated comfortably with pronated forearms, supported up to the metacarpal heads of the hand. The forearm was stabilized to restrict movement to the fingers alone. On each trial, the subject oscillated the index finger bilaterally in the transverse plane (i.e., abduction-adduction). Continuous finger displacement in the transverse and parasagittal (i.e., flexion-extension) planes was measured using a modified Selspot camera system. The electromyographic (EMG) activity of the right and left first dorsal interosseous (FDI) muscle was obtained with platinum fine-wire electrodes (see Fig. 4). All data were recorded on 12-channel FM-magnetic tape recorder for later off-line computer analysis.

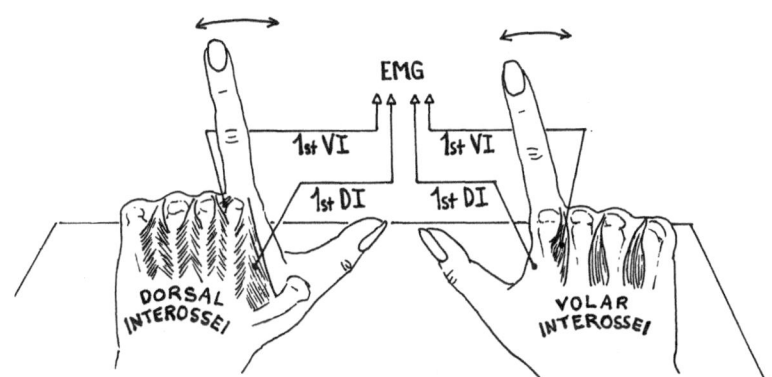

Fig. 4. General experiment set-up for recording EMG. Support splints not shown (drawing by C. Carello).

Initially, subjects were instructed to move in one of two ways: oscillation of the right (R) and left (L) index fingers in either 1) the symmetrical mode or 2) the antisymmetrical mode, at their preferred rate. The frequency of oscillation was gradually increased to a maximum of approximately 3.5 Hz. In Experiment 1, the frequency of oscillation was increased every 2-3 sec by asking the subject to slightly increase his rate. Thus, the rate of increase was not strictly controlled. In Experiment 2, the frequency of oscillation was systematically increased in 0.25 Hz steps every 4 seconds paced by a metronome. Data from trials in this experiment could therefore be averaged

in time. Averages for Experiment 1 required alignment of
trials by similar frequencies of oscillation. However, despite the lack of exact frequency equivalence, results from
the two experiments are surprisingly consistent.

6.5 Order Parameter Behavior

6.5.1 Critical Slowing Down

The time series of one trial of finger oscillation, when
the system is prepared initially in the antisymmetrical
mode, is depicted in Fig. 5a (note: the figure shows only
a portion of the trial in the vicinity of the phase transition). Here, one can clearly see the transition to the
symmetrical mode with an increase in the frequency of
oscillation. In Fig. 5b a point estimate of relative phase
for the same sample record, based upon the peak displacement of the R and L fingers, is shown. A slow oscillation
in phase, particularly before the transition, is evident.
As the transition is approached the frequency of this phase
oscillation slows; the system takes longer and longer to
return to its stationary state from a small deviation.

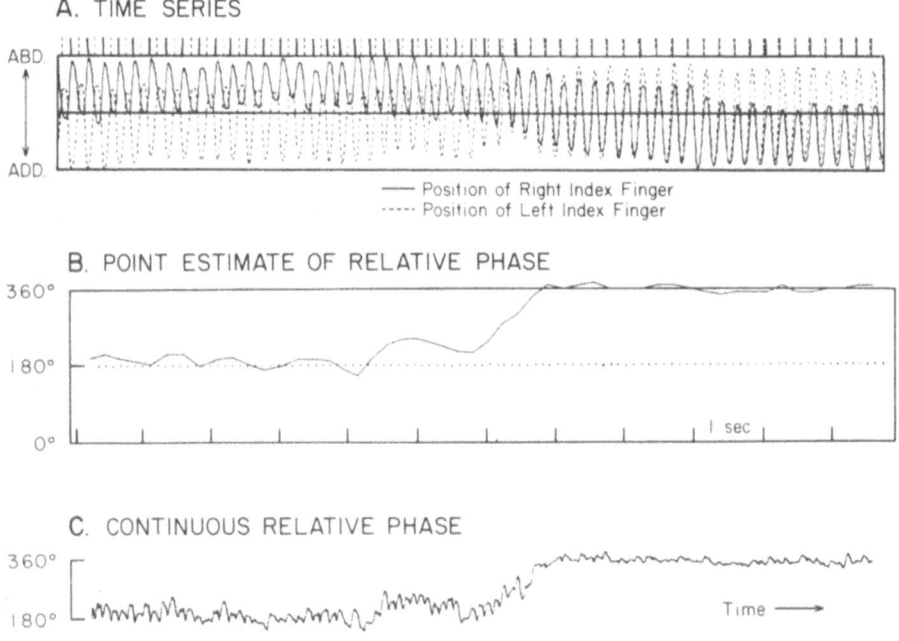

A. TIME SERIES

——— Position of Right Index Finger
····· Position of Left Index Finger

B. POINT ESTIMATE OF RELATIVE PHASE

C. CONTINUOUS RELATIVE PHASE

Fig. 5. Time series (A) and relative phase (B & C) of R and
L finger oscillation (see text for details).

This finding is a consistent feature of the experiments and is taken as preliminary evidence for the phenomenon of critical slowing down. Future work will calculate the relaxation time of the hypothesized order parameter explicitly using correlation techniques and perturbation experiments.

A continuous estimate of relative phase may be found in Fig. 5c, based upon the continuous phase angle difference between each oscillator. Note that this estimate reveals some of the microscopic details of the phase fluctuations, while preserving the slow modulations in phase described above. A clear reduction in these fluctuations occurs following the transition. All remaining data on relative phase to be reported are based upon this continuous estimate.

6.5.2   Enhancement of Fluctuations

An important feature of critical phenomena is the increase in variance of the order parameter near the phase transition. The system is said to become "soft" and thus unable to suppress critical fluctuations. The variance of the order parameter in the finger experiment is presented in Fig. 6. The SD of continuous phase was calculated in the stable regime with the transient removed, i.e., over the last 3 seconds (= 600 data points) of oscillation at each frequency. Each point on the graph represents an average of 10 trials from Experiment 2. Mean phase is presented as well.

Consideration of trials in which the system was initially prepared in the antisymmetrical mode reveals a clear increase in relative phase fluctuations as the transition is approached. The phase variance maximum at the transition is somewhat artifactual, since the phasing must change in order for a new mode to be exhibited. Note also that after the transition the variance eventually stabilizes at a lower level (corresponding to the symmetrical mode) than before the transition. So-called control trials in which the system is initially prepared in the symmetrical mode, exhibit no such fluctuational amplification with increasing driving frequency. These findings are therefore consistent with theoretical predictions and the results of the nonlinear oscillator modeling shown earlier.

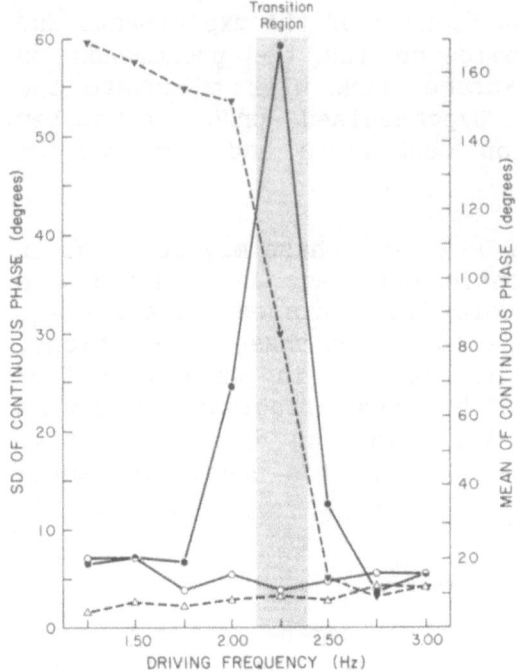

Fig. 6. Mean (▼ AMS, △ SMS) and standard deviation (● AMS,
○ SMS) of continuous relative phase at each driving frequency (n=10). AMS = antisymmetrical mode
scaled. SMS = symmetrical mode scaled.

Order parameter dynamics can be further explored by examining the spectral content of relative phase. Each sample
record of continuous relative phase was divided into eight
segments corresponding to the increments in driving frequency. The power spectral density function (PSDF) of each
segment was then determined by Fast Fourier Transform.
Average PSDFs were obtained for trials in which subjects
were initially prepared in the antisymmetrical mode, as
well as those prepared in the symmetrical mode. The results are displayed in Fig. 7. The DC component has been
removed from each plot, since it represents the mean phase
value, and overwhelms the other components, particularly in
the anti-phase mode.

Fig. 7a displays the average PSDF for trials initially
prepared in the antisymmetrical mode. Note that as the
driving frequency (ω) increases, a gradual increase in the
frequency of the dominant spectral peak occurs. This increase appears to represent, in part, the influence of the

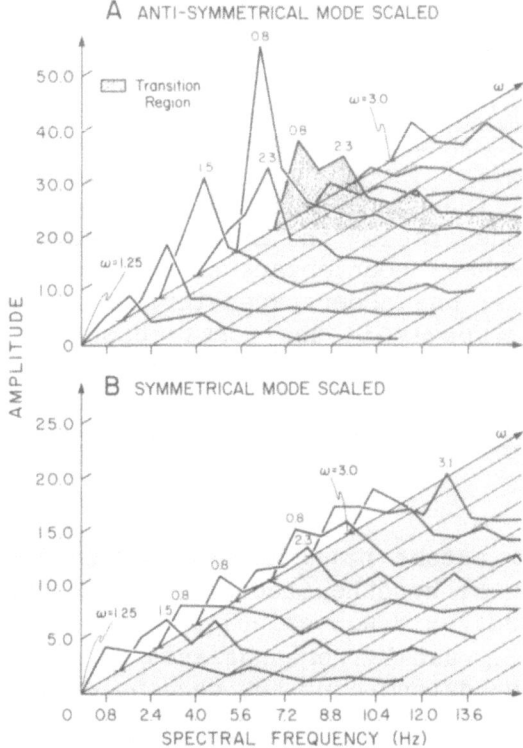

Fig. 7. Average PSDF of continuous measure of relative
phase computed at each driving frequency (ω) for
trials prepared in A. antisymmetrical and
B. symmetrical modes.

driving frequency.   Just prior to the transition, at 2.25
Hz, a dramatic increase occurs in the amplitude of the lowest frequency band, 0.8 Hz, along with the disappearance of
higher frequency components. The stippled PSDF represents
the transition region alone and reveals spectral broadening. With further increases in driving frequency the spectrum remains relatively broad and 0.8 Hz remains as a
strong harmonic.

The average PSDF of trials initially prepared in the
symmetrical mode is shown in Fig. 7b. While higher spectral components are present as the driving frequency is increased, the 0.8 Hz component is always strong, even at low
driving frequencies. Driving frequency appears to have
relatively less effect on the PSDF of the symmetrical mode
than that of the antisymmetrical mode.   The dramatic in-

crease in the amplitude of the 0.8 Hz component in the antisymmetrical mode just prior to the phase transition may represent the "swamping" of this mode's energy by that of the more stable symmetrical mode. That is, the longest lasting mode--symmetrical, in-phase--appears prominently before the transition itself. Though this interpretation is speculative at present, there does seem to be evidence that the antisymmetrical mode "feels" the driving frequency move strongly than its in-phase counterpart condition. In the language of synergetics, the order parameter is "slaving" its components less strongly in the former case than the latter.

6.6 Exploring the Neuromuscular Basis of the Transition

6.6.1 The η Parameter

In order to determine the extent to which changes in EMG activity map onto those of the hypothesized order parameter already described, the parameter η was calculated. Fig. 8a shows how this was done. R_0 and L_0 were obtained for each cycle of a sample record by determining the percent of to-

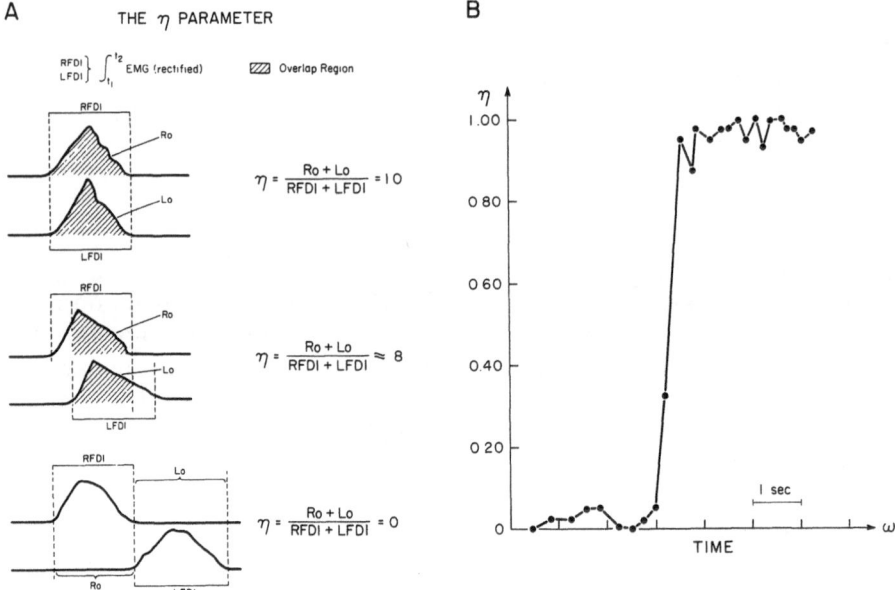

Fig. 8. The η parameter. A. Method of calculation from mean rectified, integrated EMG. B. Plot of η vs. time (and increasing oscillation frequency ω) for one representative trial.

tal mean rectified EMG of one FDI that overlapped in time with that of the contralateral FDI. Note that η is thus a sample estimate of the total energy of motor unit activity within a time interval defined by the phase between the fingers. It therefore constitutes a way of observing how the "microscopic" quantities relate to the macroscopic phasing parameter. A plot of η vs. time (and increasing frequency) for one representative trial is provided in Fig. 8b. The change in η maps quite nicely onto the change in the kinematic order parameter, as might well be expect-ed. The η parameter change appears to occur more abruptly as compared to the change in relative kinematic phase, how-ever.

6.6.2 EMG Autocorrelograms

One question concerns the nature of the neuromuscular reorganization underlying these phase transitions. In a preliminary attempt to examine this issue we looked at the autocorrelograms of mean rectified EMG for RFDI and LFDI, assuming they provide a measure of the temporal coherence of an individual muscle's activity. Two second segments of sample records prior to, during, and immediately following the transition were analyzed. The calculation of each sam-ple autocorrelogram was adjusted according to the oscilla-tion frequency of the fingers so that the same number of peaks occurred in each function. The mean value of the peaks in each function and their coefficient of variation were calculated as measures of temporal coherence. Both measures yielded similar results.

The mean peak autocorrelation of seven trials (Experi-ment 1) is presented in Fig. 9. The striking finding is the similarity between the coherence measures of the RFDI and LFDI before and after the transition, and their divergence at the transition. In the former two cases, even when the temporal coherence of one muscle is low, the contralateral FDI exhibits similar behavior. The correla-tion between the temporal coherence measures before and after the transition was above 0.90. This presumably in-dicates a tight coupling of their activity patterns, even when operating antisymmetrically. By contrast, one muscle always becomes more or less coherent in the transition re-gion. Here, correlation of the R and L coherence measure was low, negative and non-significant. Note also that the muscle showing the lowest coherence, and the direction of

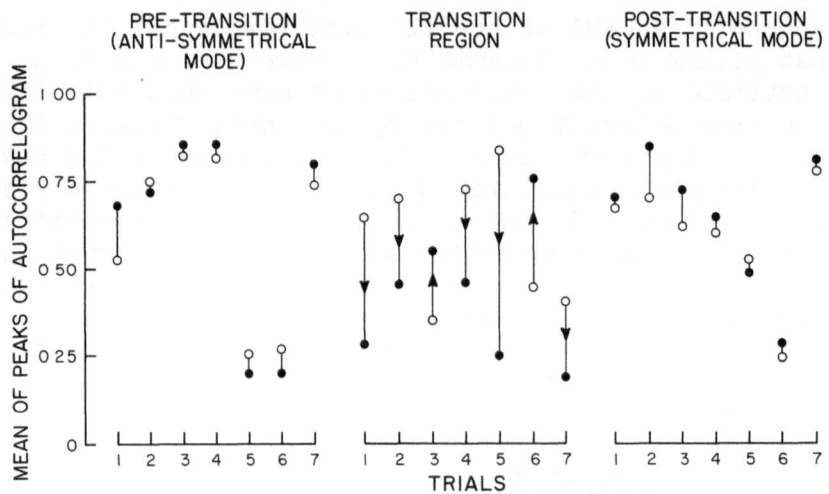

Fig. 9. Measure of temporal coherence of right FDI (●) and left FDI (O) 2 sec before, during, and 2 sec after phase transition (see text for details).

coherence change (compare with pre-transition measures) is never the same from trial to trial. Therefore, the underlying neurophysiologial mechanisms do not appear to be strictly deterministic as one might assume from a programming model of phase transitions.

6.7 Second Kinematic Phase Transition

As subjects move toward the upper extremes of oscillation frequency used in these experiments (~3.25-3.5 Hz) we have observed that a second instability occurs irrespective of the initial mode in which the subjects are prepared. In-phase modal behavior in the horizontal plane becomes unstable and gives way to a similar pattern in the vertical plane. A sample record of such an event is shown in Fig. 10 in which the displacement of each finger in both horizontal and vertical planes is plotted versus time (and, therefore, increasing oscillation frequency). Motion frequently becomes rotary in nature before simultaneous flexion-extension occurs. Further analysis, using comparable procedures to those described above, is underway.

Note that in this situation there is an additional degree of freedom available for energy dissipation. Thus a new (or different) configuration among the oscillatory com-

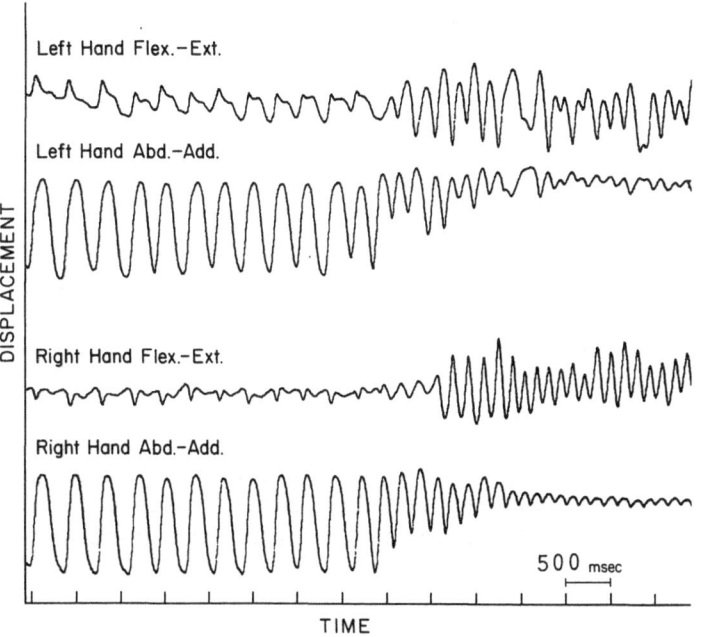

SECOND KINEMATIC PHASE TRANSITION

Left Hand Flex.-Ext.

Left Hand Abd.-Add.

DISPLACEMENT

Right Hand Flex.-Ext.

Right Hand Abd.-Add.

500 msec

TIME

Fig. 10. Time series of oscillation of R and L index finger in horizontal (abduction-adduction) and vertical (flexion-extension) planes for oscillation frequency above 3 Hz. See text for details.

ponents can occur--an additional basin of attraction appears spontaneously. The basis for this second transition is not altogether clear and requires further exploration. It may be determined, in large part, biomechanically, linked to the relaxation times of the participating muscles (i.e., FDI and first palmar interosseous, FPI). As the frequency of oscillation increases the relaxation times begin to exceed the 1/2 period of each cycle, resulting in maximum agonist-antagonist coactivity [38]. Energy can no longer be dissipated through motion in the transverse plane. However, because the experiment left open an additional degree of freedom, parasagittal motion, the system adopts this new configuration, apparently in order to dissipate the increasing energy. Both the FPI and FDI have lever arms which provide contribution to finger flexion. The extent to which the long finger flexors and extensors are also facilitated cannot be determined by the present data.

7 Concluding Remarks

Neuroscience has not looked seriously to contemporary physical theory for ways to think about brain-behavior relationships. And, with few notable exceptions (this conference being one, see also [39]), physics has made little contact with organic phenomena. Here we have shown in a very preliminary fashion--how some of the tools and concepts of nonequilibrium phase transitions may offer insight into the emergence of space-time order at a macroscopic level. In our simple experiments we have begun to identify some of the main features of non-equilibrium transitions, including symmetry breaking, critical slowing down, and enhancement of fluctuations. Further work--both theoretical and experimental--will be necessary to converge on these and other characteristics, e.g., identification of the system's time scales and especially measurement of mode relaxation times using correlation functions and perturbation techniques, classification of the stochastic nature of fluctuations, exploring the system's sensitivity to parameter change, etc.

The central thrust here, of course, is to understand coordination in the multi-degree-of-freedom motions of animals and organisms. Even if we knew all the microscopic details about the system's components we would still need a lawful description of how the components relate among themselves. An attraction of synergetics is that it deals with the formation of functional structures based on the cooperation among the system's many individual components. The theory achieves its full rigor when the system's behavior changes qualitatively, when newly emerging patterns are defined solely in terms of a few characteristic quantities, the so-called order parameters. A chief mechanism for the emergence of order lies in the competition between energy flowing into the operational components (i.e., a scaling influence) and the ability of those components to absorb the energy flow in their current configuration. As we have shown here (see e.g., Section 6.7) in the case of certain biological motions, higher bifurcations are possible if the system has available additional degrees of freedom, i.e., when a given configuration can no longer absorb the energy input. Moreover, fluctuations may permit the system's discovery of new modes or phasing structures.

If nature operates with ancient themes, as we suspect, then the same laws/strategies should appear at every level of description, and despite differences in material structure. Thus, the reductionism advocated here is not to any privileged scale of analysis, but rather to a minimum set of principles. The present treatment, preliminary though it is, may be just as pertinent to the mysteries of bacterial locomotion (see [40]) as it is to the coordinative patterns among the limbs and the abrupt transitions between them.

References

1. I. Procaccia, this volume.

2. R. Shaw: in Order and Chaos in Nature, ed. by H. Haken (Springer, Berlin, Heidelberg, New York 1981)

3. H. Haken: Rev. Mod. Phys. 47, 67 (1975)

4. H. Haken: Synergetics: An Introduction (Springer, Berlin, Heidelberg, New York 1983)

5. J.A.S. Kelso and B. Tuller: in Handbook of Cognitive Neuroscience, ed. by M. S. Gazzaniga (Plenum Press, New York 1984)

6. V. Weisskopf: Frontiers and Limits of Science (Alexander von Humboldt, Stiftung, Marz 1984)

7. N.A. Bernstein: The Coordination and Regulation of Movements (Pergamon Press, London 1967)

8. J.A.S. Kelso, B. Tuller, E. V.-Bateson and C.A. Fowler: J. Exp. Psych.: Hum. Perc. Perf. 10, 812 (1984)

9. G.M. Edelman, and V.B. Mountcastle: The Mindful Brain (MIT Press, Cambridge, MA 1978)

10. M.M. Merzenich and J.H. Kaas: Trends Neurosci. 5, 434 (1984)

11. J.A.S. Kelso: in Attention and Performance (IX), ed. by J. Long and A. Baddeley (Erlbaum, Hillsdale, NJ 1981)

12. S. Grillner: Physiol. Rev. 55, 247 (1975)

13. M.L. Shik and G.N. Orlovskii: Physiol. Rev. 56, 465 (1976)

14. D.C. Shapiro and R.A. Schmidt: in The Development of Movement Control and Coordination, ed. by J.A.S Kelso and J. Clark (Wiley, New York 1982)

15. B. Tuller, J.A.S. Kelso and K. Harris: J. Exp. Psych.: Hum. Perc. Perf. 8, 460 (1982)

16. B. Tuller and J.A.S. Kelso: J. Acoust. Soc. Am. 76, 1030 (1984)

17. J.A.S. Kelso and B. Tuller: J. Acoust. Soc. Am. Suppl. 1 77, S53 (1985)

18. J.A.S. Kelso and B. Tuller: in Sensory and Motor Processes in Language, ed. by E. Keller and M. Gopnik (Erlbaum, Hillsdale, NJ in press)

19. E. von Holst: in The Behavioral Physiology of Animals and Man: The Collected Papers of Erich von Holst (Univ. of Miami Press, Coral Gables, Fl 1937/1973)

20. M.A. Sleigh and D.I. Barlow: in Aspects of Animal Locomotion, ed. by H.Y. Elder and E.R. Trueman (Cambridge Univ. Press, Cambridge, London, New York, Sydney 1980)

21. T.A. Easton: Amer. Sci. 60, 591 (1972)

22. C.A. Fowler: Timing Control in Speech Production (Indiana University Linguistics Club, Bloomington, IN 1977)

23. J.A.S. Kelso, D.L. Southard and D. Goodman: Sci. 203, 1029 (1979)

24. M.T. Turvey: in Perceiving, Acting and Knowing: Toward an Ecological Psychology, ed. by R. Shaw and J. Bransford (Erlbaum, Hillsdale, NJ 1977)

25. W.A. MacKay: Trends Neurosci. 3, 97 (1980)

26. S. Grillner: Sci. 228, 143 (1985)

27. T.H. Bullock: in Simple Networks and Behavior, ed. by J.D. Fentress (Sinauer Assoc., Sunderland, MA 1976)

28. T.H. Bullock: in Information Processing in the Nervous System, ed. by H.M. Pinsker and W.D. Willis (Raven Press, New York 1980)

29. A.I. Selverston: Behav. Brain Sci. 3, 535 (1980)

30. G.E. Loeb and W.B. Marks: Behav. Brain Res. 3, 556 (1980)

31. J.A.S. Kelso, K.G. Holt, P.N. Kugler and M.T. Turvey: in Tutorials in Motor Behavior, ed. by G.E. Stelmach and J. Requin (North-Holland, Amsterdam, New York 1980)

32. P.N. Kugler, J.A.S. Kelso and M.T. Turvey: in Tutorials in Motor Behavior, ed. by G.E. Stelmach and J. Requin (North-Holland, Amsterdam, New York 1980)

33. M.L. Shik, F.V. Severin and G.N. Orlovskii: Biophys. 11, 1011 (1966)

34. D.F. Hoyt and C.R. Taylor: Nature 292, 239 (1981)

35. J.A.S. Kelso: Am. J. Physiol.: Reg. Integr. Comp. 246, R1000 (1984)

36. H. Haken, J.A.S. Kelso and H. Bunz: Biol. Cyber. 51, 347 (1985)

37. J. Yamanishi, M. Kawato and R. Suzuki: Biol. Cyber. 37, 219 (1980)

38. H.-J. Freund: Physiol. Rev. 63, 387 (1983)

39. Synergetics of the Brain, ed. by E. Basar, H. Flohr, H. Haken and A. J. Mandell (Springer, Berlin, Heidelberg, New York 1983)

40. L. Janos: Smithsonian 14, 127 (1983)

The Central Nervous System Utilizes a Simple Control Strategy to Generate the Synergy Used to Control Locomotion

S. Grillner

Department of Physiology III, Karolinska Institutet, Lidingövägen 1,
S-114 33 Stockholm, Sweden

The voluntary control of movement utilizes both learned and gene-
tically determined motor programs. One example of the latter
category is the neural circuitry utilized to initiate, maintain and
modify the locomotor movements. In some vertebrates this circuitry
is already functioning at birth (e.g. gazelles, horses) or after
hatching (fish, tadpoles or chickens) in other species it matures
after birth to function only at a later stage (pups, kittens or
human infants). Despite these differences it appears likely that
the control systems from cyclostomes, fish, amphibians to reptiles,
birds and mammals have essential features in common (see Grillner
1985).

There is thus a spinal network which in isolation from the
brainstem or any movement-related feedback can produce a complex
motor output identical or similar to that produced during locomotion (see Grillner and Wallén 1985). The intrinsic mechanisms of
these networks is as yet unknown even if progress has been made
both in lower vertebrates (lamprey; see Grillner and Wallén1984,
tadpole; Roberts et al. 1983). This network is referred to as the
central pattern generator. The network can be further subdivided

into smaller parts, one for each limb and further subdivision is likely to occur (Grillner, 1981, 1982).

In all species investigated from fish and cyclostomes to cat, afferents activated during the movement have a powerful effect. In fact sensory signals elicited during the activity of the CPG can reset its activity and prolong or shorten different phases of the movement, such as the support phase. These signals thus act on the CPG and the rhythm-generating mechanisms within the CPG. In addition the output of individual motor nuclei will be affected directly by signals from the different muscle receptors, to mention one category. The exact role of the input remains to be determined but there need not be any major change in the pattern of motor activity after deafferentation. Under normal conditions the actual motor pattern thus depends on a detailed central network with a powerful sensory feedback. On the spinal level, center and sensorium thus together determine the final motor output and provide a spinal pattern generator.

The spinal cat or lamprey can thus coordinate essential features of the pattern of behaviour we call locomotion. Under normal conditions these networks are under a detailed control from supraspinal structures. When neurons in the area of the caudal part of nucleus cuneiforme in mesencephalon are activated electrically or pharmacologically, locomotion is elicited via the spinal networks for locomotion. The more potent the activation, the more rapid and forceful the locomotion. A cat can even switch from one mode of interlimb coordination to another, i.e. from trot to gallop. It is worth noting that the supraspinal command signal needed to elicit stereotyped locomotor movements which will bring the animal forward is very simple, it need not contain any timing information, whatsoever. The level of activity will decide whether the speed of locomotion will be slow, moderate or fast. This simple descending signal will set the level of activity in the different spinal networks controlling the different limbs and the trunk muscles etc, i.e. a vast number of muscles will be turned on in each stepcycle each at its particular phase of the stepcycle and appropriate amplitude level.

Neurones in the mesencephalic locomotor region in all likelihood act on the gigantocellular reticulospinal neurones in the lower brainstem which descends in the ventral funiculus.

Neurones in the output nuclei of the basal ganglia (cat), i.e. the entopeduncular, and the pars reticulata of the substantia nigra project to the mesencephalic locomotor region. At least part of this projection is gabaergic and presumably inhibits MLR (Garcia Rill et al. 1985). If blockers of GABA are administered into MLR, locomotion is released, presumably indicating that these neurones are kept under tonic inhibition from substantia nigra under resting conditions, at least in the decerebrate cat. Stimulation of N. Accumbens in the rat or addition of dopaminergic agonists release locomotion (cf Grillner 1985). Thus it appears likely that the basal ganglia affect the locomotor control and take part in the initiation of locomotion and control the level of locomotor activity.

In this context it is worth noting that decorticate animals (cats) with their basal ganglia intact (Bjurstén et al. 1976) exhibit a practically unchanged motor repertoire. They initiate locomotion and with an apparent purpose such as searching for food. They avoid obstacles and perform exploratory forelimb movements and utilize this information in their motor planning. In contrast, high decerebrate animals have lost much of the adaptive control of the locomotor movement but they can still be made to walk after stimulation of the locomotor areas in the brainstem, and the coordination of the entire locomotor synergy can still be satisfactory. Such an animal will, if it meets with a wall during walking, attempt to walk through the wall rather than turn around to find another road.

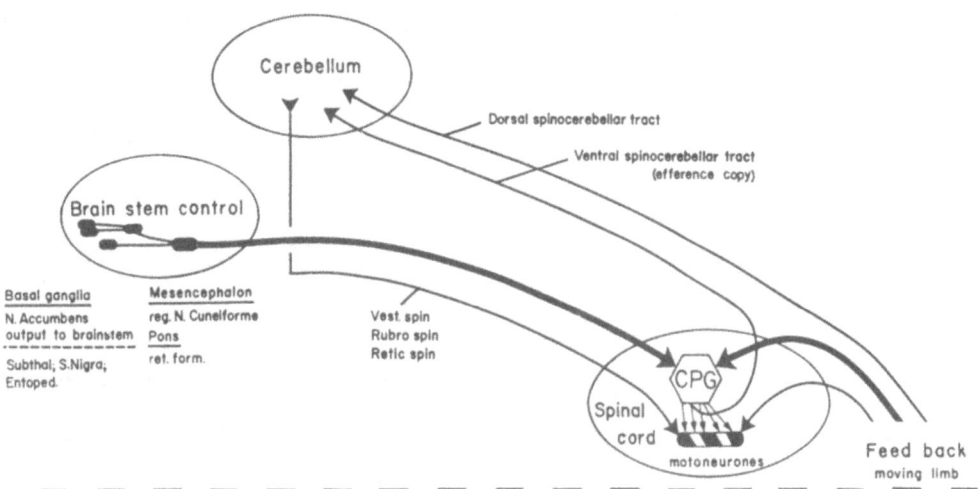

Fig. 1. Schematic diagram of the control system for locomotion in vertebrates (from Grillner and Wallén 1985).

This short contribution is only meant to give a general impression of the control strategy used by the central nervous system to control the complex pattern of motor behaviour known as locomotion. Several recent reviews have dealt with this control system in some depth (Grillner 1975, 1981, 1985, Grillner and Wallén 1984, 1985, Shik and Orlovsky 1976, Stein 1978, 1983).

ACKNOWLEDGEMENT

These studies have been supported by the Swedish Medical Research Council (project no. 3026).

1. E. Garcia Rill, R.D. Skinner and J.A. Fitzgerald: Chemical activation of the mesencephalic region. Brain Research. 330, 43-54 (1985)

2. S. Grillner: Locomotion in vertebrates: Central mechanisms and reflex interaction. Physiol. Rev. 55, 247-304 (1975)

3. S. Grillner: Possible analogies in the control of innate motor acts and the production of sound in speech. "Speech Motor Control", Eds. S. Grillner, B. Lindblom, J. Lubker and A. Persson, in Wennergren Symp., Vol 36 1982 Pergamon Press, pp 217-229

4. S. Grillner: Control of locomotion in bipeds, tetrapods and fish. "Handbook of Physiology, sec. 1. The Nervous System II. Motor Control", Ed. V. Brooks, in American Physiol. Soc. 1981 Waverly Press, Maryland, pp. 1179-1236

5. S. Grillner: Neurobiological bases of rhythmic motor acts in vertebrates, Science 228, 143-149 (1985)

6. S. Grillner and P. Wallén: How the lamprey CNS makes the lamprey swim. J. Exp. Biol. (in press)

7. S. Grillner and P. Wallén: Central pattern generators for locomotion, with special reference to vertebrates. Ann. Rev. Neurosci., 8, 233-261 (1985)

8. M.L. Shik and G.N. Orlovsky: Neurophysiology of locomotor automatism. Physiol. Rev. 56, 465-501 (1976)

9. P.S.G. Stein: Motor systems with specific reference to loco-motion. Ann. Rev. Neurosci. 1, 61-82 (1978)

10. P.S.G. Stein: The vertebrate scratch reflex. "Neural Origin of Rhythmic Movements", Eds. A. Roberts and B.L. Roberts, in Symp. Soc. Exp. Biol. 1983 Cambridge Univ. Press, pp. 383-403

Control Strategies for Complex Movements Derived from Physical Systems Theory

N. Hogan

Department of Mechanical Engineering, Massachusetts Institute of Technology, 77 Massachusetts Avenue, Room 3-449, Cambridge, MA 02139, USA

1. Introduction

The work presented here is part of an effort to develop a unified approach to the control of a system which may interact dynamically with its environment. The principal perceived problem is that when a controlled system interacts with its environment, its performance may be drastically altered. Even if the controlled system is stable in isolation, when it interacts with dynamic objects in its environment that stability may be jeopardised. This problem is particularly actue for a manipulator. Manipulation has been succintly described as a series of collisions between the manipulator and the objects in its environment [7]. Every time a manipulator grasps or releases an object, the dynamic behaviour of the physical system interfaced to the controller undergoes an abrupt change, and this change may have a profound effect on the manipulator's behaviour.

The approach discussed here is based on physical systems theory, and has been developed from an investigation of the strategies used to control the primate upper extremities, [2, 11, 15, 16, 21] and the application of similar strategies to the control of robot manipulators [12, 13, 14, 17, 18]. It may be sufficiently general to have application for controlling other complex biological systems. The ultimate goal of this work is to develop a class of controlled systems which could be dynamically coupled to or isolated from a wide variety of environments without serious degradation of performance and stability. This paper will show that the preservation of stability in the face of changing environmental dynamics can be achieved through a control strategy which ensures that a manipulator's behaviour is compatible with the physical behaviour of its environment.

2. Physical Equivalence

The basis of the approach is the concept of physical equivalence [12]. Any controlled system will consist of "hardware" components (e.g. sensors, actuators and structures) combined with controlling "software" (e.g. a neural network, brain or computer). A unified approach to the analysis and design of both the controller and the physical hardware can be developed by postulating that, taken together, the hardware and software is still a physical system in the same sense that the hardware alone is.

The value of this conjecture is its implication that no controller need be considered unless it results in a behaviour of the controlled system,which can be described as an equivalent physical system. Several well developed formalisms exist for describing physical systems, the most notable being Paynter's bond graphs [24, 27], which have been applied successfully to a broader class of systems than any other formalism. The postulate of physical equivalence justifies using the same technique to describe control systems. This provides a powerful and intuitive way of thinking about control action in physical terms, and may provide an effective vehicle for promoting communication between control system theorists and those working in other disciplines.

However, if this conjecture is to be of anything more than philosophical interest, it is necessary to clarify the definition of a physical system. What (if anything) distinguishes the differential equations used to model a physical system from any other general system of differential equations? No complete definition is attempted here, but some key issues are considered. One of the important differences lies in the structure of the equations.

3. Structure

What is meant by structure and why does it matter? Consider the differential equations for a general second-order linear system driven by a single input.

$$\begin{bmatrix} \dot{x}_1 \\ \dot{x}_2 \end{bmatrix} = \begin{bmatrix} a_{11} & a_{12} \\ a_{21} & a_{22} \end{bmatrix} \begin{bmatrix} x_1 \\ x_2 \end{bmatrix} + \begin{bmatrix} b_1 \\ b_2 \end{bmatrix} u \tag{1}$$

x_1, x_2: state variables

u: input variable

a_{11}, a_{12}, a_{21}, a_{22}, b_1, b_2: system parameters

or

$$\dot{x} = A x + B u \tag{2}$$

One important property of a system is its controllability, and this system is controllable if and only if the matrix $[B|AB]$ is of full rank.

One way of imposing structure on these equations is by restricting the values of some system parameters, and this can have a profound effect on system properties. Suppose, for example, that the parameters a_{12} and b_1 are identical to zero.

$$a_{12} \equiv 0 \tag{3}$$

$$b_1 \equiv 0 \tag{4}$$

The resulting system is structurally uncontrollable; it is always uncontrollable for all values of the remaining system parameters.

$$det[B|AB] = det \begin{bmatrix} 0 & 0 \\ b_2 & a_{22}b_2 \end{bmatrix} = 0 \tag{5}$$

If the differential equations are a mathematical model of a physical system, then that system will determine their structure. For example, dynamic interaction between a spring and a mass subject to external forces can be modelled by a second-order linear system of state equations. In this case the state variables can be given a physical meaning, and the equations may be written in phase-variable form in which one state variable is the displacement of the mass and the other its velocity.

$$\begin{bmatrix} \dot{x}_1 \\ \dot{x}_2 \end{bmatrix} = \begin{bmatrix} 0 & 1 \\ -k/m & 0 \end{bmatrix} \begin{bmatrix} x_1 \\ x_2 \end{bmatrix} + \begin{bmatrix} 0 \\ 1/m \end{bmatrix} F \tag{6}$$

x_1: displacement of mass

x_2: velocity of mass

k: spring constant

m: mass

F: external force

This system is structurally controllable. Aside from the trivial case $1/m = 0$ (corresponding to infinite mass) this system is always controllable for all values of its parameters.

$$det[\mathbf{B}|\mathbf{AB}] = det\begin{bmatrix} 0 & 1/m \\ 1/m & 0 \end{bmatrix} = -1/m^2 \tag{7}$$

This example shows that the structure imposed on the equations by the physical system they describe leads to useful restrictions on the behaviour they may exhibit.

4. Interaction

Another important characteristic of physical systems is the way they may interact. Consider two general first-order open linear systems. Each system receives an input from and delivers an output to its environment.

$$\dot{x}_1 = a_1 x_1 + b_1 u_1 \tag{8}$$

$$y_1 = c_1 x_1 \tag{9}$$

$$\dot{x}_2 = a_2 x_2 + b_2 u_2 \tag{10}$$

$$y_2 = c_2 x_2 \tag{11}$$

u_1, u_2: input variables
y_1, y_2: output variables
c_1, c_2: system parameters

The stability of each system in isolation is determined by the eigenvalues of its system matrix, in this simple case a scalar. A necessary and sufficient condition for assymptotic stability of each system is that its eigenvalue(s) be less than zero.

$$a_1 < 0 \tag{12}$$

$$a_2 < 0 \tag{13}$$

When the two systems are coupled the output of one becomes the input to the other.

$$u_1 = y_2 \tag{14}$$

$$u_2 = y_1 \tag{15}$$

The equations for the complete system are obtained by substitution.

$$\begin{bmatrix} \dot{x}_1 \\ \dot{x}_2 \end{bmatrix} = \begin{bmatrix} a_1 & b_1 c_2 \\ b_2 c_1 & a_2 \end{bmatrix} \begin{bmatrix} x_1 \\ x_2 \end{bmatrix} \tag{16}$$

A condition for stability of the coupled system is:

$$a_1 a_2 - b_2 c_1 b_1 c_2 > 0 \tag{17}$$

Although the product $a_1 a_2$ is greater than zero if the two systems are stable in isolation, stability of the coupled system requires that $a_1 a_2$ be greater than $b_2 c_1 b_1 c_2$. In general, stability of individual systems in isolation provides no guarantee of the stability of the system formed when they are dynamically coupled.

However, if the equations represent physical systems, then useful restrictions can be placed on the form of the coupling. In the formalism of bond graphs, dynamic interactions between physical systems are described (essentially by generalising Kirchoff's current and voltage laws) as an instantaneous exchange of energy without loss or storage [24]. Instantaneous energetic interaction or power flow between a physical system and its environment may always be described as a product of two variables, an effort (generalised voltage or force) and a flow (generalised current or velocity).

Energetic interaction between two systems also imposes a causal constraint on the forms of their input/output relations. One system must be an impedance, accepting flow (e.g. motion) input and producing effort (e.g. force) output while the other must be an admittance, accepting effort (e.g. force) input and producing flow (e.g. motion) output.

A mechanical spring and a frictional element experiencing a common force (i.e. in series) provides an example of an impedance; a mass and a frictional element sharing a common velocity provides an example of an admittance.

$$\dot{x}_1 = -k/b_1\, x_1 + V_1 \tag{18}$$

$$F_1 = k\, x_1 \tag{19}$$

x_1: spring displacement
V_1: input velocity
F_1: output force
k: spring constant
b_1: viscous friction constant

$$\dot{x}_2 = -b_2/m\, x_2 + 1/m\, F_2 \tag{20}$$

$$V_2 = x_2 \tag{21}$$

x_2: velocity of mass
F_2: input force
V_2: output velocity
m: mass
b_2: viscous friction constant

Assuming the usual convention [24, 27] that power is positive into a dynamic element or system imposes a sign constraint on the coupling equations. For example, if the coupling imposes a common velocity (flow) on the two systems, then to satisfy conservation of energy, the forces (efforts) must be equal but opposite (Newton's third law).

$$V_1 = V_2 \tag{22}$$

$$F_1 = -F_2 \tag{23}$$

The equations for the coupled system are again obtained by substitution:

$$\begin{bmatrix} \dot{x}_1 \\ \dot{x}_2 \end{bmatrix} = \begin{bmatrix} -k/b_1 & 1 \\ -k/m & -b_2/m \end{bmatrix} \begin{bmatrix} x_1 \\ x_2 \end{bmatrix} \tag{24}$$

A condition for stability of the coupled system is:

$$(k/b_1)(b_2/m) + k/m > 0 \tag{25}$$

In this case, if the individual systems are stable in isolation, the coupled system is also stable. Physically, this makes sense as the stability of each system in isolation guarantees that its energy is always decreasing. Coupling the two systems does not generate any energy, therefore the total energy of the coupled system is also decreasing and the coupled system is stable. Again, a knowledge of the structure of the equations for a physical system permits stronger statements about its behaviour.

5. Impedance Control

The same concepts may usefully be applied to more complex systems. If a manipulator (biological or artificial) is to interact dynamically with its environment then it is important to understand the structure of the environmental dynamics and to ensure that the behaviour of the manipulator is compatible. In the vast majority of cases, the environment a manipulator grasps consists of inertial objects, possibly kinematically constrained, and may include some elastic and frictional elements. An environment of this class can be described using Lagrange's equations in the following form.

$$L(\underline{q},\underline{\dot{q}}) = E_k^*(\underline{q},\underline{\dot{q}}) - E_p(\underline{q}) \tag{26}$$

$$\frac{d}{dt}[\partial L/\partial \underline{\dot{q}}] - \partial L/\partial \underline{q} = -\underline{P}(\underline{q},\underline{\dot{q}}) + \underline{P}(t) \tag{27}$$

$\underline{q}$: vector of generalised coordinates
$L(\underline{q},\underline{\dot{q}})$: Lagrangian
$E_k^*(\underline{q},\underline{\dot{q}})$: kinetic co-energy
$E_p(\underline{q})$: potential energy
$\underline{P}(\underline{q},\underline{\dot{q}})$: generalised frictional forces
$\underline{P}(t)$: generalised input forces

The coupling between manipulator and environment is typically such that a set of points on the manipulator have the same position and velocity as a corresponding set of points on the environmental object. These points define an interaction port. The position and velocity of the interaction port of the environment are functions of its generalised coordinates.

$$\underline{X} = \underline{L}(\underline{q}) \tag{28}$$

$$\underline{V} = \mathbf{J}(\underline{q})\underline{\dot{q}} \tag{29}$$

$\underline{X}$: interaction port coordinates
$\underline{L}(\underline{q})$: kinematic transformation equations
$\underline{V}$: interaction port velocities
$\mathbf{J}(\underline{q})$: Jacobian of kinematic transformation

As the transformation from generalised coordinates to interaction port coordinates is non-energic, the generalised input force is related to the interaction port force through the transposed Jacobian.

$$\underline{P} = \mathbf{J}(\underline{q})^t \underline{F} \tag{30}$$

$\underline{F}$: interaction port forces

Thus the input/output relation at the interaction port is:

State equations:

$$\frac{d}{dt}[\partial L/\partial \underline{\dot{q}}] - \partial L/\partial \underline{q} = -\underline{P}(\underline{q},\underline{\dot{q}}) + \mathbf{J}(\underline{q})^t \underline{F} \tag{31}$$

Output equations:

$$\underline{V} = \mathbf{J}(\underline{q})\underline{\dot{q}} \tag{32}$$

These equations show that this class of environments accepts input forces and produces output motions in response. Note that in these equations the vector of generalised coordinates may be of any order and the Jacobian need not be square. It is not, in general, possible to reformulate the equations in the dual form with velocity as the input and force as the output; this system is a generalised mechanical admittance.

Accordingly, to be compatible with this class of environments, the manipulator should be a generalised impedance, accepting motion inputs and producing force outputs in response. As the behaviour of the manipulator or the demands of the task vary, that impedance may need to be modulated or controlled, and the approach outlined in this paper and elsewhere [12, 13, 14, 16, 17, 18] has therefore been termed impedance control. The principal distinguishing feature of this approach is in the objective of the controller. Conventional controllers are usually structured so as to make some selected time function of the system state variables (e.g. position, velocity, force, etc.) converge to a desired time function. For example, almost all of present robot control technology is focused on the problem of making the robot end effector follow a desired trajectory in space [23]. An impedance controller attempts the more demanding task of making the entire dynamic behaviour of the manipulator converge to some desired dynamic function relating input motions to output forces.

The feasibility of imposing a desired impedance on a robot manipulator has been demonstrated and discussed in detail elsewhere [13, 17]. It has been shown that if a robot controller is designed with an impedance as the target behaviour some of the more prominent computational problems associated with robot control — inversion of the robot kinematic equations and computation of the inverse Jacobian in the vicinity of singular points — can be eliminated. However, it is not the intent of this paper to discuss computational techniques, as their relevance to the general problem of control of complex systems is unclear. For example, in a biological system, computational complexity may not be a major issue.

Instead a more fundamental question will be addressed: Is it useful for a manipulator to assume the behaviour of a generalised mechanical impedance? As detailed elsewhere [12, 13, 14, 17] impedance control provides a unified framework for coordinating free motions, obstacle avoidance, kinematically constrained motions, and motions involving dynamic interaction. In this paper a further benefit of impedance control is considered: the preservation of stability in the face of changes in the dynamic environment to which a manipulator is coupled. One simple (but versatile) class of impedances produces an output force as a function of only the position and velocity of the interaction port. In the following it will be shown that if the manipulator has the behaviour of this general class of impedances then a sufficient condition for the manipulator and the environment to be stable in isolation from one another is also sufficient to guarantee that the coupled system formed by dynamic interaction between manipulator and environment is also stable.

6. Preservation of Stability

To prove this result, it is convenient to express the behaviour of the environment in generalised Hamiltonian form [33]. The Hamiltonian is formed by defining the generalised momentum as the velocity gradient of the kinetic co-energy and applying a Legendre transformation.

$$\underline{p} \overset{\Delta}{=} \partial L / \partial \underline{\dot{q}} \tag{33}$$

$$E_k(\underline{p},\underline{q}) = \underline{p}^t \underline{\dot{q}} - E_k^*(\underline{q},\underline{\dot{q}}) \tag{34}$$

$$H(\underline{p},\underline{q}) = \underline{p}^t \underline{\dot{q}} - L(\underline{q},\underline{\dot{q}}) = E_k + E_p \tag{35}$$

$\underline{p}$: generalised momentum $E_k(\underline{p},\underline{q})$: kinetic energy $H(\underline{p},\underline{q})$: Hamiltonian

The displacement equations are obtained from the momentum gradient of the Hamiltonian.

$$\partial H/\partial \underline{p} = \underline{\dot{q}} \tag{36}$$

Substituting into the Lagrangian form yields the Hamiltonian form of the momentum equations.

$$\underline{\dot{p}} + \partial H/\partial \underline{q} = -\underline{P}(\underline{p},\underline{q}) + \mathbf{J}(\underline{q})^t \underline{F} \tag{37}$$

Rearranging these into the usual causal form:

$$\underline{\dot{q}} = \partial H/\partial \underline{p} \tag{38}$$

$$\underline{\dot{p}} = -\partial H/\partial \underline{q} - \underline{P}(\underline{p},\underline{q}) + \mathbf{J}(\underline{q})^t \underline{F} \tag{39}$$

This formulation has several advantages. The system equations are now in first order form (in contrast to the fundamentally second-order Lagrangian form). The structure of the Hamiltonian form of the equations is preserved under a very broad class of transformations known as canonical transformations [33]. In addition, for this system the Hamiltonian is identical to the total mechanical energy. This latter property can be used to assess system stability, as the total mechanical energy of a stable system may not grow without bound and the total mechanical energy of an assymptotically stable system must decrease. The rate of change of the mechanical energy may be expressed as follows:

$$\underline{H}_q \triangleq \partial H/\partial \underline{q} \tag{40}$$

$$\underline{H}_p \triangleq \partial H/\partial \underline{p} \tag{41}$$

$$dH/dt = \underline{H}_q^t \underline{\dot{q}} + \underline{H}_p^t \underline{\dot{p}} = \underline{H}_q^t \underline{H}_p - \underline{H}_p^t \underline{H}_q - \underline{H}_p^t \underline{P} + \underline{H}_p^t \mathbf{J}^t \underline{F} \tag{42}$$

In the absence of external forces, $\underline{F}$ is zero and this system is isolated. A sufficient condition for stability is then:

$$\underline{H}_p^t \underline{P} > 0 \qquad \text{or} \tag{43}$$

$$\underline{\dot{q}}^t \underline{P} > 0 \tag{44}$$

Now consider a manipulator with the behaviour of the following simple class of impedances:

$$\underline{F} = \underline{K}(\underline{X} - \underline{X}_o) + \underline{B}(\underline{V}) \tag{45}$$

$\underline{K}(\ )$: force-displacement relation
$\underline{B}(\ )$: force-velocity relation

$\underline{X}_o$ is the vector of desired positions of the manipulator end-effector. In the following it will be assumed to be a constant, corresponding to the maintenance of a fixed posture. If the function relating force to displacement from that posture is restricted so that it has no curl, then a potential energy function can be defined and this simple impedance can be expressed in the following Hamiltonian form:

$$\underline{q} = \underline{X} - \underline{X}_o \tag{46}$$

$$\partial E_p/\partial \underline{q} \triangleq \underline{K}(\underline{q}) \tag{47}$$

$$H(\underline{p},\underline{q}) = E_p(\underline{q}) \tag{48}$$

State equations:

$$\underline{\dot{q}} = \underline{V}(t) \tag{49}$$

$$\underline{\dot{p}} = \partial H / \partial \underline{q} + \underline{B}(\underline{V}(t)) \tag{50}$$

Output equations:

$$\underline{F} = \partial H / \partial \underline{q} + \underline{B}(\underline{V}(t)) \tag{51}$$

The rate of change of the system energy is:

$$dH/dt = \underline{H}_q^t \underline{V} \tag{52}$$

In the absence of imposed motions, $\underline{V}(t)$ is zero and this system is isolated. The rate of change of its total energy is then zero. Although the mechanical energy is non-increasing, no statement can be made about its assymptotic stability. However, one of the assumptions underlying impedance control is that the manipulator is at least capable of stably positioning an arbitrarily small unconstrained mass (i.e. a rigid body) [12]. In Hamiltonian form the equations of motion for a rigid body are:

$$H(\underline{p},\underline{q}) = E_k(\underline{p}) = 1/2\underline{p}^t \mathbf{M}^{-1} \underline{p} \tag{53}$$

$\mathbf{M}$: rigid body inertia tensor

State equations:

$$\underline{\dot{p}} = \underline{F}(t) \tag{54}$$

$$\underline{\dot{q}} = \partial H / \partial \underline{p} \tag{55}$$

Output equations:

$$\underline{V} = \partial H / \partial \underline{p} \tag{56}$$

Note that the rate of change the energy of this system is:

$$dH/dt = \underline{H}_p^t \underline{F} \tag{57}$$

Thus, in common with the simple impedance above, this environmental system has the property that when the force $\underline{F}(t)$ is zero and the system is isolated, its mechanical energy is non-increasing but no statement can be made about its assmyptotic stability.

When the rigid body and the impedance are coupled according to (22) and (23), the equations for the resulting closed system become:

$$H(\underline{p},\underline{q}) = E_k(\underline{p}) + E_p(\underline{q}) \tag{58}$$

$$\underline{\dot{p}} = -\partial H / \partial \underline{q} - \underline{B}(\underline{P}) \tag{59}$$

$$\dot{q} = \partial H / \partial p \tag{60}$$

The rate of change of the total system energy is:

$$dH/dt = \underline{H}_q^t \underline{H}_p - \underline{H}_p^t \underline{H}_q - \underline{H}_p^t \underline{B} \tag{61}$$

A sufficient condition for stability of the manipulator grasping the rigid body is:

$$\underline{H}_p^t \underline{B} > 0 \qquad \text{or} \tag{62}$$

$$\dot{q}^t \underline{B} > 0 \tag{63}$$

Now consider the stability of the system formed when the simple impedance described by (49), (50) and (51) is coupled to the more general environment described by (29), (38) and (39) through the coupling equations (22) and (23). In the following, subscript 1 refers to the manipulator and subscript 2 refers to the environment. The total system energy is:

$$H_{total} = H_1 + H_2 \tag{64}$$

its rate of change is:

$$dH_{total}/dt = \underline{H}_{1p}^t \dot{p}_1 + \underline{H}_{1q}^t \dot{q}_1 + \underline{H}_{2p}^t \dot{p}_2 + \underline{H}_{2q}^t \dot{q}_2 \tag{65}$$

$$= \underline{H}_{1q}^t \underline{J} \underline{H}_{2p} - \underline{H}_{2p}^t \underline{H}_{2q} - \underline{H}_{2p}^t \underline{P} - \underline{H}_{2p}^t \underline{J}^t \underline{H}_{1q} - \underline{H}_{2p}^t \underline{J}^t \underline{B} + \underline{H}_{2q}^t \underline{H}_{2p} \tag{66}$$

Eliminating terms:

$$dH_{total}/dt = -\underline{H}_{2p}^t \underline{P} - \underline{H}_{2p}^t \underline{J}^t \underline{B} \tag{67}$$

Using (29) and (38) the last term in (67) can be written in terms of the velocity at the interaction port.

$$dH_{total}/dt = -\dot{q}_2^t \underline{P} - \dot{q}_1^t \underline{B} \tag{68}$$

Thus the sufficient conditions (44) and (63) for stability of each of the two individual systems are also sufficient to guarantee stability of the coupled system. Intuitively, this makes physical sense because the non-energic coupling does not generate energy, thus there is no mechanism through which the total mechanical energy could grow without bound, and the frictional elements, however small, ensure that the total mechanical energy always decreases.

Summarising briefly, this discussion has shown that structuring the dynamic behaviour of a manipulator to be causally compatible with its environment has desirable stability properties. Note that the proof can be extended to more general forms of the target impedance without losing the fundamental result.

7. Application to a Biological System

Because of the generality of the physical equivalence conjecture these concepts can be applied to complex biological systems. How well do they describe observed behaviour? If the skeleton is modelled as a collection of kinematically constrained rigid bodies, then it is properly described as an admittance. Consequently, by the reasoning above, the neuromuscular system should behave as an impedance [11].

Despite the complexity of the thermodynamically non-conservative physiological processes underlying muscle contraction, under normal physiological conditions the external behaviour of a single muscle exhibits a relation between force and displacement similar to that of a spring [8, 26]. A growing body of literature in the neurosciences [1, 2, 4, 6, 15, 21, 25, 28, 31, 32] has investigated the influence of this "spring-like" behaviour on the control of movement. Indeed, one prominent and (to date) successful hypothesis [5, 10,22] explains one of the principal functions of the spinal reflex arcs (involving muscle spindles and Golgi tendon organs) as preserving the spring-like behaviour of an individual muscle in the face of perturbing effects. Furthermore, the relation between force and velocity produces a behaviour similar to that of a frictional element [3, 9, 19, 20], and thus a single muscle does, in fact, exhibit the behaviour of an impedance of the form of (45).

When muscles act in coordinated synergy, there is no guarantee that the behaviour of the complete neuromuscular system will be equally simple. For example, the presence of intermuscular spinal reflex arcs could introduce a relation between force and displacement with non-zero curl, or a relation between force and velocity with non-zero curl [11]. Such a system would still be an impedance, but would not enjoy the stability properties discussed above. However, recent experiments by the author and colleagues [21] have investigated the patterns of postural stiffness of the human upper extremity. Under steady state postural conditons, the anti-symmetric component of the stiffness was negligible in comparison to the symmetric component of the stiffness (see fig. 1) verifying that under these conditions the entire neuromuscular system

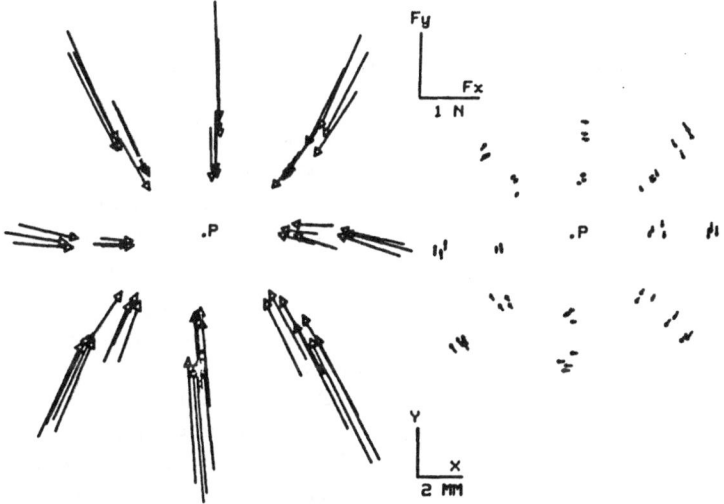

a) CONSERVATIVE COMPONENT   b) ROTATIONAL COMPONENT

Figure 1: While human subjects maintained a fixed posture of the upper extremity, a series of small (approximately 4 to 8 mm in magnitude) displacements were imposed on the hand and the steady state postural restoring force generated by the neuromuscular system in response was measured. The postural stiffness matrix was estimated by multivariable regression of between 50 and 60 observations of the force vectors onto the corresponding displacement vectors. This figure shows graphical representations of the symmetric (conservative or spring-like) component and the antisymmetric (rotational or curl) component of the postural stiffness. In these diagrams the two components are represented by drawing the force vectors obtained by multiplying each of the imposed displacement vectors by the symmetric (part a) and antisymmetric (part b) components of the postural stiffness. Each force vector is drawn with its tip at the tip of the corresponding displacement vector. For clarity, the displacement vectors are not shown. The nominal hand posture is at point P in each diagram.

of the upper extremity behaves as a simple impedance of the form of (45). Note that this behaviour requires that either the intermuscular reflex feedback is non-existent or that it is exquisitely balanced [11]. For example, the gain of the reflex pathways relating torque about the elbow to rotation of the shoulder must be identical to that relating torque about the shoulder to rotation of the elbow. These results suggest that despite the evident complexity of the neuromuscular system, coordinative structures in the central nervous system go to some lengths to preserve the simple "spring-like" behaviour of the single muscle at the level of the complete neuromuscular system.

If such finely tuned coordinative structures exist, what is their purpose? The analysis presented in this paper offers one explanation of the benefits of imposing the behaviour of a generalised spring on the neuromuscular system. If the curl of the force displacement relation is zero then the stability of the isolated limb is guaranteed with even the most modest frictional effect. Furthermore, when the limb grasps an external object -- even an object as complicated as another limb on another human -- then if that object is stable in the sense described above, the stability of the coupled system is again guaranteed.

8. Conclusion

The approach outlined in this paper offers a new perspective on the control of complex systems such as the primate upper extremity. An unique feature of the approach is that it is firmly based in physical systems theory. One important aspect of the dynamic equations of a physical system is their structure. If a manipulator is to be physically compatible with its dynamic environment then its behaviour should complement that of the environment. In the most common case in which the environment has the behaviour of a generalised mechanical admittance, the manipulator must have the behaviour of a generalised mechanical impedance, and its controller should not attempt to impose any other behaviour.

Imposing appropriate structure on the dynamic behaviour of a manipulator can result in superior stability properties. It must again be stressed that in general the stability of a dynamic system is jeopardised when it is coupled to a stable dynamic environment. In contrast, in this paper it was shown that if the force displacement behaviour of a manipulator has the structure of a generalised spring then the stability of the manipulator is preserved when it is coupled to a stable environment. Experiments to date indicate that the behaviour of the neuromuscular system of the human upper extremity has precisely this structure.

Control of a complex system is not exclusively a matter of preserving stability; acceptable performance must also be achieved. A clear definition of "acceptable performance" may prove to be elusive, but one desirable feature is that the manipulator should have a sufficiently rich repertoire of behaviour. In that context it is interesting to note that the impedance control strategies discussed in this paper give the manipulator the behaviour of a set of coupled nonlinear oscillators. Coupled nonlinear oscillators exhibit a prodigious richness of behaviour, and recent research has shown that some of their behavioural peculiarities are qualitatively similar to aspects of coordinated human movement [29, 30].

ACKNOWLEDGEMENT

This work was supported in part by National Science Foundation Research Grant ECS 8307641, National Institute of Neurological Disease and Stroke Research Grant NS 09343, and National Institute of Handicapped Research Grant 900 820 0048.

REFERENCES

1. E. Bizzi, A. Polit and P. Morasso: "Mechanisms Underlying Achievment of Final Head Position". Journal of Neurophysiology, Vol. 39, pp. 435-444, 1976.

2. E. Bizzi, N. Accornero, W. Chapple and N. Hogan: "Posture Control and Trajectory Formation During Arm Movement". Journal of Neuroscience, Vol. 4, pp. 2738-2744, November 1984.

3. B. Bigland and O.C.J. Lippold: "The relation between force, velocity and integrated electrical activity in human muscles". Journal of Physiology, Vol. 123, pp. 214-224, 1954.

4. J.D. Cooke: "Dependence of Human Arm Movements on on Limb Mechanical Properties". Brain Research, Vol. 165, pp. 366-369, 1979.

5. P.E. Crago, J.C. Houk and Z. Hasan: "Regulatory Actions of the Human Stretch Reflex". Journal of Neurophysiology, Vol. 39, pp. 925-935, 1976.

6. A.G. Feldman: "Functional Tuning of the Nervous System with Control of Movement or Maintenance of a Steady Posture. III. Mechanographic Analysis of the Execution by Man of the Simplest Motor Tasks". Biophysics, Vol. 11, pp. 766-775, 1966.

7. R.C. Goertz: "Manipulators used for Handling Radioactive Materials". chapter 27 in Human Factors in Technology, edited by E.M. Bennett, McGraw-Hill, 1963.

8. A.M. Gordon, A.F. Huxley and F.J. Julian: "The Variation in Isometric Tension with Sarcomere Length in Vertebrate Muscle Fibers". Journal of Physiology, Vol. 184, pp. 170-192, 1966.

9. A.V. Hill: "Heat of Shortening and the Dynamic Constants of Muscle". Proceedings of the Royal Society (London), Vol. B-126, pp. 136-195, 1938.

10. J.A. Hoffer and S. Andreassen: "Regulation of Soleus Muscle Stiffness in Premammillary Cats: Intrinsic and Reflex Components". Journal of Neurophysiology, Vol. 45, pp. 267-285, 1981.

11. N. Hogan: "The Mechanics of Multi-Joint Posture and Movement Control" submitted to Biological Cybernetics, 1985.

12. N. Hogan: "Impedance Control: An Approach to Manipulation: Part I - Theory". ASME Journal of Dynamic Systems, Measurement and Control, Vol. 107, pp. 1-7, March 1985.

13. N. Hogan: "Impedance Control: An Approach to Manipulation: Part II - Implementation". ASME Journal of Dynamic Systems, Measurement and Control, Vol. 107, pp. 8-16, March 1985.

14. N. Hogan: "Impedance Control: An Approach to Manipulation: Part III - Applications". ASME Journal of Dynamic Systems, Measurement and Control, Vol. 107, pp. 17-24, March 1985.

15. N. Hogan: "An Organising Principle for a Class of Voluntary Movements". Journal of Neuroscience, Vol. 4, pp. 2745-2754, November 1984.

16. N. Hogan: "Adaptive Control of Mechanical Impedance by Coactivation of Antagonist Muscles". IEEE Transactions on Automatic Control, Vol. AC-29, pp. 681-690, August 1984.

17. N. Hogan: "Impedance Control of Industrial Robots". Robotics and Computer Integrated Manufacturing, Vol. 1, pp. 97-113, 1984.

18. N. Hogan: "Mechanical Impedance Control in Assistive Devices and Manipulators". paper TA 10-B in Proceedings of the 1980 Joint Automatic Control Conference, edited by B. Friedland and H.A. Spang, American Automatic Control Council, 1980.

19. G. Joyce, P.M.H. Rack and D.R. Westbury: "The Mechanical Properties of Cat Soleus Muscle During Controlled Lengthening and Shortening Movements". Journal of Physiology, Vol. 204, pp. 461-474, 1969.

20. B. Katz: "The Relation between Force and Speed in Muscular Contraction", Journal of Physiology, Vol. 96, pp. 45-64, 1939.

21. F.A. Mussa-Ivaldi, N. Hogan and E. Bizzi: "Neural, Mechanical and Geometric Factors Subserving Arm Posture in Humans". Journal of Neuroscience (in press) 1985.

22. T.R. Nichols and J.C. Houk: "Improvement in Linearity and Regulation of Stiffness that Results from Actions of Stretch Reflex". Journal of Neurophysiology, Vol. 39, pp. 119-142, 1976.

23. R.P.C. Paul: Robot Manipulators: Mathematics, Programming, and Control. M.I.T. Press, Cambridge, 1981.

24. H.M. Paynter: Analysis and Design of Engineering Systems. M.I.T. Press, Cambridge, 1961.

25. A. Polit and E. Bizzi: "Characteristics of the Motor Programs Underlying Arm Movements in Monkeys". Journal of Neurophysiology, Vol. 42, pp. 183-194, 1979.

26. P.M.H. Rack and D.R. Westbury: "The Effects of Length and Stimulus Rate on Tension in the Isometric Cat Soleus Muscle". Journal of Physiology, Vol. 204, pp. 443-460, 1969.

27. R.C. Rosenberg and D.C. Karnopp: Introduction to Physical System Dynamics. McGraw Hill, New York, 1983.

28. R.A. Schmidt and C. McGown: "Terminal Accuracy of Unexpectedly Loaded Rapid Movements: Evidence for a Mass-Spring Mechanism in Programming". Journal of Motor Behaviour, Vol. 12, pp. 149-161, 1980.

29. J.A. Scott Kelso: "Phase-Transitions and Critical Behaviour in Human Bimanual Coordination". American Journal of Physiology, Vol. 246, pp. R1000-R1004, 1984.

30. J.A. Scott Kelso and B. Tuller: "Converging Evidence in Support of Common Dynamical Principles for Speech and Movement Production". American Journal of Physiology, Vol. 246, pp. R928-R925, 1984.

31. J.A. Scott Kelso and K.G. Holt: "Exploring a Vibratory System Analysis of Human Movement Production". Journal of Neurophysiology, Vol. 43, pp. 1183-1196, 1980.

32. J.A. Scott Kelso: "Motor Control Mechanisms Underlying Human Movement Reproduction". Journal of Experimental Psychology, Vol.3, pp. 529-543, 1977.

33. E.L. Stiefel and G. Scheifele: Linear and Regular Celestial Mechanics. Springer-Verlag, 1971.

Parallel Processes in Oculomotor Control

K. Hepp

Physics Department, E.T.H., CH-8093 Zürich, Switzerland

With the availability of VLSI technology to build supercomputers
with a large number of interacting processors, the brain-computer
analogy has found a revived interest. The RAM machine by von Neumann, [1] with a global memory for data and programs and with control by a single program counter (which holds the address of the
next instruction and is updated in a regular cycle), cannot serve
as a useful model for the brain. Much more interesting are parallel
computers, where the architecture depends strongly on the algorithms to be performed. The ultracomputer model [2] uses a large
number N of identical processors which are locally connected to a
few nearest neighbor processors, e.g. in a higher dimensional cubic
lattice. Such an architecture has a programming style, in which
periods of independent parallel computations alternate with periods
of interprocessor communication and data shuffling. For certain
problems algorithms can be developed where an N-processor ultracomputer is N times faster than a RAM machine. Such ultracomputers are
also structurally very different from the brain. Closest to the
brain's mode of operation is the data flow concept [3] . Here a
large number of processors with local memory are - for a given algorithm - connected in such a way as to exploit the maximal parallelism within the constraints imposed by the data dependence of the
computation. A program on a data flow computer is a graph, where
the nodes represent functions, which are implemented by the processors, and the arcs dependencies between functions. The control in

such a machine is "distributive". An instruction can be executed
once all required input values have been computed. Enabled processors transform input values into output values sent to other processors, which all generate functions on the data without any "side effects". Data flow computers can exploit the highest parallelism in
any algorithm without major control problems. However, the communication problem is immense for millions of processors arranged in
two-dimensional layers.

The data flow computer is an interesting conceptual model for the
brain. In the nervous system the processors are neurons or local
circuits built from axo-dendritic and dendro-dendritic interconnections [4]. The brain has solved the connectivity problem by sophisticated 3-dimensional arrangements which are unraveled by modern neuro-
anatomy. In the sensory and motor periphery many algorithms are implemented by permanent neuronal circuits: no reprogramming has to
convert acoustic to visual pathways or the oculomotor system to the
control of jaw movements. However, on the cortical level modules are
switched together in multiple patterns dependent on the data flow in
different tasks [5] .

Studies of the oculomotor system, where eye movements are generated from visual and vestibular inputs, have disclosed a number of
interesting parallel processes in the central nervous system of
higher mammals. The oculomotor system generates a number of distinguishable programs which can be classified as fast eye movements
(FEM), slow conjugate eye movements (SEM) and vergent eye movements.
In the 1983 symposium on synergetics of the brain I have reviewed
results in oculomotor neurophysiology obtained in collaboration with
V. Henn, and I have characterized parallel operations in the generation of FEM [6]. I shall use the same notation in this lecture and

refer to [6] for references to original contributions and review articles.

The best understood parallel operation in the saccadic system is "foveation": optical images of the world in retinal coordinates are transformed into motor error maps, e.g. in the deep layers of the superior colliculus (DSC), filtered by target selection processes. In Fig. 1 a target point P has been selected for foveation. Then in the DSC a population of vectorial long lead burst neurons $V_1, \dots V_n$ increase their firing rate beyond threshold for a synchronized burst. The strength[*] of the burst in V_i depends on the position of the saccade vector $\overrightarrow{OP}$ (from the fovea O to P) relative to the center $\overrightarrow{OV_i}$ of the movement field of V_i. One assumes that V_i is connected to the horizontal and vertical burst generators in the brainstem by synaptic strength dependent on the component $(\Delta h_i, \Delta v_i)$ of $\overrightarrow{OV_i}$. Then this circuit provides a rapid acceleration of the eye into the direction of the target, in which the horizontal and verti-cal eye velocity components are synchronized in such a way that the eye trajectory is straight even for oblique saccades. Recently, Hikosaka and Wurtz [7] have reversibly inactivated parts of the DSC

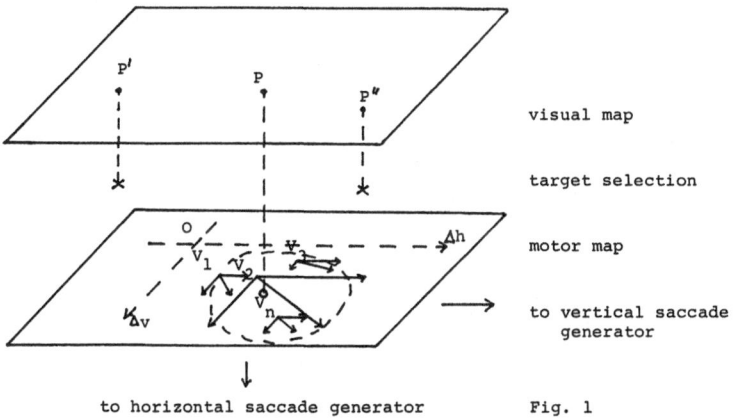

visual map

target selection

motor map

to vertical saccade generator

to horizontal saccade generator

Fig. 1

*) represented by arrow length

burster map in the alert monkey. After the lesion saccades were difficult to initiate and slowed down, as predicted by the functional interpretation of the DSC output as target selective saccade trigger.

It is puzzling that the movement fields for vector bursters everywhere in the brain are about 10 times larger than the visual receptive fields in the primary visual cortex (VIX) . In many situations the saccadic system first approximately "grasps" a novel target and brings it to the center of the fovea by a second corrective saccade. This is the typical mode of operation for large saccades. It is also found in a situation, where a subject is instructed to look as fast and as accurately as possible at two target points which unexpectedly appear in the dark. If both target points P_1 and P_2 have the same separation from the fovea, $|\overrightarrow{OP_1}| = |\overrightarrow{OP_2}|$, then a gradual transition between a monomodal and a bimodal saccade pattern occurs, when the angle between $\overrightarrow{OP_1}$ and $\overrightarrow{OP_2}$ increases [8] . The size of the movement fields of the vector bursters in the DSC could explain the neuronal implementation of this choice process, if one assumes that each excitatory movement field has an inhibitory surround. Then the target selection process can be localized in the DSC: For angles smaller than 30 deg between $\overrightarrow{OP_1}$ and $\overrightarrow{OP_2}$ the excitatory centers of the activated neurons in Fig. 1 for targets in P_1 and P_2 overlap strongly, and the first saccade will be made to a point between P_1 and P_2 . For angular separations larger than 60 deg a bimodal distribution is generated by the dominance of either the population around $\overrightarrow{OP_1}$ or around $\overrightarrow{OP_2}$.

It is clear that the rapid visuo-motor pathway from VIX to DSC has to be kept under control, so that we are not permanently distracted by potential saccadic targets during fixation or smooth pursuit. By single-cell recording in alert monkeys and reversible chemical lesions Hikosaka and Wurtz [9] have identified a population of neurons in the substantia nigra pars reticulata (SNR), which inhibit the output neurons of the DSC in exactly such situations. Furthermore Wurtz et al. (see e.g. [10]) have identified many cortical regions, where neurons with visual receptive fields show an enhanced premotor response, if the visual stimulus is used as a saccade target. These investigations on "visual attention" are the first step in the ana- lysis of higher cortical circuits in the parallel and hierarchical control of FEM.

The stereotypical pattern of normal saccades can be explained by the ballistic input into the saccadic burst generator from structures like the DSC. Since dysmetric saccades impair visual acuity, the pulse generator has to be under permanent adaptive control. There is a strong projection from the DSC to the vermal cortex (VER) of the cerebellum. One mossy fiber input of VER is from vectorial long lead bursters which code saccade vectors independently of eye position, and another input codes the presaccadic eye position by their firing rate similar to motoneurons. Mc Elligott and Keller [11] have identified output Purkinje cells from VER with a saccade-related burst and pause pattern which differs for centrifugal and centripetal saccades with identical eye displacement vectors. Lesions of the vermis or of the medial cerebellar nuclei (which contain interneurons from VER to the burst generators in the PPRF and RMRF with a complex saccade related firing pattern [12]) lead to saccadic dysmetria [13], [14], [15] .

Therefore another parallel circuit in the control of saccadic eye movement is implemented as a transcerebellar adaptive gain control. Fig. 2 summarizes the data flow in a very incomplete block diagram , in which other FEM generating pathways have been omitted.

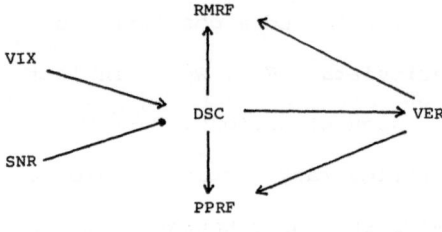

Fig. 2

Several mechanisms have developed to generate compensatory SEM in order to prevent blurring of images on the retina during movements (see e.g. [16], [17]). Head movements induce compensatory SEM into the opposite direction via the vestibulo-ocular reflex (VOR). Movements of the visual surround induce SEM in the direction of the moving pattern by the optokinetic reflex (OKR) and movements of selected target objects by the smooth pursuit system. The VOR is an open loop control system in the sense that the induced eye movement is not reducing its sensory input from the vestibular organ. OKR and pursuit act in a closed loop by reducing retinal slip. Continuous rotation of the head or of the visual surround induces nystagmus, a periodic pattern of compensatory "slow phase" eye movements reset by FEM. In Fig. 3A the slow phase velocity pattern of nystagmus is described under vestibular (VE) and visual stimulation (VI). In the first case, the monkey is sitting in the dark with his head fixed and the chair rapidly accelerated to 80 $^{\circ}$/s , rotated for 2 min with constant velocity and then stopped. In the second case the monkey is stationary in the dark, while a drum with black and white stripes is rotated around him with 80 $^{\circ}$/s in the opposite direction. The light is switched on for 2 min and then off again.

Under vestibular stimulation the slow phase velocity is compensatory with an initial value close to 80 $^{\circ}$/s which decreases exponentially with a time constant T_V of 10-40 sec. After deceleration the same pattern is induced in the opposite direction. During visual stimulation there is a rapid increase (a) in compensatory slow phase velocity to about 50 $^\circ$/s, followed to a slower increase with time constant T_V to a steady state level (b) of almost 80 $^\circ$/s . With lights off eye velocity decreases sharply to about 60 $^\circ$/s (c) and then the optokinetic afternystagmus (OKAN) decreases with time constant T_V to zero (d).

Under the same conditions the firing rates of vestibular ganglion cells (VG) and of an identified population of central vestibular neurons (VN) and of Purkinje cells in the flocculus division of the cerebellum (FLO) are also described. In Fig. 1B VG neurons, the primary afferents from the horizontal canals (see Fig. 4 in [6]), are unmodulated by visual input. They show under vestibular stimulation a pattern similar to the VOR, which decays more rapidly with a time constant T_C of about 5 sec. In Fig 1C VN neurons describe faithfully the time course T_V of the VOR in the dark. Under visual stimulation they show the time course T_V of the slow build-up (b) and the decay of the OKAN (d), but they lack the rapid eye velocity changes (a) and (c). Furthermore, the steady-state level would have been the same for all drum velocities above 60 $^\circ$/s (saturation), which is also the velocity level after the drop c . The firing pattern in Fig. 1D of FLO neurons is complementary to that of the VN pattern. Finally, the firing pattern of horizontal motoneurons (MN) and of burst-tonic neurons (BT) in the eye position integrator is a faithful image of the motor output.

Büttner, Henn and Waespe [17] have carefully analyzed the firing patterns in these neuronal populations in the full range of physiological stimuli, in the normal alert monkey and after selective inactivation of the VN- and FLO-neurons. Their results can be consistently described by a complementary operation of VN and FLO in the visuo-vestibular interactions generating compensatory SEM. The brain-

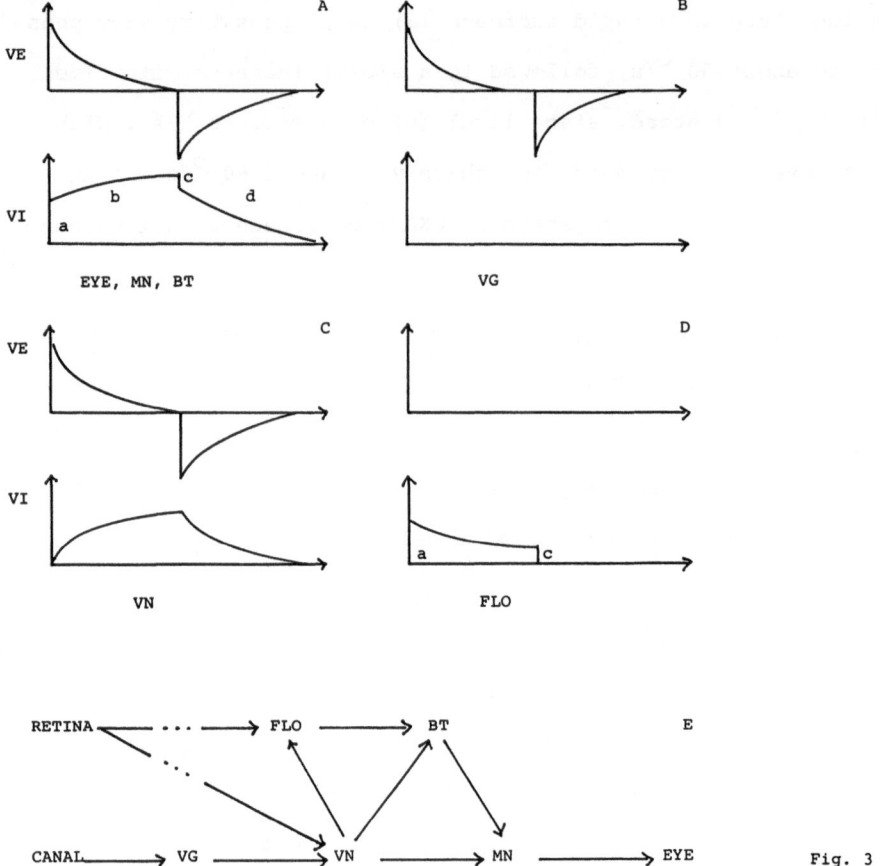

Fig. 3

stem pathway involves VN as a velocity storage, which increases T_c to T_v and maintains the OKAN. The velocity storage is a short term memory shared by the vestibular and the visual sensory systems. In certain parameter ranges, as typically in Fig. 3, the FLO Purkinje cells generate a motor output which complements the VN output for optokinetic stimuli of high velocity or acceleration. For this purpose they are informed by mossy fiber input about the state of the velocity storage VN, about the input BT to the motoneurons and they receive cortical visual inputs, since these FLO Purkinje cells are an important pathway in the smooth pursuit system [18] . The visuo-vestibular interactions with the parallel pathways through VN and FLO have been successfully modelled by Raphan and Robinson (see

176

[16]). However, there are many unsolved problems. The smooth pursuit system is poorly understood even on the algorithmic level. Although the excitatory and inhibitory connections between the neuronal populations in the cerebellar cortex are well understood by the work of Eccles, Ito and Szentágothai (see e.g. [19]) , nobody can compute the neurodynamic processes in the flocculus of the alert monkey. And finally there is the disturbing finding that the two short-term memories in the brainstem, the velocity storage and the position integrator, are not disjoint. There are many neurons which can be vestibularly modulated, even when the VOR is suppressed, and which carry an eye position and velocity signal without vestibular stimulation.Which are the target neurons for the floccular pathway in the monkey?

Although our understanding of parallel processing in the brain has significantly increased in the last decades, we cannot yet observe the data flow in the alert monkey through all interacting processors simultaneously. With the advent of powerful computers there might be the hope of combining a more extensive data analysis and a realistic simulation of unobservable processes to confront theoretical concepts of robotics and artificial intelligence with the real brain, our "most important possession which didn't come with an owner's manual" [20] .

[1] J. von Neumann, "Collected Works", Vol. 5, Pergamon London, 1963

[2] J.T. Schwartz, ACM Trans Prog Lang Sys 2, 484 (1980)

[3] J.B. Dennis, IEEE Computer, Nov 1980, p. 48

[4] G.M. Shepherd, "The Synaptic Organization of the Brain", Oxford, 1978

[5] V.B. Mountcastle, "An organizing principle for cerebral function: the unit module and the distributed system". In

"The Mindful Brain" (G.M. Edelman, V.B. Mountcastle eds)
MIT Press Cambridge (1978)

[6] K. Hepp, V. Henn, "Neurodynamics of the oculomotor system:
     space-time recording and a non-equilibrium phase transition".
     In "Synergetics of the Brain" (E. Başar, H. Flohr, H. Haken,
     A.J. Mandell eds) Springer, Berlin (1983)

[7] O. Hikosaka, R.H. Wurtz, J. Neurophysiol. $\underline{52}$, 266 (1984)

[8] F.P. Ottes, J.A.M. van Gisbergen, J.J. Eggermont, Vision Res
     $\underline{24}$, 1169 (1984)

[9] O. Hikosaka, R.H. Wurtz, J. Neurophysiol. $\underline{49}$, 1230, 1254, 1268,
     1285 (1983); $\underline{53}$, 292 (1985)

[10] R.H. Wurtz, M.E. Goldberg, D.L. Robinson, Prog. Psychobiol
     Physiol Psychol $\underline{9}$, 43 (1980)

[11] J.G. Mc Elligott, E. Keller "Neuronal discharge in the posterior
     cerebellum: its relationship to saccadic eye movement genera-
     tion" in "Functional Basis of Ocular Motility Control" (G.
     Lennerstrand, D.S. Zee, E.L. Keller eds), Pergamon Press,
     Oxford (1982)

[12] K. Hepp, V. Henn, J. Jaeger, Exp. Brain Res. $\underline{45}$, 253 (1982)

[13] L. Ritchie, J. Neurophysiol. $\underline{39}$, 1246 (1976)

[14] L.M. Optican, D.A. Robinson, J. Neurophysiol. $\underline{44}$, 1058 (1980)

[15] T. Vilis, R. Snow, J. Hore, Exp. Brain Res. $\underline{51}$, 343 (1983)

[16] V. Henn, B. Cohen, L.R. Young, Neurosci Res Prog Bull $\underline{18}$, 459
     (1980)

[17] W. Waespe, V. Henn, "Cooperative functions of vestibular nu-
     clei neurons and floccular Purkinje cells in the control of
     nystagmus slow-phase velocity: Single-cell recordings and
     lesion studies in the monkey". In "Adaptive Mechanisms in
     Gaze Control - Facts and Theories" (A. Berthoz and G. Melvill
     Jones, eds) Elsevier, Amsterdam (1985)

[18] S.G. Lisberger, A.F. Fuchs, J. Neurophysiol. $\underline{41}$, 733, 764
     (1978)

[19] M. Ito, "The Cerebellum and Neural Control", Raven Press, New
     York (1984)

[20] Advertisement in the Scientific American, Sept 1979, p. 93

Dynamics and Cooperativity in the Organization of Cytoplasmic Structures and Flows

D.G. Weiss

Zoologisches Institut, Universität München, Luisenstraße 14,
D-8000 München 2, Fed. Rep. of Germany

1.What is Cytoplasm ?

The cytoplasm of all animal and plant cells is a complex system consisting of an aqueous solution of a large number of charged and uncharged low molecular weight compounds including ions, sugars, amino acids amongst others. In addition,charged polymers of various kinds such as polypeptides, polysaccharides and copolymers of these are present at a relatively high concentration.

Furthermore, a large variety of structures are present in the form of organelles which are surrounded by a boundary of apolar lipids and form compartments of a chemical composition which is in principle similar to the surrounding but highly specialized,due to the presence of specific sets of enzymes. Some organelles are packages of building material (vesicles) or other material needed only at specific sites in the cell (e.g. secretory and synaptic vesicles), while other organelles provide the energy for the cell's metabolism (plastides and mitochondria) or are part of a waste disposal system (lysosomal organelles). Organelles carry a large number of positive and negative charges, but have an external negative net charge at physiological pH.

The cytoplasm is in addition internally structured by the presence of various proteinaceous filaments,which form a meshwork extending throughout the cell [1]. Its biochemical composition has been analyzed and its universal presence has been recognized only during the last decade. The various types of filaments occurring in all kinds of animal cells are listed in Table I together with their presumed functions. These include maintenance of cell shape and form by providing an internal scaffold,

Table I   Cytoskeletal Filaments Typically Present in an Animal Cell

name	structure	function
microfilaments	6 nm, double helix; actin	contraction of the cell; cytoplasmic streaming in plants; skeleton (stress fibers); anchoring other filaments
myosin filaments	12 nm, short rods; myosin[a]	contraction of the cell; cytoplasmic streaming in plants
intermediate filaments	10 nm, solid rod; cell type-specific proteins such as desmin, vimentin, keratin, neurofilament proteins	skeleton (highly crosslinked)
microtubules	25 nm, long hollow tubes; α and ß tubulin	saltatory organelle movement in animal cells; cytoplasmic streaming in some plants; cilia; mitosis; skeleton; no contraction

[a]it should be noted that myosin is present in most if not all cells, although only in muscle cells they form readily discernable long filaments.

Abbreviations: MT, microtubule; FGE, force generating enzyme.

locomotion of cells by reorganizing this scaffold, production of motion of parts of the cell (by filaments sliding against each other as in muscle or cilia) and production of a permanent motion of the internal structures of the cell (cytoplasmic streaming and organelle movement).

Thus, understanding cytoplasm means not only understanding the bewildering complexity on the molecular and ultrastructural level, but also the amazing world of movements the cellular components are able to perform. Some of these movements are prominent enough to be seen with ordinary light microscopes. However, the generality of this phenomenon has been recognized only recently,since its study requires advanced techniques such as fluorescence microscopy and especially video-enhanced contrast light microscopy [2]. Our present knowledge of the molecular composition of cytoplasm is comparably good, whereas we are only beginning to understand some of the complex motile phenomena and their mechanisms [1, 3, 4] although they have long been recognized to be an integral part of "living cytoplasm" [5].

2.The Nerve Cell Paradigm

The cytoplasm of the nerve cell is organized very similarly to these principles. The fact that neurons have extremely long extensions (axons and dendrites) and the fact that both synthesis and catabolism of proteins are absent in these long projections, being restricted only to the cell body, requires a most powerful intracellular transport system to supply materials to the cell periphery and to retrieve the material bound to be degraded. Transport in neuronal processes, i.e. axoplasmic transport, can indeed be regarded as a variant of cytoplasmic transport in general, which is, however, organized to work more efficiently. The axonal cytoplasm has, therefore, been used as a well-suited model system to study cytoplasmic structures and their movements [6-9]. The knowledge obtained on the dynamic aspects in the neuronal cytoplasm has not only helped to understand the complex system "cytoplasm" but also the formation, maintenance, modification and function of the likewise complex but highly ordered nervous system itself.

3. Self-Assembly of the Cytoskeleton

The ordered construction of a three-dimensional structure as complex as the cytoskeleton is difficult to explain in biochemical terms. Regulation of protein synthesis and enzyme activities would be required in an extremely complicated manner. However, some intrinsic properties of the cytoskeletal subunit proteins, already explain part of the process. The subunits of these proteins, usually as nucleotide di- or triphosphate complexes, assemble into their filament forms until an equilibrium is reached. The local concentration of subunits,together with some regulatory proteins,are therefore assumed to determine the average length of the filaments [10].

The cytoskeleton originates on the plasma membrane,where specific anchoring sites for actin filaments are provided [11]. It was shown recently that only the first protein [12] of a series has to be inserted into the membrane in a controlled way. Forming a "receptor cascade",several additional proteins are added until the sites which determine the stoichiometry are occupied, while surplus amounts of these additional proteins remain in solution and are subject to proteolytic digestion [12], i.e. excess production followed by catabolism of the surplus. The degree and mode of crosslinkage of the filaments themselves is an unsettled issue [13, 14].

In axons,the filaments retain their polarity with e.g. the microtubules (MTs) pointing with their faster growing (+) ends away from the cell body [15]. Since vigorous organelle and possibly solute traffic takes place along MTs, and in other cells along actin (see below), it can be assumed that these dynamic processes influence the parallel alignment of filaments and perhaps even the assembly kinetics at free ends. In vitro-experiments on the activity of an actin filament-associated enzyme show that its activity can be modulated by changing the streaming velocity along the filaments [16]. Similar experiments indicating flow-dependence also of

180

actin assembly have been performed [17, 18]. These are probably examples where dynamic cooperativity in the sense of Shimizu and Haken [19, 20], by influencing the protein molecules' conformation or alignment, acts to regulate the formation, three-dimensional arrangement and enzyme activity of the filaments .

4.The Microstream Concept of Cytoplasmic Transport

As an example of cytoplasmic transport,the mechanism of axoplasmic transport has been studied by several groups. The many experimental reports allowed us to summarize the properties of this phenomenon as well as to compile a list of properties which the transport mechanism must have (Tables I and II in [21]). The problems of cyto-plasmic viscosity and energy requirement could also be adressed [21-23]. We concluded from all the available information including biochemical, pharmacological and ultra-structural data that transport is organized as microstreams along the outside of MTs [22-24]. That bulk water-flow is taking place was inferred from the fact that all kinds of components of the axonal cytoplasm including soluble molecules seem to be conveyed in a non-specific manner. This was further supported by the finding that outflow profiles of radioactive material through axons can be described by the same equations applied in chromatography [25] and their changes of shape and position can be simulated by the principles of chromatography [26].

We suggested that this kind of movement can only be brought about by the coopera-tive action of as yet unknown MT-associated force-generating enzymes (FGEs) [22, 24]. By cooperativity we mean that a mechano-chemical enzyme reaction which is usually scalar is forced by a vectorial matrix, such as a MT to which the FGEs are as-sociated, to work in an oriented mode,so that the whole set of individual enzyme molecules would cooperate to produce a vectorial effect without violating Curie's

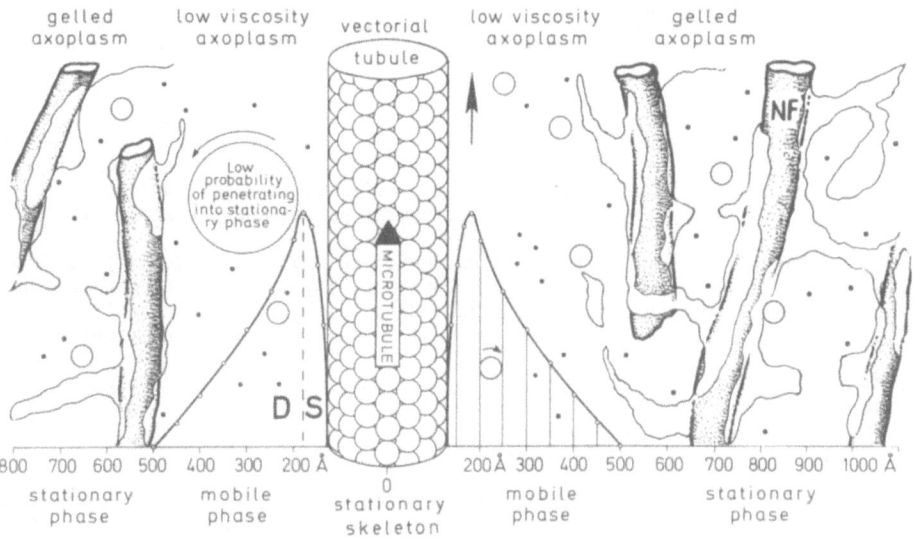

Fig.1 Schematic representation of the velocity profile around a microtubule (MT) as proposed by the microstream hypothesis. The streaming region is depicted here to extend about 40 nm from the microtubule surface. Stationary axoplasm consists of neurofilaments (NF) with additional proteins which together are assumed to give rise to the high macroviscosity of cytoplasm. The force generating enzymes (FGEs), which are proposed to be located near the MT surface, are not depicted. Material participating in rapid transport is shown as particles (spinning and excluded from the stationary phase), as protein molecules (small rings) and as low molecular weight molecules (dots, with free access to both stationary and moving phase). Reproduced from [22]

symmetry principle [27]. Today and in light of recent work [20, 28] we can even consider the possibility and test it experimentally, whether the movement itself, by producing flow along the MTs, acts to improve or stabilize the orientation of the FGEs by hydrodynamic means. This is especially important,since the FGEs seem to be soluble and not permanently MT-associated proteins [29].

The microstream hypothesis as summarized schematically in Fig. 1 is compatible with the experimental findings [22] and proposes a transport mechanism which is simpler than those suggested by other models. However, many biologists consider ratchet-type mechanisms more likely, since they feel that interactions of organelles with ratchet-like structures are more easily acceptable and more persuasive than streams which are invisible. Therefore, most competitive hypotheses propose carriers (filaments or organelles) which are suggested to bind the material to be transported (see [6, 21] for some).

5.Cooperativity

Rubinow and Blum [30] provided us with a theoretical model of axoplasmic transport which also presupposed a carrier system and reversible association of proteins and organelles with the moving carriers. Mathematical equations representing the concentrations of the moving carriers and the transported material at any position and time were developed.

When these authors tried to denote the structural entity corresponding to their theoretical "carrier", they had to rule out some of the generally discussed candidates and finally "recognized the possibility that 'carrier' considerations mediated by chemical reaction might simply be a wrong concept". We, however, want to draw again attention to the fact that a "<u>carrier</u> stream" concept is capable of fitting both the experimental and Rubinow and Blum's theoretical demands. The conclusions at which Rubinow and Blum [30] arrived from their calculations are all intrinsic to the microstream hypothesis [21, 24] and directly resulted from the visualization of axoplasmic transport as a carrier stream phenomenon,and treating it in terms of chromatography theory [23, 26, 31].

From their analysis these authors concluded that a travelling wave solution for the concentration of the transported material, which had been shown experimentally to exist [25], is possible provided that the interaction, assumed to be chemical in nature, between carriers and transported material exhibits positive cooperativity. The following processes can be suggested as having positive cooperativity:

a. The non-Newtonian behavior of cytoplasm, which resembles in some properties a Bingham body and in others a thixotropic fluid [32, 33], may lead to catastrophic network breakage upon mechanical stress [22, 24]. As a result of the activity of FGEs this may lead to the self-enhancing opening of channels through the cytoplasm.

b. Structural failure at loci of minimal cross-linkage of structural elements (presumably mainly in MT regions [32]) also leads to a self-enhancing formation of channels. These may finally bridge even longitudinal intertubule gaps or, in front of an obstruction, lateral intertubule gaps.

c. Particulate material, if available in increased amounts, has clearly been shown to increase the transport velocity [34, 35]. In terms of the microstream hypothesis this is explained by the hydrodynamic interactions between transported organelles and the structure of the MT domains. The spatial distributions of MTs may not always allow the passage of an organelle, despite streaming in that region. A moving organelle may form optimal channels through the cytoskeleton which may accommodate other particles which follow,and channels may be kept open by the repeated passage of organelles, e.g. when the concentration of organelles is increased. This so-called snow-plough effect [36] may, therefore, be explained as the combined effects of disruption of anchoring filaments, widening of converging microtubular regions and MT displacement leading to effective channels.

d. Another type of cooperative effect is the association of a number of flow fields each around a single MT. One or two MTs may be incapable of moving larger organelles protruding partially into gelled regions. Movement may become possible, however, if such organelles are surrounded by several MTs. Such situations have often been depicted in the electron micrographs of axoplasm (e.g. [37] Fig 5b). Using Rubinow and Blum's equations and terminology, it was found [35] that for synaptic vesicles the best fit with the experimental data for the number of theoretical attachment sites is n=4, which appears reasonable when taking into account the geometry in MT domains and the size of vesicles.

It is clear that cooperativity is abundant in living systems, while it remains to be clarified whether all these possible cooperative effects turn out to be those which have been modeled by the equations of Rubinow and Blum, and what their relative contributions may be.

6.Dynamic Cooperativity in Actin Systems

Further support for a MT-based microstream model came from reports showing experimentally that streaming phenomena occur along actin filaments [19, 38]. These authors were not only able, by applying the principles of synergetics, to explain this phenomenon on a theoretical basis,but they also invoked a different type of cooperativity, the so-called dynamic cooperativity [19, 20]. They consider the macroscopic streaming as being a truly self-organizing phenomenon,since the microscopic dynamics of holonic elements (here the myosin heads) become linked to each other through a feedback loop, i.e. a stream produced by the elements and slaving them to work in a cooperative manner [20]. The cooperative dynamics of the actin-myosin system [19] are not only useful to explain the stream cell developed by these authors (Fig 2a) but also cytoplasmic streaming in plant cells (Fig 2b) and muscle contraction (Fig 2c) [19]. Streaming in plant cells has been suggested to be due to the movement of myosin attached to organelles [39]. This in turn can be assumed to produce bulk water-flow along the actin bundles present in these plants by viscous coupling,as was shown previously [40] (Fig 2b). It is important to note that the actin-myosin system is capable of producing the streaming both in the absence or presence of organelles (c.f. Fig. 2a and b). It may therefore be premature to discuss only mechanisms of the latter type when treating either actin-based [41] or MT-based organelle movement [28, 42].

An actin-based motile system, namely the rotating polygons in the extruded cytoplasm of the alga Nitella [41], develops movement irrespective of the number of vesicles attached to or moving along them (Fig. 3). Their rotation may therefore be due to forces produced by streaming along the circular actin bundles consisting of many actin filaments [43] and not necessarily by myosin-coated organelles moving along them. However, why these rings rotate as polygons, mainly hexagons, remains a mystery.

It should be noted that the work of Shimizu shows not only that streaming occurs in actin-myosin systems, but also that streaming is indeed necessary for the production of the macroscopic dynamics, i.e. bulk water-streaming or muscle contraction, since it forces the elements to cooperate.

7.Analogous Behavior of Microtubule Systems

It was shown very recently that individual MTs from squid giant axons show essentially analogous behavior to that of actin bundles [28, 42]. The gliding movement of MTs had been earlier predicted from our microstream hypothesis [21] as being due to the reaction forces from the forward propulsion of fluid produced by MT-associated FGEs.

In fresh preparations, or when ATP was added, organelles and particles which, according to their size and abundancy, were classified as mitochondria, small vesic-

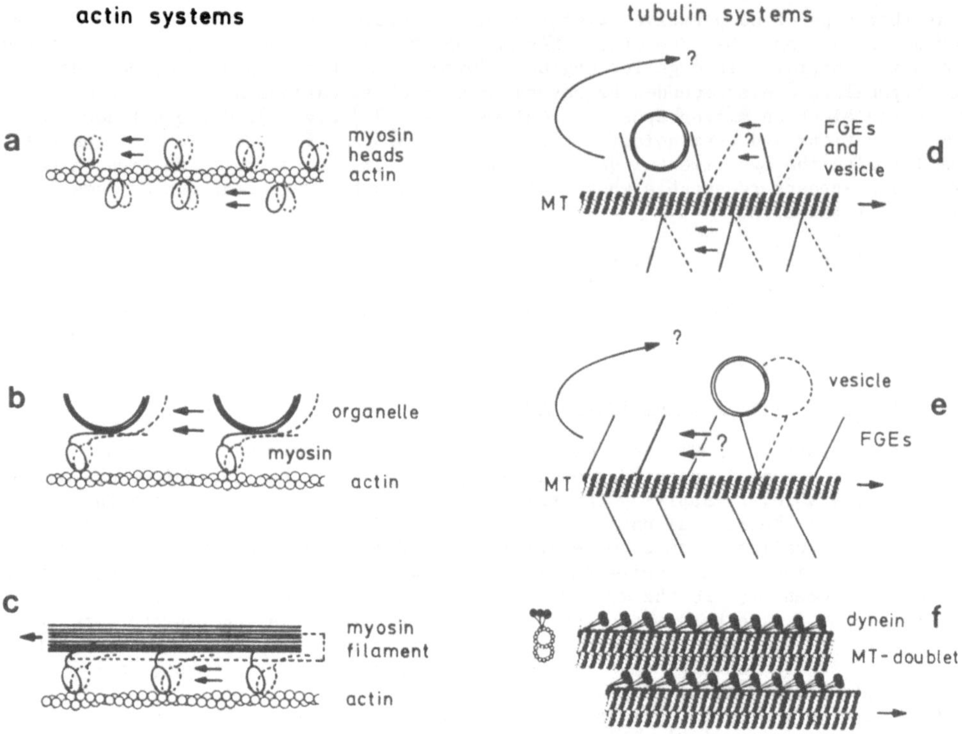

actin systems   **tubulin systems**

Fig.2   Hydrodynamic versus solid state mechanisms of intracellular motility. (a) Represents the mechanism which was shown to work in the stream cell of Shimizu and Yano[19,38].(b) Organelle movement and cytoplasmic streaming in the plant cell, e.g. in Nitella[39,40].(c) Muscle contraction. (d) The hypothetical microstream concept of axoplasmic transport[22,24]. FGEs are analogous to the myosin heads in (a), but are depicted as rods undergoing cycles of conformational changes, although nothing is known about their form or mode of action. (e) Hypothetical solid state mechanism whereby the FGE transfers the material via a direct binding e.g. [28,46](f) Sliding of MT which is the basis of movement of cilia. (a,d) are purely hydrodynamic mechanisms, (c,f) are solid state mechanisms, while (b,e) are intermediate since (b) makes use of both principles and in (e) one of the two directions might be realized on a hydrodynamic basis (curved arrow). Fluid streams (double arrows) are probably re-quired for proper functioning in (a) and (c)[19], and they are evident in (b). They are proposed as being the primary event in (d), but only secondary, although probably indispensable for proper functioning, in (e) and (b). They can be expected to exist also in (f). The mechanisms in the top row can be considered simple and may have emerged early in evolution, while (c) and (f) are very specialized and highly deve-loped motile systems. MT, microtubule; dashed lines represent one-unit movement caused e.g. by one cycle of conformational changes. The FGEs are known to be myosin in (a–c), dynein in (f), but are unknown in (d,e).

les (50–100 nm) and larger round organelles (200 nm), were seen to attach to the MTs and to be transported to their ends (Fig. 4). They can easily switch from one MT to an intersecting one. The particles were transported unidirectionally in fresh, ATP-containing preparations at velocities of 1.3 - 2 µm/s. In older preparations, or without added ATP, an increase of MTs transporting bidirectionally and a decrease in velocity were observed. Fluorescently labelled polystyrene particles were also trans-ported [28].

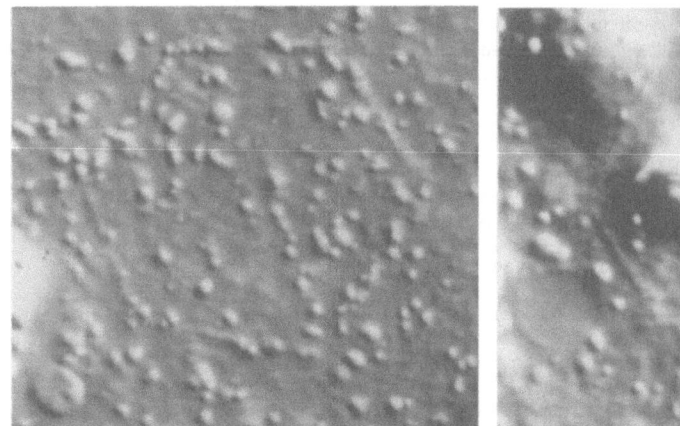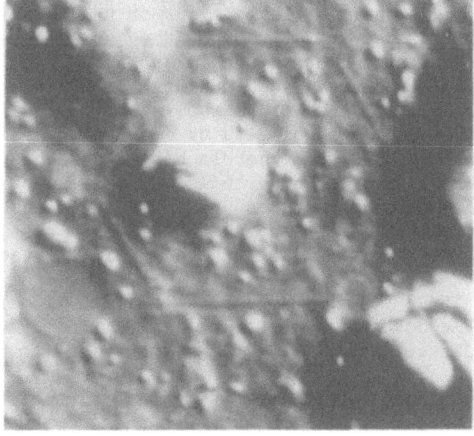

Fig.3   Rotating hexagons,consisting of actin bundles form when the cytoplasm of
Nitella cells is extruded. Movement of individual organelles is either in the same
direction as the movement of the corners or in the opposite direction. The two
hexagons shown here (20 and 17 µm) are from the same preparation,and were observed
only a few minutes apart. They rotate irrespectively of the presence of moving
organelles. The many organelles in the surrounding are in Brownian motion. AVEC-DIC
microscopy according to[2].

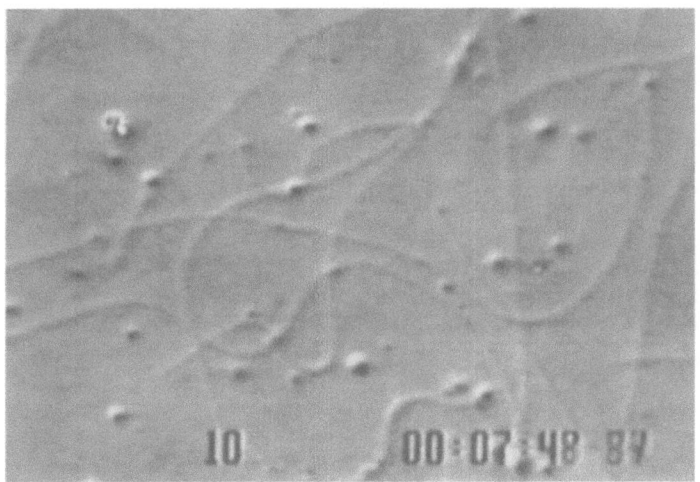

Fig.4   A single frame from a video recording of the movement of synaptic vesicles
and larger organelles along single native microtubules from the extruded cytoplasm of
squid giant axon. Particles move in both directions and can attach to more than one
microtubule. The larger, presumably lysosomal organelles move predominantly in one
direction (0.1 - 0.5 µm/s) while the majority of the small vesicles moves more
rapidly (0.4 - 1.0 µm/s) and in the other direction. The diameters of all structures
are inflated by diffraction to 0.1 - 0.2 µm. AVEC-DIC microscopy according to [2].
Field width 21 µm. For methods and quantitative results see[28].

Additional motile behavior of the relatively stiff MTs themselves could be observed,irrespective of whether they transported organelles along themselves or not.
Movements of protruding MTs appeared as gentle lateral motions resembling sea-
weed. MTs often glided out of the bulk of the filamentous meshwork,while one end
remained attached, thus forming a growing loop. MTs which fell off the axoplasm
showed longitudinal gliding over the glass surface at velocities between 0.2 and 0.6
μm/s(Fig. 5). When the MTs hit an obstacle,or when one end became entangled, the free
rear end started a rhythmic fishtailing or a circling motion. Long MTs may form rings
or U-shaped structures which rotate in the presence of ATP. Although during all these
movements MTs appear as stiff elastic rods, they can be bent by the pushing forces
exerted by their tail ends to a minimal radius of curvature of about 0.2 μm [28].

These findings also show that the properties of bidirectional intracellular organelle movement, i.e. cytoplasmic transport, can to a great extent be brought about by
a single, native, cytoplasmic MT with its associated proteins.

In analogy to actin-based particle motion [41] and filament motion [41, 43],which
can both be explained as being due to streaming (Fig. 2a) and/or ratchet-type mecha-
sms (Fig. 2b), (both thought to depend on dynamic cooperativity[19]) we proposed that
MT-associated FGEs, are similarly slaved to cooperate by the streaming they produce,
thereby eliciting axoplasmic transport [28].

The updated microstream hypothesis,which explains all three types of movement
(anterograde and retrograde transport as well as MT gliding) is shown in Fig. 6. The

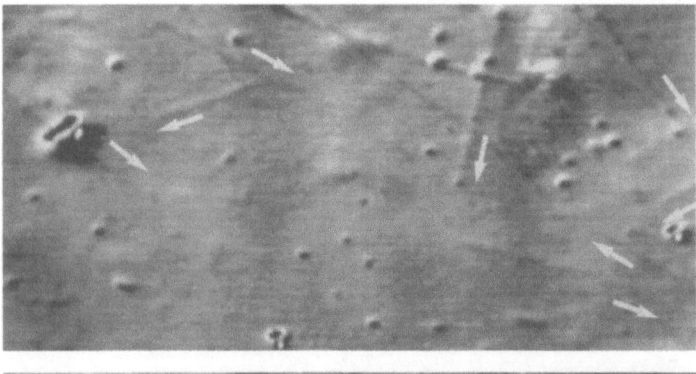

Fig.5 Short segments of microtubules from sheared extruded squid axoplasm are shown
by AVEC-DIC microscopy to glide over the surface of a cover glass [28]. The time
interval between the two frames is 11 sec. Some microtubules are stuck to the glass,
while most of them are motile, with the directions of their movement being random.
Size and magnification as Fig.4.

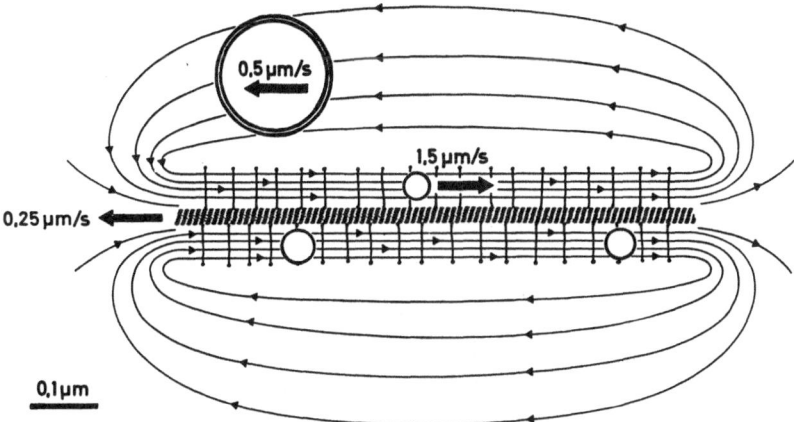

Fig.6   Schematic representation of the modified microstream hypothesis, which accounts for the motile phenomena observed in intact and extruded squid axoplasm. Small
organelles move rapidly anterogradely (to the right), while larger organelles and the
microtubules move more slowly and in the retrograde direction (to the left). Movement
of the microtubules themselves is probably prohibited in intact axoplasm. Whether the
bidirectional streams can indeed be stabilized at such extremely small Reynolds
number conditions is presently being investigated.

outer stream, suggested by this model to exist further apart from the surface of the
MT, should allow for backflow of fluid which may be stable at the exceptionally small
Reynolds number of the system (Re=10^{-9}). The original theory of transport of particles in microstreams assumes that forces that maintain the particle in the stream are
present,and so strong that Brownian bombardment would not interfere. So far we do not
know what such a force could be. If a further biophysical analysis of the movement
showed that such attraction would be possible, e.g. by electrostatic or hydrodynamic
forces, then the scenario of the microstream hypothesis would become acceptable [21,
24]. Bidirectional flow is also reported to occur at various distances from the
surfaces of swimming microorganisms [44]. Much remains to be learned from a quantitative treatment of the phenomenon that would take into account the laws of hydrodynamics at very low Reynolds numbers [45].

   If such forces,which might attract unbound organelles to the MT-associated mechanism,are not to be found, then transient binding of the organelles to e.g. the FGEs
would be necessary [28], as was already suggested as a possibility earlier [46].

   A different model recently suggested for MT-movement [29],proposing the interaction of MTs with FGEs which are randomly associated to a surface, is not suited for
the explanation of the highly ordered MT movements observed,and fails completely to
address the above mentioned cooperativity inherent to axoplasmic transport. Similarly, an association of FGEs to the organelles,which was suggested by the same authors
[29], misses the point that organelles have an intrinsic direction of movement [8, 37,
47]. Directionality cannot be brought about by association of FGEs to a spherical
organelle,but only by association to a vectorial MT see [21, 22, 48].

   The similarity of actin- and MT-based systems of cytoplasmic movements can also be
shown by another example. Cytoplasmic streaming,which occurs along the inner side of
the plant cell wall (cyclosis),is usually an actin-myosin system (Table I). Some
plants, however, produce the same macroscopic streaming phenomenon by a MT-based
mechanism, which shows the same pharmacological properties as cytoplasmic transport
in animal cells, such as neurons [49]. Interestingly, there are even plants having
both systems working simultaneously in the same cell but conveying different types of
organelles [50]. This further underlines the close analogy between the mechanisms
depicted in Fig. 2 a and b for actin-based systems with those in Fig. 2 d and e
representing MT-based systems.

8.Modes of Force Generation

The experimental findings on MT-based motility in the axon are compatible with both ratchet or microstream mechanisms (Fig. 2d and e) or with a combination of both [28] while several other hypotheses could already be ruled out [28]. There is, however, nothing known regarding the mode of mechano-chemical conversion of the potential energy from the ATP to kinetic energy. As yet we still do not know whether this is achieved directly or through an additional form of potential energy such as a chemical or electrical gradient (c.f. [17, 48]).

Biochemists tend to prefer direct energy conversion, such as a cycle of conformational changes of the FGE which would either move the transported material specifically [28, 46] or the water and other molecules in its surrounding by unspecific action [22, 24]. Another possible mechanism especially suited to convey fluids would be to make use of the fact that the ATP-cleavage is a vectorial reaction if the enzymes are all oriented on a vectorial structure. Directed release of charged reaction products would, by virtue of electroosmosis, cause the surrounding fluid to stream, especially if the efficiency is increased by cooperative action. This has been proposed as a theoretically possible means of force-generation for actin-based systems [17] and for axoplasmic transport [22, 24, 48].

Another possibility, still more abstract and therefore not considered seriously by many biologists, is the production of "structured flows" by means of dissipative structures [51]. In concentrated multicomponent solutions of macromolecules, one observes diffusive-convective phenomena which develop streaming velocities which are in the order of those observed intracellularly, and therefore are several orders of magnitude faster than diffusion. They do not depend on metabolic energy, but can be driven by gravity or very shallow concentration or temperature gradients, caused e.g. by MT-associated ATPase activity [51, 52, 53]. Such "structured flows" could be combined with MT-associated ATP-consuming processes which add directionality and possibly increase the stability of the flows. In this case we would be dealing with a different kind of self-organizing phenomenon, responsible for the generation of the motive force.

9.The Organization of Many Microstreams in the Cell

The actin bundles in *Nitella* cells are all aligned in parallel close beneath the cell wall, thus producing a stable cycling stream of cytoplasm of relatively high speed (70 μm/s) in the cell [54, 55]. The situation in animal cells is surprisingly similar, although their internal movement appears at first sight to be random. In axons we find anterograde transport (towards the synapse) and retrograde transport (back to the cell body) with "turnaround" at the synapse [8]. The MTs are known to be organized in bundles, but it is not known whether different bundles transport only in one direction [22] or whether they maintain the bidirectionality which single MTs show outside the cell [28]. We had proposed, on the basis of the microstream hypothesis, that the many individual streams coalesce to form a system of flows in the cytoskeletal matrix, which eventually stabilizes hydrodynamically [22, 23]. This would, in addition, have many aspects in common with the "structured flows" produced in model systems by means of dissipative structures [52, 53].

The individual MTs cannot apparently move inside the cell, as they do on microscope slides, instead we have to assume that they are well anchored in the cytoskeleton. In cross-sections they are often seen to be organized in an almost hexagonal array [c.f. 10]. Should each one indeed maintain its bidirectional flow at different distances from the surface (see Fig. 6), then the flows in the axon would form a very elongated hexagonally ordered pattern, somewhat resembling the Benard instability. From our model, we would suggest the anterograde movement to be centrally located around the MTs, while retrograde movement would occur in the intervening space in the hexagonal array.

10.Bidirectional Transport of Molecular Information

This bidirectional axoplasmic transport system provides a means to supply material required for the turnover of all structures and for metabolism to all parts of the cell. Here again,the working principle appears to be: production in excess, lack of enzymatic regulation of the amounts transported [8, 48] , and catabolism of the surplus material after its return from the cell's periphery.

It is well known that the neuron also uses this intracellular transport system to exchange molecular information between the cell body and the cell periphery (axon, synapse, dendrites) on their respective biochemical and physiological states [6, 7, 9]. Since the synapse is not only capable of releasing small molecules and proteins by exocytosis,but also to sample and retrieve the extracellular microenvironment by endocytosis, as well as conveying the material taken up back to the cell body's synthetizing machinery, it becomes clear that axoplasmic transport also provides a reciprocal and continuous system of molecular communication between cells.

There are molecules, called trophic substances, which are produced by neurons as well as by neighboring cells, such as glial or muscle cells, whose presence is mutually required for the survival and maintenance of the different cell types (survival factors) [57, 58]. For example, during development,sympathetic neurons emerge in surplus numbers. Those neurons whose axons find their proper target cells, e.g. gland or muscle cells, take up the trophic substance called nerve growth factor (NGF) which is transported retrogradely towards the cell body [57]. Neurons receiving NGF in this way survive, while those missing their target cells die (excess production followed by destruction of the surplus). Similar situations have been found for muscle cells,which require a neuron-derived survival factor [56]. Although only a few of these reciprocally active survival factors have been characterized, such a system is capable of generating and maintaining the complicated wiring patterns which are necessary for any functioning neural network [c.f. 59]. In other words, the self-assembly of the extremely complex nervous systems is controlled by this reciprocal system of intra- and extracellular transfer of molecular information, the mathematical treatment of which may well give new insights into the formation of ordered neural networks.

11.Conclusion

With the aim of qualitatively discussing problems of cooperativity which may be present during the formation and functioning of the cytoplasm, a wide range of phenomena have been described,which depend on cooperativity of various kinds. Further assessment of these phenomena is evidently required to delineate which processes are of static and which of dynamic cooperativity. We can now start to test these questions experimentally,as well as theoretically using the criteria given by Shimizu [19].

The present qualitative discussion,together with previous experimental and theoretical work by Shimizu [19]and our group [22, 48] strongly suggests that in several instances it is a hydrodynamic flow which gives rise to dynamic cooperativity. Therefore, this concept was applied to several levels of cellular and intercellular material and information flows (Table II). As a result, a better interpretation can be provided for cellular phenomena,such as the formation of the cytoskeleton, the mechanisms of intracellular motility, the arrangement of flows in a cell, as well as the control of the formation of multicellular networks. These phenomena are amenable to experimental and theoretical treatment,such as studies of their phase-transitions, instabilities, stochastic properties, or dimensionalities. It is to be hoped that a new view of these old problems from this angle will provide us with working hypotheses suitable to better understand the complex structures and flows in living cytoplasm.

Table II   Cytoplasmic flow viewed in the overall context of cellular processes illustrating how this concept may help to understand neuronal function

selforganization of the cytoskeleton
↓
organization of bidirectional transport/flow along single filaments or microtubules
↓
organization of bundles of filaments/microtubules to a stable system with bidirectional streaming
↓
continuous bidirectional system of molecular communication in the axon (anterograde, retrograde, and turnaround)
↓
continuous bidirectional system of intercellular molecular communication by uptake and release of trophic substances at the axon terminals
↓
organization, maintenance, and modification of ordered neural networks regulated by uptake and release of trophic substances
↓
ordered neural networks allow for ordered information flow and processing which are prerequisite for all higher nervous system functions

Acknowledgment

The experimental work of the author was supported by Deutsche Forschungsgemeinschaft and Boehringer Ingelheim Fonds. The author thanks R.D. Allen, G.W. Gross, R. Jarosch and H. Shimizu for many fruitful discussions.

References

1   Organization of the Cytoplasm, Cold Spring Harbor Symposia on Quantitative Biology, Volume XLVI (Cold Spring Harbor, 1982)
2   R.D.Allen, N.S. Allen: J.Microsc. 129 3–17 (1983)
3   B. Alberts, D. Bray, J. Lewis, M. Raff, K. Roberts, J.D. Watson (eds.): Molecular Biology of The Cell (Garland, New York 1983)
4   M. Schliwa: In Cell and Muscle Motility, J.W. Shay (ed.) (Raven, New York 1984) Vol. 5 1–84
5   O. Bütschli: Investigations on Microscopic Foams and on Protoplasm (Black, London 1894)
6   D.G. Weiss (ed.): Axoplasmic Transport (Springer, Berlin 1982)
7   D.G. Weiss, A. Gorio (eds.): Axoplasmic Transport in Physiology and Pathology (Springer, Berlin 1982)
8   D.G. Weiss: In Axoplasmic Transport in Physiology and Pathology D.G. Weiss, A. Gorio (eds.) (Springer, Berlin 1982) 1–14
9   S. Ochs: Axoplasmic Transport and its Relation to Other Nerve Functions (New York 1982)
10   A. Wegner: J. Mol. Biol. 108 139–152 (1976)
11   D. Branton, C.M. Cohen, J. Tyler: Cell 24 24–32 (1981)
12   E. Lazarides: Eur. J. Cell Biol. Suppl. 7, 36 77 (1985)
13   K.R. Porter, M. Beckerle, M. McNiven: Mod. Cell Biol. 2 259–302 (1983)
14   H. Ris: J. Cell Biol. 100 1474–1487 (1985)
15   P.R. Burton, J.L. Paige: Proc. Natl. Acad. Sci. USA 78 3269–3273 (1981)
16   M. Yano, H. Mioh, H. Shimizu: In International Cell Biology 1984, S. Seno, Y. Okada (eds.), (Academic Press Japan, Tokyo 1984) 502
17   A. Oplatka: In Biological Structures and Coupled Flows, A. Oplatka, M. Balaban (eds.) (Academic Press New York 1983) 207–222
18   E.A. Cerven: In International Cell Biology 1984, S. Seno, Y. Okada (eds.), (Academic Press Japan, Tokyo 1984) 479
19   H. Shimizu: Adv. Biophys. 13 195–278 (1979)
20   H. Shimizu, H. Haken: J.theor. Biol. 104 261–273 (1983)

21   D.G. Weiss, G.W. Gross: Protoplasma 114 179-197 (1983)
22   D.G. Weiss, G.W. Gross: In Axoplasmic Transport, D.G. Weiss (ed.) (Springer, Berlin 1982) 362-383
23   G.W. Gross, D.G. Weiss: Protoplasma 114 198-209 (1983)
24   G.W. Gross: Adv. Neurol. 12 283-296 (1975)
25   G.W. Gross, L.M. Beidler: J. Neurobiol. 6 213-232 (1975)
26   G.H. Stewart, B. Horwitz, G.W. Gross: In Axoplasmic Transport D.G. Weiss (ed.) (Springer, Berin 1982) 414-422
27   P. Curie: J. de Physique, 3 serie III 393-398 (1894)
28   R.D. Allen, D.G. Weiss, J.H. Hayden, D.T. Brown, H. Fujiwake, M. Simpson: J. Cell Biol. 100 1736-1752 (1985)
29   R.D. Vale, B.J.Schnapp, T.S.Reese, M.P.Sheetz: Cell 40 559-569 (1985)
30   S.I. Rubinow, J.J. Blum: Biophys. J. 30 137-148 (1980)
31   G.W. Gross, G.H.Stewart, B.Horwitz: Brain Res. 216 215-218 (1981)
32   G.W. Gross, D.G. Weiss: In Axoplasmic Transport, D.G. Weiss (ed.) (Springer, Berlin 1982) 330-341
33   K.A. Rubinson, P.F. Baker: Proc. Roy. Soc.,Ser.B 205 323-345 (1978)
34   D.J. Goldberg, J.E.Goldman, J.H.Schwartz: J. Physiol. 259 473-490 (1976)
35   S. Mackey, G. Schuessler, D.J. Goldberg, J.H. Schwartz: Biophys. J. 36 455-459 (1981)
36   J.H. Schwartz, D.J. Goldberg: In Axoplasmic Transport, D.G. Weiss (ed.) (Springer, Berlin 1982) 351-361
37   R.S. Smith: J. Neurocytol. 9 39-65 (1980)
38   M. Yano, Y. Yamamoto, H. Shimizu: Nature 299 557-559
39   M.P. Sheetz, J.A. Spudich: Nature 303 31-35 (1983)
40   E.A. Nothnagel, W.W. Webb: J. Cell Biol. 94 444-454 (1982)
41   R. Jarosch: Biochem. Physiol. Pflanzen 170 111-131 (1976)
42   R.D. Vale, B.J.Schnapp, T.S.Reese, M.P.Sheetz: Cell 40 449-454 (1985)
43   S. Higashi-Fujime: J. Cell Biol. 87 569-578 (1978)
44   T.Y. Wu: Fortschritte Zoologie 24 (2-3) 149-169 (1977)
45   E.M. Purcell: Am. J. Physiol. 45 3-11 (1977)
46   J.P. Heslop: Soc. Exp. Biol. Symp. XXVIII 209-227 (1974)
47   S. Tsukita, H. Ishikawa: J. Cell Biol. 84 513-530 (1980)
48   D.G. Weiss: In Axoplasmic Transport, Z. Iqbal (ed.) (CRC Press, Boca Raton Fl. 1985) in press
49   K. Kuroda, E. Manabe: Proc. Japan. Acad. 59 Ser.B, 131-134 (1983)
50   H.-U. Koop, O. Kiermayer: Eur. J. Cell Biol. 22 355 (1980)
51   W.D. Comper, B.N. Preston: Adv. Polymer Sci. 55 105-151 (1984)
52   B.N. Preston, W.D. Comper, L. Austin: J. Neurochem. 44 Suppl. S7 and S123 (1985)
53   W.D. Comper, N.P. Barry, L. Austin: Neurochem. Res. 8 943-953 (1983)
54   B. Corti: J. de Physique (Rosier) 8 232 ff (1776)
55   N.S. Allen, R.D. Allen: Annu. Rev. Biophys. Bioeng. 7 497-526 (1978)
56   B.H. Smith, G.W. Kreutzberg: Neurosci. Res. Prog. Bull. 14 (1976)
57   Y.A: Barde, D. Edgar, H. Thoenen: Annu. Rev. Physiol. 45 601-612 (1983)
58   M. Schwab: In Synergetics of the Brain, E. Basar, H. Flohr, H. Haken, A.J. Mandell (eds.),(Springer Series in Synergetics Vol. 23, Heidelberg 1983)

Part IV

Computers and Computing

Parallel Processes and Parallel Algorithms

F. Hossfeld

Zentralinstitut für Angewandte Mathematik, Kernforschungsanlage Jülich,
D-5170 Jülich, Fed. Rep. of Germany

1 Introduction

Whereas many processes are naturally parallel, our perception of processes involved in the dynamics of physical systems, and the design of algorithms in order
to model and compute the behaviour of such systems, has been dominated by the paradigm of the sequential flow of operations and data on which the breakthrough of
computational science has fundamentally been built, thus enforcing the appearance
- and the tremendous success - of the serial von Neumann type computer architecture.

Today, however, the novel category of parallel algorithms is becoming more and
more important in computing theory and practice [1], since due to the growing
pressure from advanced scientific and technical applications requiring computer
power far beyond the capacity of today's large-scale general purpose - i.e. serial
- computers [2], the increasing interest in parallel processing has enhanced the
developments in the design, analysis, and implementation of parallel algorithms
significantly; the promising progress in VLSI technology is favouring highly parallel computer structures additionally [3].

2 Parallel versus Sequential Processing

The expectations concerning parallel processing are based, on the one hand, on the
potentiality that a computer system consisting of p parallel processors, instead
of one, might be able to solve a problem with p-fold speed, and, on the other hand,
on the technological perspectives that the progress in VLSI will provide the means
to build parallel computers with many thousands, if not millions, of processing
elements which can be technically operated and economically justified.

A necessary prerequisite of these potential enhancements in computer power is the
parallelization of the problems in question. While, for instance, the addition of
two vectors in linear algebra can be easily transformed into parallel structure,
since the additions of the vector components are independent with respect to the
individual operands, there are computations which cannot be speeded up by any number of parallel processors; a well-known example is the computation of a (high)
power of x, say x^{2^k}. With a serial computer, k steps of successive squaring yield

194

the result; since within one step only the power x^2, in two steps only x^4, and so on, can be computed, this computation principally cannot be speeded up on a parallel processor system [4].

Therefore, the central issue of parallel processing aims at the design of originarily parallel algorithms whose performance can be related with optimal sequential algorithms. A parallel algorithm which solves a given problem of size n with P(n) parallel processors within the time T(n), yields a corresponding sequential algorithm which solves the same problem on an equivalent serial computer within the time P(n)·T(n). The product P(n)·T(n), therefore, has the lower bound in the time complexity of the best sequential algorithm known to solve that problem. A reasonable objective in designing parallel algorithms is to reach this target.

A major difficulty in designing and evaluating parallel algorithms is the strong dependency on the underlying architecture of the parallel processor system, as compared to sequential algorithms. On the one hand, the utmost abstraction from the specific computer structure is necessary to analyze the complexity of the algorithms, on the other hand the space of possible efficient parallel algorithms is bounded by the parallel processor even if it is described by an abstract model only. Although the body of theoretical knowledge about parallel algorithms is still fragmentary, this research field has already achieved a plethora of important results. However, many of these results have been of only academic interest, due to the idealistic computational models on which the design and analysis of these parallel algorithms have been built, ignoring widely the characteristics and restrictions in real parallel computer structures and, in particular, the implications of parallel algorithms with respect to the intrinsic data-transfer requirements. While in the sequential case it is sufficient to treat the time (and space) complexity of an algorithm, the performance of parallel algorithms is, in addition, essentially determined by the novel phenomena involved with the data communication complexity which takes into account the interaction between the algorithmic dynamics and the intrinsic structure of the flow of data with respect to specific parallel processor architectures.

2.1 Model of Parallel Computations

In order to design and analyze parallel algorithms, some model of computation has to be specified. In the literature, however, the model is based very often on idealistic or unrealistic conceptual characteristics including the assumption of the so-called free parallelism, which means that the number of available parallel processor elements is unlimited in the context of the parallel algorithm under consideration.

In many cases, the specifications of the parallel computer model are defined by the following characteristics:

(I)   The parallel processor system consists of p individual processors (where p might be infinite).

(II)   The computational problem to be solved by the parallel processor system can be represented as sequences of binary arithmetic and logic operations.

(III)   Each of the p processors can execute any of these operations at any time.

(IV)   Every operation can be executed within one unit of time (time step).

(V)   Other processes involved with the computational solution of the problem on the parallel processor (e.g. data transfers, control etc.) do not require any time to be executed.

(VI)   There are no data access conflicts.

To characterize the gain due to the parallel computation as compared with the sequential algorithm for the same problem, the "speedup" and "efficiency" are approved measures. Let $T_p(n)$ be the time complexity of an algorithm for a certain problem of size n on p processors (where n, as usual in complexity theory, is an adequate measure of the size of the input data). Then the speedup S_p and the efficiency E_p are defined as

$$S_p = T_1/T_p , \qquad E_p = S_p/p ,$$

with the trivial bounds $1 \le S_p \le p$ and $1/p \le E_p \le 1$.

Principally based on this model, Munro and Paterson [5] achieved the following result:

<u>Theorem:</u>

If the sequential computation of a single result consists of q binary arithmetic operations: $T_1 = q \ge 1$, the parallel computation utilizing p processors requires

$$T_p \ge \begin{cases} \lceil \log(q+1) \rceil , & \text{if } q < 2^{\lceil \log p \rceil}, \\ \lceil (q+1-2^{\lceil \log p \rceil})/p \rceil + \lceil \log p \rceil , & \text{if } q \ge 2^{\lceil \log p \rceil}. \end{cases}$$

(The logarithms are to base 2 throughout this paper.)

If the computation with binary operations is represented by binary trees where the leaves are the operands and the nodes correspond to the operations, the theorem of Munro and Paterson becomes plausible immediately.

Also utilizing binary tree structures, an effective design principle for parallel algorithms can be illustrated: the recursive doubling (Fig.1), which is a special case of tree-height reduction as a general design method [6]. Recursive doubling is equivalent to the principle of divide-and-conquer widely used in designing efficient sequential algorithms: a computation is recursively divided into two independent subtasks of identical complexity; but contrary to the sequential algorithm, the subtasks on each level of recursive decomposition are executed in parallel on

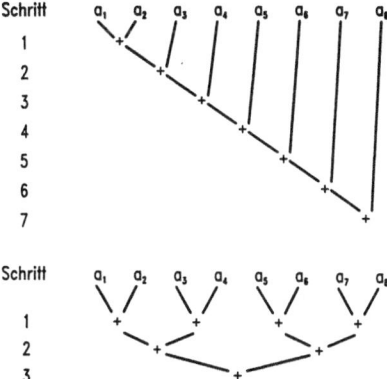

Schritt

1 2 3 4 5 6 7

a_1 a_2 a_3 a_4 a_5 a_6 a_7 a_8

Fig. 1 The Principle of Recursive Doubling
(Example: Summing up n=8 numbers)

Schritt

1 2 3

a_1 a_2 a_3 a_4 a_5 a_6 a_7 a_8

individual processing elements, where parallel processing is starting on the lowest level of independent (elementary) subtasks after feeding the input operands into the leaves of the equivalent binary tree [7].

If we consider the ordinary sequential multiplication of matrices,

C = A • B,

where A, B, and C are (nxn) square matrices, just for simplicity, the computation of each element c_{ij} requires n multiplications and (n-1) additions, thus leading to the cubic time complexity $n^2(2n-1)$:

$$T_1(n) = O(n^3).$$

For the optimal parallel algorithm we need n^3 parallel processors in order to execute the n multiplications in one time step for all inner products

$$c_{ij} = \sum_{k=1}^{n} a_{ik} b_{kj} \qquad , 1 \le i,j \le n,$$

simultaneously; based on the recursive-doubling principle, the n^2 summations of the individual products can be performed with $n^2 \cdot n/2$ processors by adding them pairwise in parallel, continuing with the parallel pairwise addition of the intermediate sums recursively, thus yielding each inner product in $\lceil \log n \rceil$ time steps. The time complexity of this parallel algorithm is therefore

$$T_p(n) = \lceil \log n \rceil + 1 = O(\log n)$$

where, in total, $P=n^3$ parallel processor elements are required. Comparing this result with the sequential algorithm of the ordinary type outlined above (or even with Strassen's algorithm /8/), the speedup by the parallel algorithm is tremendous:

$$S_p(n) = O(n^3/\log n).$$

Similar results hold true for many other computational problems, thus leading to very optimistic views on parallel computing.

2.2 Computation versus Communication

Referring to the computational model of a parallel processor system as described by the characteristic features (I) - (VI), the aspects of data access and data transfer within such a system have been ignored. Considering the principal structure of the parallel processor architecture which is illustrated in Fig. 2, however, the data communication requirements inherent in the parallel algorithms turn out to be crucial for the performance of parallel computers and parallel algorithms if implemented in hardware and software.

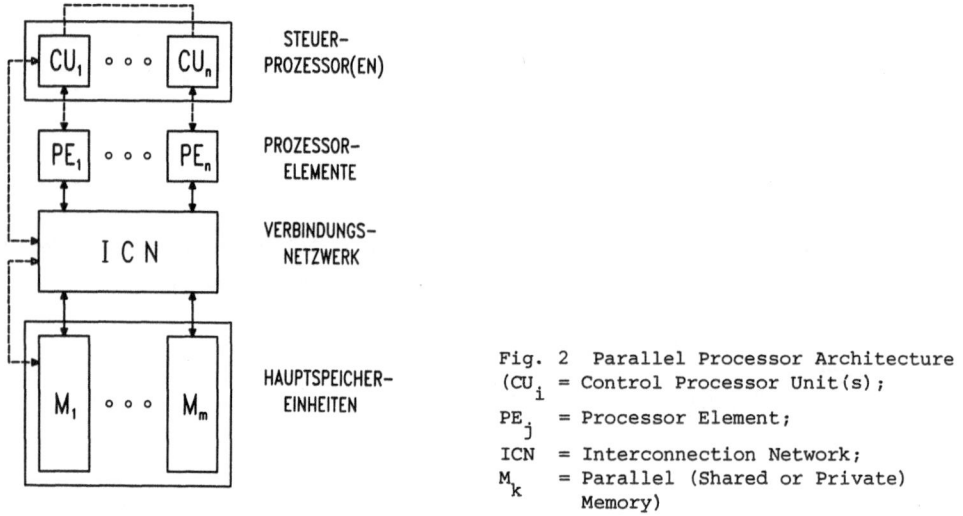

STEUER-
PROZESSOR(EN)

PROZESSOR-
ELEMENTE

VERBINDUNGS-
NETZWERK

HAUPTSPEICHER-
EINHEITEN

Fig. 2 Parallel Processor Architecture
(CU_i = Control Processor Unit(s);
PE_j = Processor Element;
ICN = Interconnection Network;
M_k = Parallel (Shared or Private) Memory)

If we ignore the architecture of the vectorprocessors,which is based on the pipeline principle [9] and which is dominating the field of high-speed computing today by systems like CRAY-1, CRAY X-MP, CDC CYBER 205, and the forthcoming supercomputers from Fujitsu, Hitachi, and NEC, not to forget the thousands of "array" processors installed from Floating Point Systems Inc., there are two main concepts of "genuine" parallel processors:

1) parallel processor with shared memory,
2) parallel processor with private memory;

198

in both concepts, the data communication requirements have to be satisfied by the novel architectural component of the so-called interconnection network. The two concepts, however, are characterized by the organizational structure of their main memory.

While already in general-purpose computers the organization of the main memory turned out to be a challenging design area because of the technologically determined gap between processor and memory cycle time which had to be partially, at least, bridged by the technique of memory interleaving [9], the organization of main memory in parallel processor architectures is of qualitatively and quantitatively new dimensions.

In the shared-memory concept, where all parallel processing elements utilize one common memory, the simultaneous access to this shared memory from different processors leads to access conflicts which deteriorate the performance of the parallel computers and algorithms, since the orderly provision of the processor elements with the data cannot be guaranteed. Thus, the access conflicts depend on the organizational structure of the shared memory, but also on the data structure utilized in the algorithm. The requirements of conflict-free access of parallel processors to memory has led to the theory - and practice - of skewing schemata, which are based on the mapping of "d-ordered N-vectors modulo M", well known in basic number theory (see [10] for a review of the present state of the art). The results can be implemented by structuring the shared main memory into parallel units, where the number of units is determined by selection rules from skewing theory.

In order to avoid these access conflicts, the shared memory can be distributed over the parallel processor elements by providing them with "private" memories each, thus leading to an alternative parallel processor architecture. In this concept, however, data communication is crucial, since the access by one processor to data produced by another processor and stored in its private memory has to be achieved by transmitting these data through the interconnection network, which now is the only data transfer vehicle in the system.

From a technical point of view, the interconnection network has to be as simple as possible, since the number of processors is intended to be very large in this architecture, in order to exploit the inherent parallelism. Since during execution of a parallel algorithm the necessary operands, which have to be available to a parallel processor element, are typically results of operations by other processors, they must be transferred to the processor requesting the operands according to a data-flow structure determined by the underlying algorithm.

Obviously, the models of parallel computing must include the costs of synchronization and data transfers, in addition to the operation counts for the computational steps of a parallel algorithm. It is necessary to define, in addition to the time (and space) complexity, a measure which takes into account these data communication costs: the communication complexity $C_p(n)$, which is counting the number of data transfer steps necessary for a parallel processor system. Thus, $C_p(n)$ depends on the algorithm, the problem size n and the interconnect structure of the parallel computer as well; hence, the design and analysis of parallel algorithms is getting even more complicated.

Hence, we have two options in order to design and implement parallel algorithms effectively: (1) We have to look for more powerful interconnections yielding sub-linear communication complexity; (2) We have to search for parallel algorithms which balance the computation time and the communication effort by the same com-plexity order.

3 Parallel Algorithms with Balanced Computation and Communication Processes

Many investigations have been dedicated to the complexity of parallel algorithms in various application fields (see, for instance, [11-13]). An extraordinary example of adapting the interconnection network to the computational and data-flow struc-ture of an algorithm (and vice versa) is represented by the parallel Fast Fourier Transform.

3.1 The Fast Fourier Transform

Cooley and Tukey have outlined already in 1965 in their original paper on the sequential FFT algorithm that it can be translated into a parallel algorithm [14]. The question was, however, whether the complicated data flow of the sequential FFT could be mapped into an efficient interconnection network coupling the parallel processors.

We will consider the case $N=2^n$. As is well known, the (complex) Discrete Fourier Transform (DFT) $Q(j)$ is computed from the (complex) Fourier coefficients $C_0(k)$ by the Fourier series

$$Q(j) = \sum_{k=0}^{N-1} C_0(k) \cdot w^{jk} \qquad , 0 \leq j \leq N - 1,$$

where w is the N-th root of unity: $w=\exp(2\pi i/N)$, $i=(-1)^{1/2}$. Thus, considering only the complex multiplication, the DFT yields the time complexity of the sequential algorithm:

$$T_1(N; DFT) = O(N^2).$$

If the indices j and k are represented as binary numbers

$$j = (j_{n-1}, \ldots, j_0),$$
$$k = (k_{n-1}, \ldots, k_0),$$

the vector Q of the Fourier Transform can be computed as an n-fold summation over the binary k_r, $0 \leq r \leq n - 1$:

200

$$Q(j_{n-1}, \ldots, j_0) = \sum_{k_0} \sum_{k_1} \cdots \sum_{k_{n-1}} C_0(k_{n-1}, \ldots, k_0) \cdot w^{jk}.$$

Based on this relation, the recursive Cooley-Tukey algorithm can be established:

$$C_1(j_0, k_{n-2}, \ldots, k_0) = \sum_{k_{n-1}} C_0(k_{n-1}, \ldots, k_0) \cdot w^{jk_{n-1}2^{n-1}} \tag{1}$$

$$C_s(j_0, \ldots, j_{s-1}, k_{n-s-1}, \ldots, k_0) = \tag{2}$$

$$\sum_{k_{n-s}} C_{s-1}(j_0, \ldots, j_{s-2}, k_{n-s}, \ldots, k_0) \cdot w^{(j_{s-1}2^{s-1} + \ldots + j_0)k_{n-s}2^{n-s}},$$

for $2 \leq s \leq n$.

This special case of the sequential FFT yields the time complexity

$$T_1(N; \text{FFT}) = O(N \log N).$$

The logical order of the tranform is obtained by bit reversion:

$$Q(j_{n-1}, \ldots, j_0) = C_n(j_0, \ldots, j_{n-1}).$$

This recursive algorithm is easily translated into a parallel algorithm. Introducing the binary representations

$$r = (r_{n-1}, \ldots, r_0)$$
$$f(r, k) = (r_{n-1}, \ldots, 0, r_{n-k-1}, \ldots, r_0)$$
$$h(r, k) = (r_{n-1}, \ldots, 1, r_{n-k-1}, \ldots, r_0)$$
$$g(r, k) = (r_{n-k}, \ldots, r_{n-1}, 0, \ldots, 0)$$
$$\text{rev}(r) = (r_0, \ldots\ldots\ldots, r_{n-1}) = g(r, n)$$

and defining the operations as to be executed in parallel for the set of indices in brackets, the parallel FFT for the case $N=2^n$ can be written in the following form:

$$z_j = w^j, \quad (0 \leq j \leq N-1);$$
$$c_j = C_0(j), \quad (0 \leq j \leq N-1);$$

for k=1 step 1 until n do

$$C_j = C_{f(j,k)} + Z_{g(j,k)} \cdot C_{h(j,k)} \qquad , \quad (0 \le j \le N-1);$$

$$Q_j = C_{rev(j)} \quad , \quad (0 \le j \le N-1).$$

This parallel algorithm yields the time complexity

$$T_p(N; \text{FFT}) = O(\log N)$$

utilizing P=N processor elements.

If we consider the data-flow graph corresponding to the sequential Cooley-Tukey algorithm, as illustrated in Fig. 3, which in each recursive step s brings together pairs of operands (according to the "butterfly" operation [15]) which are initially separated by a displacement 2^{n-s}, we can map this data-flow graph into an isomorphous data-flow graph (Fig. 4) which consists of n successive perfect-shuffle

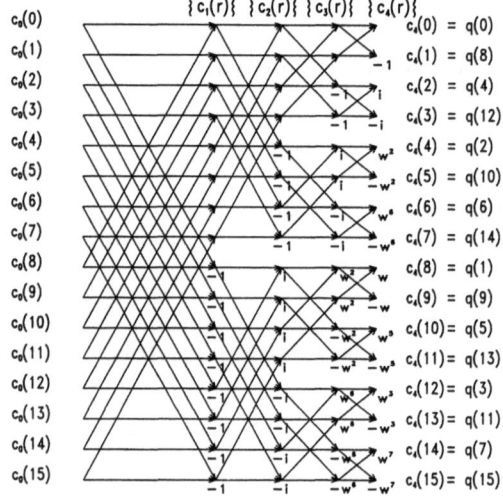

Fig. 3 Data-Flow Graph of the sequential FFT Algorithm ($N=2^4$)

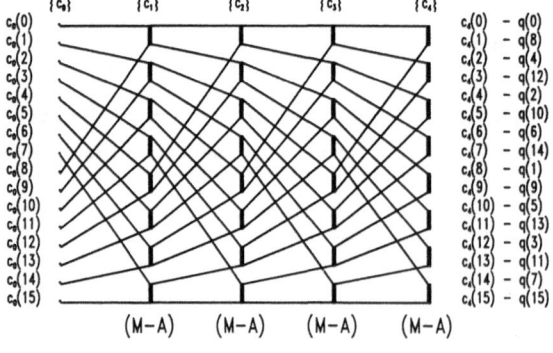

Fig. 4 Isomorphous Data-Flow Graph of the FFT with $n=\log N$ Perfect-Shuffle Stages ($N=2^4$)

stages with N/2 "butterfly" processors on each stage which perform the complex multiplication of one operand with the adequate power of w as well as the addition,
and, respectively, subtraction of this product to or, respectively, from the other
operand fed into the processor element [16].

The perfect-shuffle interconnection realizes the following connections between P =
2^q processor elements PE(i), where $q \geq 2$:

$$PE(i) \quad \longrightarrow \quad PE(PS(i)),$$

where the operation PS(i) performs a left-shift with cyclic closure on the bit
pattern of the address i:

$$PS(i) \quad = \quad (i_{q-2}, \ldots, i_0, i_{q-1}).$$

In this parallel FFT utilizing the perfect-shuffle interconnection, the structure
of the parallel computations and the parallel flow of data are optimally balanced
yielding O(logN) each. Therefore, this parallel procedure can easily be mapped onto
an integrated parallel FFT computer (Fig. 5).

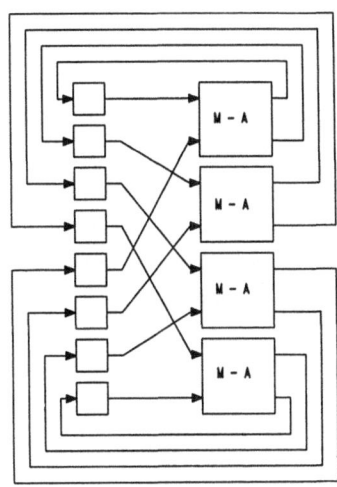

Fig. 5   Integrated Parallel FFT Processor
$(N=2^3)$

3.2 Linear-Time Systolic Algorithms

Due to the alternating processes of arithmetic operations and synchronous data
transfers, in analogy to the rhythm of heart function and blood flow through the
veins, algorithms like the FFT on a shuffle-connected parallel processor are called
"systolic" algorithms [16].

Since this initiating paper of Kung and Leiserson, an important progress is seen in
the development of linear-complexity systolic algorithms which can be mapped onto
regular "systolic arrays" in VLSI technology, thus harmonizing the arithmetic compu-

tational steps and the data-transfers. This leads to parallel hardware algorithms with already wide spread applications [16-18].

The origin of such algorithms may be dated well back in 1969 when Cannon designed a parallel multiplication algorithm for (NXN)-square matrices C = A $\cdot$ B on a two-dimensional (NXN)-mesh-connected computer which, compared with the optimal algorithm yielding O(logN) time complexity, shows a linear dependency on the matrix size N. The Cannon algorithm, thus, provides a comfortable compromise: $T_p(N)$ = O(N) with P = N^2 processors.

If we consider again the inner-product form of matrix multiplication

$$c_{ij} = \sum_{k=0}^{N-1} a_{ik} b_{kj} \quad , \quad 0 \le i, j \le N-1,$$

and if we introduce the modulo function, we can substitute k = (i+j-s) modN yielding

$$c_{ij} = \sum_{s=0}^{N-1} a_{i,\ (i+j-s)\bmod N} \cdot b_{(i+j-s)\bmod N,j}$$

$$= \sum_{r=0}^{N-1} a^*_{i,(j-r)\bmod N} \cdot b^*_{(i-r)\bmod N,j} \quad .$$

We store the elements of the square matrices A and B into the registers of the N^2 processor elements of such a nearest-neighbours interconnected array (Fig. 6) in such a skewed way that row i of A is shifted to the left by i columns, column j of B is shifted upward by j rows, while simultaneously closing rows and columns cyclicly (Fig. 7); the multiplication algorithm proceeds rhythmically by multiplying first the element of A stored in register AREG(i,j) with the element of B stored in

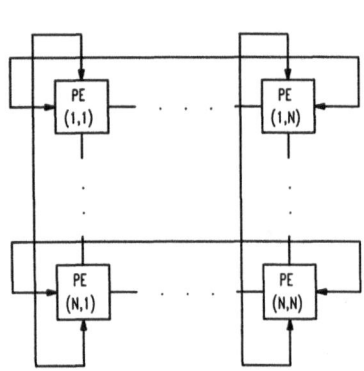

Fig. 6 (NxN)-Array of Nearest-Neighbour Connected Processor Elements (with Cyclic Boundary Connections)

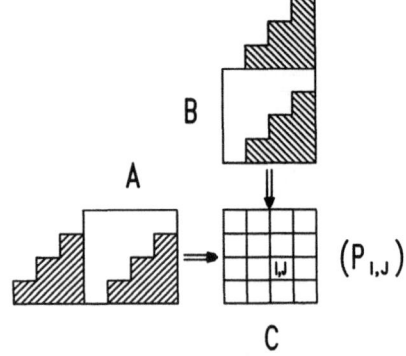

Fig. 7 Initial Skewed Arrangement of the Square Matrices A and B in the Cannon Algorithm of (Systolic) Matrix Multiplication ($P_{i,j}$= Processor Element of the (NxN)-Array yielding C=A$\cdot$B)

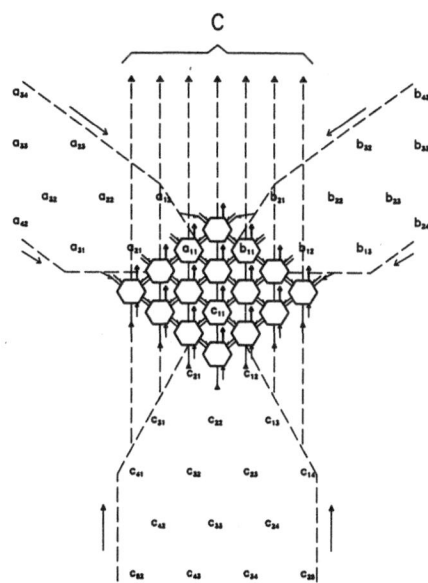

Fig. 8 Hexagonal Systolic Array for
Banded-Matrix Multiplication

register BREG(i,j) of processor PE(i,j) and adding the product to the accumulated
intermediate result in register CREG(i,j), simultaneously in all PE(i,j), $0 \leq i, j$
$\leq N-1$, and then shifting the elements of A to their neighbours on the right-hand
side and the elements of B to their lower neighbours and repeating the arithmetic
operations of multiplication and addition.

After N rhythmic steps alternating data-shifts and arithmetic operation, the product
matrix C is obtained, c_{ij} being stored in PE(i,j), $0 \leq i, j \leq N-1$; thus, T_p(N;Cannon)
= O(N), balancing computation and communication in this systolic algorithm and array.
Modifying the processor array to a hexagonal structure, the multiplication of banded
matrices can be mapped into a systolic algorithm (Fig. 8). By a minor change of the
connection paths,this hexagonal array can be utilized for the LU-decomposition of the
matrix C if C is shifted through the processor array as indicated in Fig. 8 [16].

This systolic LU-decomposition of an (nxn)-matrix C is based on the assumption that
pivoting is not required; in this case, the LU-process can be written in the recurrent way:

$$c_{ij}^{(1)} = c_{ij} , \qquad 1 \leq i, j \leq n;$$

$$c_{ij}^{(k+1)} = c_{ij}^{(k)} - l_{ik} u_{kj}, \qquad 1 \leq i,j \leq n, \ 1 \leq k \leq n;$$

$$l_{ik} = \begin{cases} 0 & \text{if } i < k, \\ 1 & \text{if } i = k, \\ c_{ik}^{(k)} u_{kk}^{-1} & \text{if } i > k, \end{cases}$$

$$u_{kj} = \begin{cases} 0 & \text{if } j < k, \\ c_{kj}^{(k)} & \text{if } j \geq k, \end{cases}$$

$$1 \leq i,j,k \leq n.$$

The recurrence system can be derived from the sequential algorithm:

```
LU-ALGORITHM:
    for k = 1  to n - 1 do

            find l so that
```
$$\left| C(l,k) \right| \; = \; \max \; (\left| C(k,k) \right|, \; \ldots, \; \left| C(n,k) \right|)$$

$(T_k^k):$

$PIV(k) = l$	pivot row
$C(PIV(k),k) \;\; = \;\; C(k,k)$	exchange
$r \; = \; 1/C(k,k)$	
for i = k+1 to n do	
$\quad C(i,k) \; = \; r * C(i,k)$	L-elements

```
            for  j = k+1  to n do
```

$(T_k^j):$

$C(PIV(k),j) \;\; = \;\; C(k,j)$	exchange
for i = k+1 to n do	
$C(i,j) = C(i,j) - C(i,k)*C(k,j)$	U-elements

This algorithm,which includes pivoting,can be structured into the task sequence

$$(T_k^j \; , \;\; 1 \le k \le j \le n),$$

which can be mapped onto a task-dependency graph (Fig. 9) which illustrates that the tasks

$$(T_k^{k+1}, \; T_k^{k+2}, \; \ldots\ldots, \; T_k^n \; ; \;\; 1 \le k < n)$$

are mutually independent [19].

Therefore, a parallel LU-algorithm can be designed which will efficiently run on an MIMD parallel computer (MIMD = <u>M</u>ultiple <u>I</u>nstruction streams – <u>M</u>ultiple <u>D</u>ata streams [20]). With $P = \lceil n/2 \rceil$ processors, the tasks can be distributed in a way that processor j will execute the task sequence

$$T_1^{2j-1}, \; T_1^{2j}, \; T_2^{2j}, \; T_2^{2j+1}, \; \ldots\ldots, \; T_{n-2(j-1)}^n \; ;$$

thus yielding the time complexity of this parallel algorithm from the longest path in the dependency graph [19]

$$T_p(n;LU) \;\; = \;\; O(n^2).$$

The sequential algorithm has the well-known cubic complexity:

$$T_1(n;LU) \;\; = \;\; O(n^3),$$

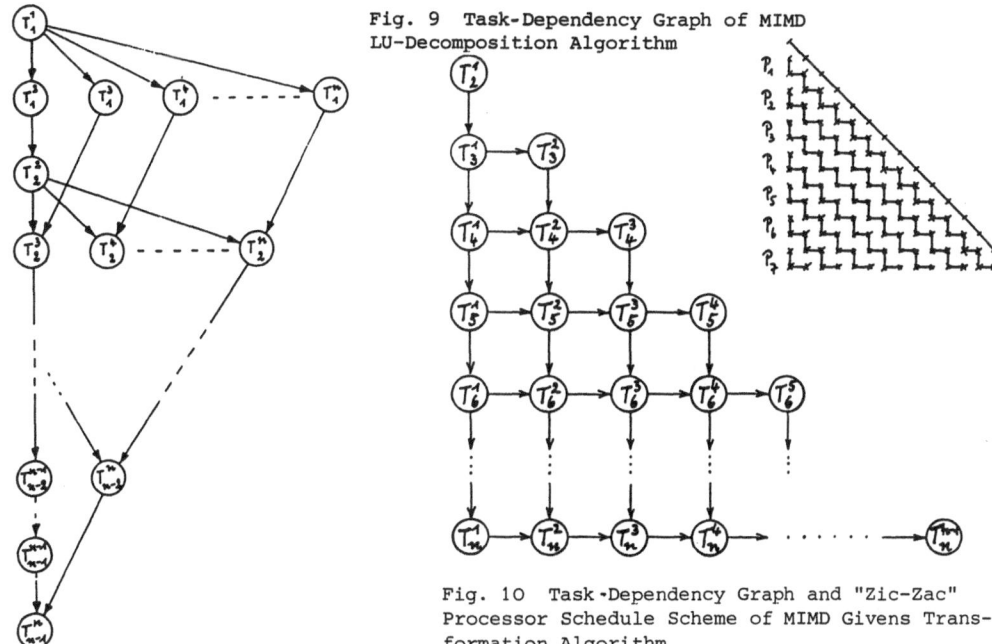

Fig. 9 Task-Dependency Graph of MIMD
LU-Decomposition Algorithm

Fig. 10 Task-Dependency Graph and "Zic-Zac"
Processor Schedule Scheme of MIMD Givens Transformation Algorithm

while the systolic algorithm is a linear-time algorithm [16].

For stability reasons, usually, pivoting cannot be omitted in LU-decomposition;
therefore, the systolic LU-algorithm will be limited to well-defined problems. This
disadvantage of the systolic LU-algorithm can be overcome by applying the Givens
transformation in order to reduce a matrix A into (upper) triangular form [18]. The
Givens algorithm is based on the orthogonal factorization of the real, non-singular
(nxn)-matrix $A = (a_{ij})$:

$$Q \cdot A = R,$$

where Q is an orthogonal matrix which is built as the product of elementary rotations; R is the resulting upper-triangular matrix.

From the task structure according to a maximum parallel dependency graph (Fig. 10),
a parallel MIMD-algorithm has been designed for $P = \lceil (n-1)/2 \rceil$ parallel processors
[19] yielding

$$T_p(n; \text{ Givens}) = O(n^2),$$

compared to the sequential Givens transformation algorithm with

$$T_1(n; \text{ Givens}) = O(n^3).$$

From a detailed analysis of the Givens transformation and the data flow structure
involved, a nearest-neighbour connected triangular systolic array of processing
elements $PE(j,k)$, $1 \le k < i \le n$, has been constructed for a systolic algorithm

207

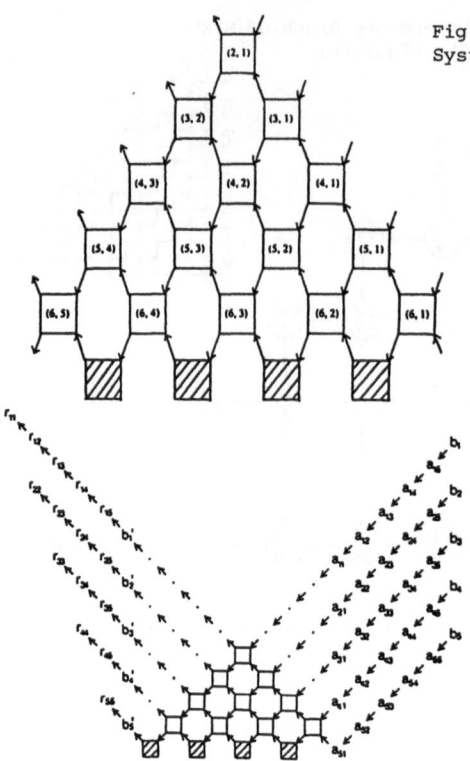

Fig. 11 Processor Array and Data Flow of
Systolic Givens Transformation Algorithm

(Fig. 11). Although it realizes the Givens rotations and the data communication
streams in linear time, it yields a numerically stable solution of dense systems of
linear equations [18].

There is strong scientific and technological emphasis on the development of systolic
systems. So far, these efforts have resulted in an already broad spectrum of systolic
algorithms and arrays:

- matrix multiplication of dense and banded matrices,
- LU decomposition of dense and banded matrices,
- Givens rotation of (banded) matrices,
- inversion of triangular and symmetric matrices,
- (triangular) systems of linear equations,
- Toeplitz systems of equations,
- eigenvalues of symmetric matrices,
- least squares problems,
- convolution, filtering; DFT,
- geometric problems (convex hull),
- greatest common divisor of polynomials and binary numbers; polynomial
  multiplication and division,
- interpolation,
- dynamic programming.

4 Parallel Processes and Cellular Automata

In a sense, the systolic arrays and algorithms balancing the parallel flow of computation and communication on a lattice of processing elements are in structural
accordance with the traditional strategy which builds the models of natural systems
on the discretization of time and space in order to analyze the dynamics and evolution of these systems; the space-time discretization in the numerical treatment of
differential equations, which form the mathematical basis for most current models of
natural systems, is the most prominent category.

On the other hand, systolic structures are quite similar to a category of mathematical models which just recently seems to gain new scientific interest in the investigation of physical systems, showing complex, self-organizing and "chaotic" behaviour: the cellular automata [21].

Cellular automata have five fundamental defining characteristics:

(1) They consist of a discrete (k-dimensional) lattice of sites.

(2) They evolve in discrete time steps.

(3) Each site takes on a finite set of possible values.

(4) The value of each site evolves according to the same deterministic rules.

(5) The rules for the evolution of a site depend only on a local neighbourhood of
    sites around it.

Cellular automata provide rather general discrete models for the evolution of parallel processes in homogeneous systems with local interactions [22]. Some generalizations introducing randomness even make cellular automata analogous to lattice
spin systems, which are widely studied with respect to phase-transitions by high-
speed computers [2]. As originally introduced by von Neumann and Ulam, cellular
automata are also used as models for biological systems and, again with extensions
for certain randomness, for neural networks [21].

Whereas part of the growth of interest in cellular automata today may be due to the
availability of high-speed computing facilities, which allow for extensive simulations of cellular automata, the study of these parallel processes may also influence the design of future parallel processing computers far beyond the concept of
systolic arrays; the progress in the high-integration technologies may, on the one
hand, provide necessary prerequisites, while, on the other hand, theoretical
equivalences between certain cellular automata and universal computers may contribute the appropriate conceptual basis for this purpose.

Whereas most parallel computer concepts which have been pursued so far involve a relatively small number of high-level processors, the cellular automata, as well as the systolic arrays, suggest the design of parallel computer systems with very large numbers of relatively primitive processing elements. Such parallel processing systems seem to have better capabilities to map natural systems and their parallel space and time evolution into computer architecture. However, while the algorithmic analysis has initiated the design of systolic structures, the programming and control of parallel processing systems which may imitate cellular automata will be a significant challenge.

5 Conclusion

While parallel algorithms were originally designed for linear algebra problems which are dominating the field of scientific-technical applications, new areas are of growing interest. Increasing research activities are focussing on parallel graph algorithms (see [13] for a survey); also systolic algorithms have already been designed for graph problems [23, 24] and even for relational data base operations [25, 26], thus entering the field of important non-numerical applications. These research areas will expand further due to the impact of artificial intelligence and the computer projects of the next generations, where the issue of parallel processes and algorithms will become crucial.

In addition, the progress in new circuit technologies will give impact to the design of systolic algorithms with rather higher than two-dimensional operation and data flow giving rise to innovative 3D-systolic hardware structures. (This may also influence the further role of cellular automata.)

Unfortunately, we must admit that, compared to high-speed computer architecture and hardware as well as parallel algorithm design, relatively limited efforts have been dedicated so far to develop adequate "parallel" languages which are capable to principally enhance the exploitation of the potential of parallel processing. In addition, there is, of course, still a lack in parallel-computing methodology. Progress in this field is also required with respect to parallel numerical mathematics in order to investigate, for instance, the stability of parallel algorithms.

Literature:

1   Hossfeld, F.; Weidner, P.: Parallele Algorithmen, Informatik-Spektrum 6 (1983), 142.

2   Hossfeld, F.: Nonlinear Dynamics: A Challenge on High-Speed Computation, Proc. Int. Conf. Parallel Computing 83 (M. Feilmeier et al., eds.), North-Holland 1984, pp. 67.

3   Uhr, L.: Algorithm-Structured Computer Arrays and Networks, Academic Press, 1984.

4   Borodin, A.; Munro. I.: The Computational Complexity of Algebraic and Numeric Problems, Elsevier, 1975.

5   Munro, I.; Paterson, M.: Optimal Algorithms for Parallel Polynomial Evaluation, J. Computer and Syst. Sci. 7 (1973), 189.

6   Dekel, E.; Sahni, S.: Binary Trees and Parallel Scheduling Algorithms, IEEE Trans. Computers C-32 (1983), 307.

7   Horowitz, E.; Zorat, A.: Divide-and-Conquer for Parallel Processing, IEEE Trans. Computers C-32 (1983), 582.

8   Aho, A.V.; Hopcroft, J.E.; Ullman, J.D.: The Design and Analysis of Computer Algorithms, Addison-Wesley, 1974.

9   Hockney, R.W.; Jesshope, C.R.: Parallel Computers, Hilger, 1981.

10 van Leeuwen, J.; Wijshoff, H.A.G.: Data Mappings in Large Parallel Computers, Proceedings GI-13. Jahrestagung, Informatik-Fachberichte Bd. 73, Springer, 1983, pp. 8.

11 Heller, D.: A Survey of Parallel Algorithms in Numerical Linear Algebra, SIAM Rev. 20 (1978), 740.

12 Bitton, D.; De Witt, D.J.; Hsiao, D.K.; Menon, J.: A Taxonomy of Parallel Sorting, ACM Computing Surveys 16 (1984), 287.

13 Quinn, M.J.; Deo, N.: Parallel Graph Algorithms, ACM Computing Surveys 16 (1984), 319.

14 Cooley, J.W.; Tukey, J.W.: An Algorithm for the Machine Calculation of Complex Fourier Series, Math. Comp. 19 (1965), 297.

15 Nussbaumer H.J.: Fast Fourier Transform and Convolution Algorithms, Springer, 1981.

16 Kung, H.T.; Leiserson, C.E.: Systolic Arrays (for VLSI), Sparse Matrix Proceedings 1978, SIAM 1979, p. 256.

17 Brent, R.P.; Kung, H.T.; Luk, F.T.: Some Linear-Time Algorithms for Systolic Arrays, IFIP 83, Elsevier/North-Holland, 1983, p. 865.

18 Bojanczyk, A.; Brent, R.P.; Kung, H.T.: Numerically Stable Solution of Dense Systems of Linear Equations Using Mesh-Connected Processors, SIAM J. Sci. Stat. Comput. 5 (1984), 95.

19 Lord, R.E.; Kowalik, J.S.; Kumar, S.P.: Solving Linear Algebraic Equations on an MIMD Computer, J. ACM 30 (1983), 103.

20 Hwang, K.; Briggs, F.A.: Computer Architecture and Parallel Processing, McGraw-Hill, 1984.

21 Farmer, D.; Toffoli, T.; Wolfram, S. (eds.): Cellular Automata, Proceedings of an Interdisciplinary Workshop , Los Alamos. Special Issue of Physica 10D, North-Holland, 1984.

22 Wolfram, S.: Statistical Mechanics of Cellular Automata, Rev. Mod. Phys. 55 (1983), 601.

23 Nassimi, D.; Sahni, S.: Finding Connected Components and Connected Ones on a Mesh-Connected Parallel Computer, SIAM J. Computing 9 (1980), 744.

24 Hambrusch, S.E.: VLSI Algorithms for the Connected Component Problem, SIAM J. Computing 12 (1983), 354.

25 Kung, H.T.; Lehman, P.L.: Systolic (VLSI) Arrays for Relational Data Base Operations, Proc. ACM-SIGMOD 1980 International Conference on Management of Data, ACM, 1980, p. 105.

26 Lehman, P.L.: A Systolic (VLSI) Array for Processing Simple Relational Queries, in: H.T. Kung et al. (eds.), VLSI Systems and Computations, Carnegie-Mellon University, Computer Science Press, 1981, p. 285.

Massively Parallel Multi-Computer
Hardware=Software Structures for Learning

L. Uhr

Department of Computer Sciences, University of Wisconsin,
Madison, WI 53706, USA

1. Abstract

This paper suggests how appropriately structured hardware/software multi-computers might be built and used to explore ways that intelligent systems can evolve, learn and grow. It examines what computers are, the great variety of topologies that can be used to join large numbers of computers together into massively parallel multi-computer networks, and the great sizes that today's and tomorrow's micro-electronic VLSI ("very large scale integration") technologies make feasible. Then it describes several multi-computer structures that appear to be especially appropriate as the substrate for systems that evolve, learn and grow, and begins to sketch out a system of this sort.

2. Introduction

How can "mind" arise in a biological/chemical/physical world? At some points in some evolutionary processes special nodes appear that, more or less loosely, might be characterized as "intelligent," "percipient," "thoughtful," "conscious," "mind." What might such nodes look like? Can we define systems where such nodes tend to form? Can we define systems that continue to form and re-form such nodes, tending to build successively more complex and more powerful nodes of this sort, using the same construction principles, or construction principles that themselves have been formed in this self-organizing evolving/learning process?

2.1 Multi-Computer Networks, Programs, and Intelligence

Computers - both the basic concept and the massively parallel multi-computers that are necessary for realistic attempts to use them to achieve intelligence - may well throw important light on these vital issues.

A basic "universal," "general-purpose" computer is an extremely and elegantly simple device. Even the very smallest and slowest general-purpose computer (and virtually all of today's computers are general-purpose) is capable, given enough time, of computing anything describable, hence anything that any other computer, no matter how powerful, can compute.

2.2. The Working Hypothesis that Computers/Programs Can Be Intelligent

There is good reason to believe - that is, to accept as a working hypothesis well worth pursuing - that among all the different processes that a computer can potentially accomplish (when given the necessary program and/or hardware structure) are perceiving, thinking, remembering, understanding, learning, ...; that is, processes people are inclined to characterize as "intelligent." It is important to emphasize that this is an hypothesis, to be confirmed or denied by empirical evidence. It will be confirmed when we achieve a hardware/software structure that really learns, thinks and understands. It will be denied when we achieve a barrier to such a sensational positive achievement, plus a proof that this barrier is absolute.

2.3. The Powerful Multi-Computer Structures Now Becoming Feasible

There appear to be a number of potentially extremely powerful topologies that might be used to construct multi-computers with many thousands, millions, or even billions of individual processors. For example, today (ignoring the major but one-time costs of the research needed to achieve the original discoveries, inventions and wise choices and of the design and development of the hardware and software), a network of a million (small) computers could be built today for less than a million dollars [1]. And the technologies involved are advancing so rapidly that we can double the size (or cut the price in half) roughly every 12 to 20 months into the foreseeable future (that is, for the next 10 to 20 years)!

The real problem is to develop a good enough understanding of the problem to create good hardware designs and good programs to run and use this hardware.

3. *Universal Turing Machines, Finite State Devices, General-Purpose Computers*

There have been a number of independent inventions of the underlying theoretical concept of computers, proofs as to their generality, and proofs as to their equivalence. Especially interesting is the relatively little known fact that Emil POST, in 1936 [2], and Alan TURING, also in 1936 [3], independently proposed the theoretical construct that has come to be known as the "Turing machine" (a very simple kind of computer with a read-shift-write "head" that can move over, input from, and output to a potentially infinite "tape") *that is capable of doing anything that any other computer might conceivably do - given enough time.* All it does is have the "head" execute a sequence of instructions (stored on its tape) from the simple repertoire:

> READ the current symbol;
> SHIFT to the next symbol;
> WRITE the current symbol onto the current memory/tape location;
> IF the current symbol is x, THEN DO instruction i, ELSE DO j;

Rather than read, process, write and store everything one bit at a time, a system can be defined to work with B-bit entities (encoding numbers and symbols). The system can further be extended to use 2, 3, or any number of read-process-write heads. All might work with one common tape; or each might have its own tape; or several might share the same tape. All such systems have been proved to be equivalent. That is, the very simplest single-head Turing machine that uses the very simplest 1-bit processor working on 1-bit pieces of information is capable of executing any program that any other N-bit Turing machine, with no matter how many processors or tapes, can execute. This is the basis for using a 1-processor serial computer - it is capable of doing anything that any multi-computer, no matter how large, can do.

3.1. *Finite State Machines, McCulloch-Pitts Neurons, Petri Nets*

No living brain, indeed no actual computer, can have an arbitrarily large "potentially infinite" memory. The Turing machine is important as a theoretical construct that allows us to explore the capabilities of computers and prove their generality. But every actual multi-computer that is built, and every program that it executes, will be finite, hence equivalent to a finite-state automaton rather than a universal Turing machine.

A Turing machine's tape memory can be replaced by a set of IF-THEN rules - which can be embedded in a set of processors that actually execute these rules. This embodies a particular program into a particular equivalent automaton, replacing the Turing machine's memory (which contains the program's instructions and intermediate results). It is no longer a general-purpose system, since it does not have a potentially infinite memory, and it executes only its own program.

Warren MCCULLOCH and Walter PITTS [4] developed a calculus of networks built from extremely simple neuron-like components. A neuron either fires a +1 or an absolute inhibition into another neuron's synapse, whose threshold is a positive integer value. A neuron fires whenever enough neurons fire into it to exceed its threshold, but no inhibitory neurons fire into it. Steven KLEENE [5] proved that McCulloch-Pitts neurons are equivalent to finite-state automata. More efficient variants can be defined, by assigning values other than 1 and absolute inhibition to firings - e.g., real or integer numbers, either positive plus inhibition, or positive and negative. These serve to give simpler and more realistic nets; but an equivalent McCulloch-Pitts net can always be constructed.

C. A. PETRI [6] formulated a system that specifies more about the actual flow of information (data-flow). A Petri net is a directed bipartite graph (a graph is a set of nodes joined by links) built from "places" linked to "transitions." A "token" is put on a place, "filling" it. A transition "fires" as soon as all its input places are full; the result is that one token is removed from each of its input places and one token is put on each of its output places. A token can be interpreted as a piece of information, the result of input or a previous process. A transition can be interpreted as a process that is executed as soon as all its necessary inputs (tokens) have been received. Thus the Petri net actually specifies how information flows into a process, possibly forming queues, and how the processor actually processes that information.

213

3.2. The Basic Structure of a Single-CPU General-Purpose Serial Computer

The conventional 1-CPU serial computer has three basic parts:
  A) The Central Processing Unit (CPU);
  B) Memory storage;
  C) Input-and-Output (I-O) Devices.
It has the same structure as a single-head Turing machine, with modifications:

. ALL information, including data to be operated upon AND ALSO the actual program that specifies the set of operations to be executed, are input to and stored in the Memory.

. The CPU typically contains a processing unit, that actually executes the program's instructions, and a controller, that gets and decodes the program's instructions and, in general, controls and runs the show, telling the computer exactly what to do.

. The use of a "random-access memory" ("RAM"), where any location can be accessed by a single "fetch" or "store" instruction, means that the system does not need to shift one location at a time along a linear tape. Any piece of data, and any instruction, can be fetched immediately. The use of an 8-bit, 16-bit, 32-bit, or 64-bit "word" of information that is (in parallel) fetched, stored, and operated on means that the system need not do everything "bit-serially" (that is, one bit at a time). These two improvements (plus electronic technologies that run at nanosecond speeds) turn the excessively slow single-head Turing machine into very fast computing engines.

A computer can be built with any number of CPUs. But almost all computers built to date have been of the very simplest sort, with only one single processor. This kind of "general-purpose single-CPU serial stored-program digital computer" executes only one instruction at a time, sequentially.

4. State-Driven, Rather than Programmed, Devices

The salient difference between a Turing machine and what are often thought to be the much more limited finite state machine, McCulloch-Pitts neuron network, and Petri (data-flow) net, is the Turing machine's "potentially infinite memory" on which both the program's operations and the information to be operated on are stored. But a program can either be a structure of symbols, to be interpreted by hardware, or an actual hardware structure. There are two ways (these can be mixed together) that the execution of a program can be handled:

A) The program can be stored in memory (much as though written on paper). The computer then looks at and follows each instruction: it fetches it; decodes it into signals; sends these signals to set its hardware switches to fetch the required pieces of information; pumps this information through the indicated structure of processing switches; stores the results where indicated; and gets the next instruction.

B) The program can be built into the hardware. That is, all the switches can be positioned exactly as the program would have positioned them, and the information pumped directly through them. Now each instruction specifies a particular small graph-structured piece of hardware - the logic gates (built from switches) needed to execute that instruction. The instruction's structure of processes (encoded in symbols) is thus crystalized into a quite similar structure of (hardware) switches linked by wires.

4.1. Information-Flow Hardware/Software Systems

Now, rather than fetch information from memory, operate on it, and store it back into memory, we can do something much more elegant. We can link the structure that executes the next instruction's operation directly to this instruction's structure. Thus a total structure is built where the outputs from each operation input directly to the gates that execute the next operation. This gives a hardware structure almost identical to the structure of the Petri net that describes the particular structure of operations that it executes.

We can build this kind of direct hardware embodiment for any specific program. Now data are actually input not to a general input device that immediately stores them on a disk or some other suitably large general memory, but to the actual structure of gates that will execute the program's first process. The results are then similarly input to the next structure(s) of gates, and the process continues....

The close resemblance to networks of neurons through which impulses that code information flow is not at all coincidental, and should be taken very seriously. Such an information-flow "data-flow" system is an optimal design for the algorithm it reflects. It uses less hardware and less energy than a general-purpose computer, and it is substantially faster. It is not capable of executing any possible program, because it has no "potentially infinite" tape on which all programs are stored, to be decoded and executed. But the ability to add to and change its structure would be equivalent. This building and rebuilding might actually change hardware, or take place within the limits imposed by an appropriate pre-built multi-computer topology.

4.2. *Perceiving and Learning from Environments Result in Potentially Infinitude*

We human beings, along with all other living, hence finite, animals, do not have within the shell of our skin a potentially infinite memory. But we may well have a potentiality to learn, remember and know so much more than we can ever achieve that our memory is effectively potentially infinite with respect to the use we will ever make of it.

We also have a second resource that may be even more powerful. We have the external world around us. Our "input" devices are our eyes, ears, skin and other sensors; our "output" devices are our legs, arms, fingers, mouth, etc. The Turing machine's "head" serves both as its "input" and "output" devices, and also its "fetch" and "store" device. The tape serves both as internal directly accessible memory and also as all auxiliary memories (disks, magnetic tapes, punched cards, or whatever), and - an unspecified and slightly mysterious mechanical detail - as the transducer phases of the input and output devices.

In this same somewhat mysterious sense, it seems fruitful to view the whole real world, or more exactly the immediate and potentially reachable sensible environment around the system - the aspects of the real world with which the system interacts and can, potentially, interact - as part of the system's "tape." This is still finite, but it is certainly large enough.

Thus it seems illuminating and fruitful to think of the surrounding environment as a vital part of the intelligent system's stores of information. We human beings make obvious use of this memory all the time, when we look around, go to the library or the Galapagos, read books, write books, jot down notes or intermediate results. What may well be an even more important use is central to perception and learning - simply looking again; that is, being able to re-input, observe and study information from the environment which, although it will inevitably be changed at least a tiny bit, will usually be substantially the same.

5. *The Need for, and the Possibility of, Massively Parallel Multi-Computer Structures*

A number of programs have been coded for conventional computers that model large networks of neuron-like components. But even the largest single-CPU serial computer can handle (taking hours or days of computer time) only a few minutes of activity of only a few hundred or a few thousand neurons. A true nerve net-like data-flow multi-computer could be built to handle millions of (simplified models of) neurons millions of times faster. Such a system may well be quite suited to today's VLSI technologies. Each neuron-like processor would be quite small and simple compared to the processors in conventional computers.

But unfortunately there appears to be virtually no work at present directed toward designing and building such architectures. Possibly the main problem is that nobody is investigating whether there are relatively general-purpose topologies that might be used. Until it becomes clear what specific operations to build into the individual neuron-like nodes, and how to link nodes together, it seems more reasonable to build each node as a general-purpose computer. But this is a far more costly node to build (chiefly because of the relatively great expense of the controller needed to decode instructions and control processing, compared to the relatively simple processor needed to compute neuron-like threshold operations).

5.1. *Fabricating Increasingly Dense VLSI Chips*

Today's basic computer component is a rectangular chip (usually made of silicon crystal) roughly 4 to 6 mm. on each side. The largest chips in commercial production contain 100,000 to 500,000 transistors. Computers are used to help design and to draw the intricate chip layouts of transistors linked into logic gates, which in turn are linked together into more complex structures. Typically a half dozen or so enormous floor-plan-like designs will be output, for the different layers of the circuit - much as a number of different master masks are used to make colored silkscreen reproductions. These are photographed (today using visible light; tomorrow, in order to increase resolution, with x-rays or electron beams), and drastically reduced in size. Then several hundred copies are spread out like tiles, each mask laying down its layer in turn, over a round wafer. Finally the wafer is very carefully baked, much like a pot, and, when done, the separate copies are diced into separate chips.

Chips are designed in terms of a minimal unit of measurement, lambda, such that the minimal distance between transistors or wires is at least 2 lambda. In current commercial technology lambda is typically 2 microns. A transistor might be as small as 2 or 3 microns in length or breadth, a wire even narrower. These sizes have been growing steadily smaller, and will continue to grow smaller, at least until lambda=.25 micron. As a result, chips have been doubling in size (that is, in terms of the number of devices) every 12 to 18 months for the past 20 to 30 years. This doubling should continue into the foreseeable future. Chips with a million transistors have already been fabricated experimentally. It seems likely that million transistor chips will be available commercially by 1987, ten million transistor chips by the early 1990s, and, possibly, billion transistor chips by 2001. And new technologies - e.g., wafer scale integration, 3-dimensional chips or stacked wafers, optical multi-computers, and biological computers - may well vault substantially beyond these figures.

5.2. *The Sizes of Today's Computers and Their Components*

The processor in a general-purpose 32-bit computer typically has 10^4 to 10^6 transistors; the controller needs as much again; the main high-speed memory needs ten to hundreds of times as much. The general-purpose 1-bit processors in the large parallel array multi-computers that have recently been built with over 4,000 (Stewart REDDAWAY's 1978 DAP [7]), 10,000 (Michael DUFF's 1976 CLIP4 [8]) and 16,000 (Kenneth BATCHER's 1980 MPP [9]) computers need 10^2 to 100^3. Their controllers still need 10^4 to 10^6 (as a result only one is used). Their memories (at least to date) have been relatively small per processor (10^2 to 10^4) but reasonably large *in toto* (10^6 to 10^8). A simple threshold processor (to model a simple neuron) might need only 10^2 or even 10^1 transistors *in toto*, since it needs neither memory nor controller. Almost all its gates would be dedicated to processors - which are the crucial structures that actually do the work [10].

A reasonable rule of thumb is that the feasible limit to the size of a computer is the number of separate basic components (today these are the chips) that must be handled individually and wired together. One to one hundred components gives a micro-computer. One thousand components gives a large mini-computer; 10,000 gives a very large computer; 100,000 to 300,000 components gives a buildable but extremely large super-computer. Therefore networks with 10^7 to 10^{10} neuron-like processors, each with 10^1 to 10^2 transistors, could be built today using 10^4 to 10^5 chips with 10^5 to 10^6 transistors each.

6. *Topologies of Multi-Computer Networks: Used, Proposed, Possible*

Several of the people building large multi-computer networks have been interested in using them to model very large neuron networks. Some (e.g., Larry WITTIE [11]) have thought of assigning a relatively large sub-net of neurons to each of the general-purpose computers in the hardware structure. Others (e.g., Danny HILLIS [12]) appear to be considering assigning each more or less neuron-like process to a different processor. But the topologies of their systems (arrays, N-cubes) do not appear appropriate for efficient mapping and execution of the more complex structures that seem most interesting.

6.1 *Actual Topologies Used for Multi-Computers*

It is instructive to look at some of the graph structures that have been used, and might be used, for multi-computer networks. First, it is important to mention that any possible N-node graph might be used as the underlying topology of an N-computer network. So the possibilities are astronomical. The combinatorial problems of dealing with graphs of even a few dozen or a few hundred, much less many thousands or millions, of nodes are so great that discoveries of good graphs are difficult, and few.

The topologies used in most multi-computers actually built or designed to date have been line (bus), ring, crossbar, star, array, N-cube and tree (see [1], [13]).

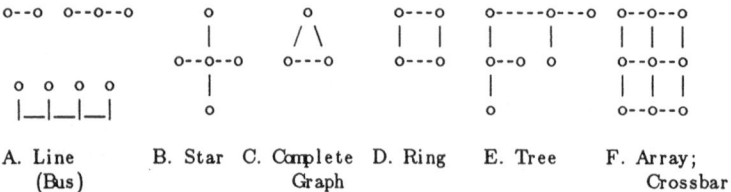

A. Line   B. Star   C. Complete   D. Ring   E. Tree   F. Array;
   (Bus)   Graph   Crossbar

Fig. 1. Small examples of several basic topologies for multi-computer structures

Buses, rings, crossbars and stars cannot be used for more than 8, 16, or possibly, 32 computers (or with heroic efforts perhaps a few more). The complete graph is often suggested as the desired ideal, since it would appear to allow every computer to communicate directly with every other computer. (Actually, this would not be the case; efficient messages passing is a major, unsolved problem.) But an N-node complete graph needs N-1 links and interfaces, and this number almost immediately grows beyond any practical bound.

Pipelines, 1-dimensional arrays, and 2-dimensional arrays are all attractive topologies (though each has its problems). Data can be flowed (as in the Cray super-computers and in the image processors like the Cyto-computer, deAnza and Vicom) through a pipeline of processors much like an assembly-line of workers. A large array can execute large numbers of local window operations everywhere in one parallel instruction.

N-Cubes and trees can be made ever-larger:
. Link two N-cubes together into an N+1-cube.
. Raise the degree (the number of links to each node) of the tree - that is, increase the number of branch-ings at each node. Raise the diameter (the longest shortest distance between any pair of nodes) of the tree - that is, add links to the leaves (those nodes with only one link).

Three other major types of topologies have been used occasionally:
. Trees are augmented with extra links, usually with the hope of drawing otherwise quite distant nodes closer together.
. A number of small, simple clusters are linked together (usually via a bus, ring or crossbar).
. A special kind of NlogN reconfiguring network is used, one that allows N processors to access M memories, or up to N processors to shuffle messages to one another.

6.2. *Dense, Complex, Intricate Graphs*

How should we go about choosing among these, and the potentially infinite number of other possible graphs?

Part of the problem is that we don't have a very firm set of criteria for judging graphs. This is a large is-sue in itself. To keep things simple, I will only mention one possible criterion, although the one that is prob-ably most often suggested: the packing of as many nodes as possible within a given diameter for a given de-gree. I shall use the somewhat unsatisfactory term "density" that we coined at Wisconsin (when we discovered a number of new, denser graphs a few years ago) for this measure.

The densest possible graphs would be those that can be achieved by taking a degree N tree - whose radius (from leaf to root) is 1/2 its diameter (the distance between a pair of most distant leaves) - and judiciously adding N-1 links to each leaf node, until its diameter has been reduced to equal its radius. That is, the de-gree N tree has been turned into a degree N graph with half the diameter. Obviously, no further reduction in diameter could possibly be made, since no extra links can be added to any node other than the leaves. The following is the formula for computing this upper bound to the number of nodes in a graph of given de-gree and diameter.

$$N_{(d,k)} \quad = \quad \frac{d(d-1)^k - 2}{(d-2)} \qquad (d>2) \qquad [N=\text{nodes}, d=\text{degree}, k=\text{diameter}]$$

A. J. HOFFMAN and R. R. SINGLETON [14] proved that at most three graphs (other than the for our purposes trivial degree=2 polygons and diameter=1 complete graphs) can possibly achieve what they named the Moore bound (after E. F. Moore, who pointed it out to them at a meeting): (3,2) with 10 nodes, (7,2) with 50 nodes, and the not yet discovered (57,2) with 3250 nodes.

Figure 2 contains the 10-node (3,2) Petersen graph. It is drawn in a number of different ways, to show how differently exactly the same graph can look.

 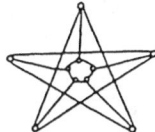 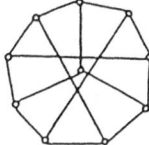 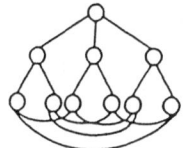

Fig. 2. The 10-node optimal Petersen graph drawn in several different ways)

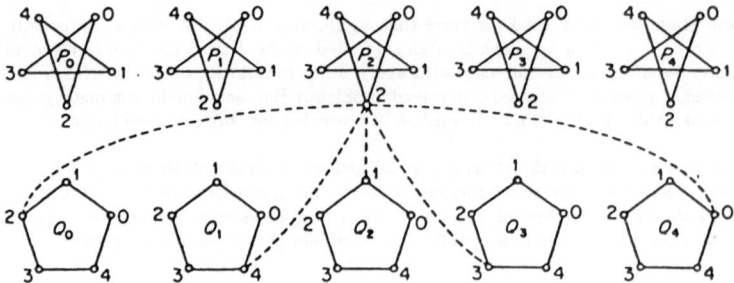

Fig. 3. The 50-node optimal Singleton graph sketched out

Figure 3 shows how the 50-node (7,2) Singleton graph can be constructed. (An actual drawing would, so far as anybody can tell, look like a mess of spaghetti.)

Figure 4 shows a small example of the DE BRUIJN graph [15], which, although producing graphs far less dense than the Moore bound, produces graphs that are among the densest known for even degrees beyond 15,000 or so nodes. The extremely small De Bruijn graph drawn is not at all dense. But it gives a feeling for how an N degree tree can be augmented to an N+1 degree De Bruijn graph, by adding N links to each leaf and 1 link to each non-leaf, judiciously joined together.

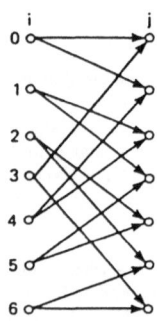
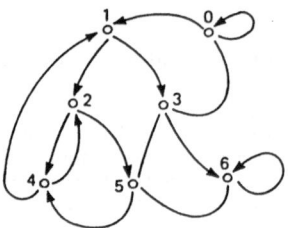

Fig. 4. A small De Bruijn graph (which is relatively dense when large)

Figure 5 shows a few more small graphs with interesting properties. In particular, the first graph shows how to compound N+1 N-node graphs with degree d, diameter k into a new graph with N(N+1) nodes, degree d+1 and diameter 2k+1. A good example of such a compound is the graph with 2550 nodes constructed from 51 50-node Singleton graphs, which is the densest degree 8 diameter 5 graph discovered to date [16]. Many of the densest graphs discovered to date combine a number of good dense graphs together using one or another of several newly discovered compounding operations (see BERMOND *et al.* [17] for the most recent discoveries of dense graphs).

It is important to emphasize that all the graphs pictured in this paper's figures are extremely small. The actual topologies that a multi-computer built from millions or billions of individual general-purpose comput-

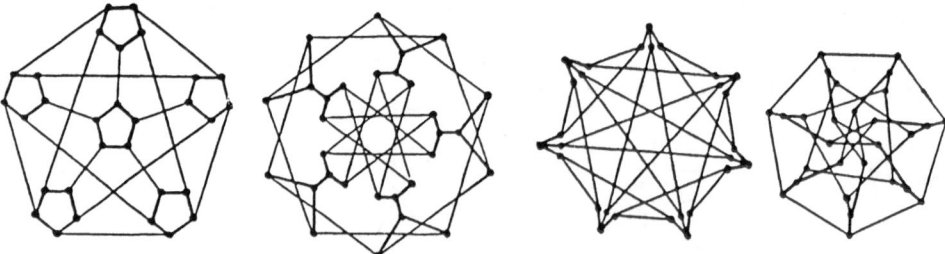

Fig. 5. A few very small graphs of interest

ers or neuron-like processors, whether compounds of compounds, N-dimensional arrays, augmented-tree graphs, or some entirely different topology, are far too complex to be pictured.

7. Graphs that Appear to Offer the Most Promise of Being Evolvable/Learnable

An especially attractive structure for a system that needs to perceive objects, remember facts and thoughts, and reason and make inferences, would appear to be a network that, from almost any node, locally fanned out (and in) logarithmically in all directions - call this a fan-out/fan-in structure. Such a system would minimize the distance from any node to any other node.

7.1. Arrays for Fan-Out/Fan-In Networks

Probably the simplest fan-out/fan-in network structure is a D-dimensional array. A D-dimensional array with only right angle square links has 2D links to each interior node. Thus a 5-dimensional square array has two links in each dimension and therefore 10 links to each node; an 8-dimensional array has 16; a 2-dimensional array has 4. But the array's locally dense linkage means that many fan-out paths fan in to the same node, and the total number of nodes fanned out to in S steps is only $O(S^D)$.

Consider an array linked at its edges to form a torus: Each node is now the central root of a fan-out/fan-in structure. Such a system will have N^D nodes (N=length in each dimension, D=number of dimensions) and 2D links to each node. E.g., in a 3-dimensional array each node has 6 square links. Compare the following examples: A 10^{24} array has 10^8 nodes, each with 8 links, and a greatest distance (diameter) of 196. A 10^8 array has 10^8 nodes, each with 16 links, and a greatest distance of 72.

7.2. Graphs that Have the Desirable Property of Logarithmic Fan-Out

Because an array has many local cycles far fewer nodes are fanned out to than would be the case if every new link connected to a new node (as shown for a very simple example in Fig. 6). This gives much denser graphs, with $O(d^S)$ nodes in S steps with d links from each node to new nodes.

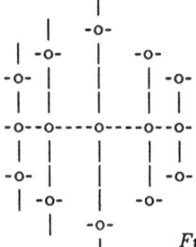

Fig. 6. Logarithmic fan-out in a degree 4 graph, every link connecting to a new node

Optimal fan-out can be achieved with a graph of this sort, where each of a degree d node's links goes to a different node.

A tree is actually such an optimal structure from the point of view of its root, and also of all other interior nodes that are relatively distant from any leaf nodes. Picking up Fig. 6 by its center node gives a tree. Trees are reasonably dense - better than N-dimensional arrays, including the 2^N arrays (often called [binary] N-cubes) that are widely used in today's multi-computers. For example, a degree 20, diameter 20 2^{20}-cube has 2^{20} (roughly a million) nodes. A degree 20, diameter 20 tree has $20(19^9)$ nodes in its leaves alone. However, since a leaf of a tree has only one link there is no fan out from it, and little fan out from nodes near to it. Since a degree d tree has roughly d-1 times as many leaf nodes as interior nodes, this means that most of the nodes in a tree will not be at the center of a fan-out/fan-in structure.

7.3. Augmented Trees for Relatively Dense Fan-Out/Fan-In Structures

Trees can be judiciously augmented with additional links, giving graphs many of whose nodes have good fan-out/fan-in properties to reasonably great distances. The leaves are obvious nodes at which to add links. They are on the periphery and hence more distant from other nodes than are interior nodes, and they have room for more links, since they have only one whereas the interior nodes have d. Such augmentations will substantially increase the density of a tree. But there may well be reasons to link a leaf to an interior node (as in the De Bruijn graphs). Or links might be added between pairs of internal nodes (as is done in graphs that have been .called X-tree constructions [18]).

Indeed, one might want to add links between pairs of nodes simply because they are related, from whatever point of view. This suggests that, in general, there should be a physical link, a physical pathway, wherever there might be good reasons to move directly from one node to the other. But that is in general impossible, since the requirements of whatever hardware we are using will restrict the number of links that can be connected directly to a single node. For today's hardware this is usually a small number like 4, 8, or possibly 16.

When a tree is judiciously augmented with new links at its leaves, to pull specific nodes or regions of the tree closer together, more and more of its interior nodes have increasingly uniform logarithmic fan-outs. A Moore graph is entirely uniform; it has this property everywhere, since every node is the root of an identical tree. It seems likely that carefully augmented tree graphs can come reasonably close.

Augmented binary trees may well be good structures for our purposes. But it appears that at least a few significant improvements can be made. First, there are no compelling reasons not to use trees of higher degree. Binary trees are actually general, in that any other tree could be mapped into a binary tree. But they can be wasteful. For when more than three nodes need to be linked to a single node extra intervening nodes are needed to handle this, by fanning out to them. Binary trees may be desirable for technical reasons, for reasons much like those that compel the building of computers from binary rather than 3-state or N-state switches. But more physical links would appear to make more room for, hence simplify, learning.

An interesting rule would assign an increasing cost to each link added to a node, reflecting how difficult it is to squeeze in new links. An important physical interpretation underlying such an assumption comes when a node is treated as a physical entity with a surface bounding skin (in a 2-dimensional space this becomes the bounding line that designates the perimeter). Note that the surface may be quite irregular and extensive, as is the case with living neurons and their many dendritic fibers. We can conceive of treating a VLSI processor as an irregular design on a chip, where the possibility exists of linking to many spots on its perimeter. For example, the processor's memory might meander around it, with input ports scattered through it.

In tandem to this would be a (possibly quite complex) rule that computes the expected value that might be gained by linking some pair of nodes.

8. Toward Graphs that Might Be Evolvable/Learnable/Growable

What kinds of topologies might be suitable substrates to evolve, learn and grow potentially very large and powerful software/hardware data-flow structures? (The physical hardware might add, modify and delete components - if a technology that made this feasible were available. Or the hardware might first be built, and the software processes might then grow and change within the hardware's limits.)

There appear to be two different types of graphs that a learning system might grow:
. First, a good graph that may be, or may (from a certain point of view or because of the way it is drawn) appear to be, very complex might turn out to be quite simple in that some very simple procedure can be used to generate it.
. Second, it may be possible to continually make small additions to a graph, following some simple and plausible procedure, and thus gradually build it into successively more complex graphs, such that these incremented graphs continue to be good, useful graphs.

8.1. Problems that Can Arise in Attempting to Grow Globally Structured Graphs

A good example of the first type is the De Bruijn family of graphs. A degree d De Bruijn graph can be generated by the very simple algorithm:
  **Create N nodes, ordered.
    Starting with the first node,
      link each node to the next d/2 other nodes in order
      (returning to the first node after the last node)
      until all nodes have been so linked.**

The rule for creating a De Bruijn graph might well be simple enough to be generated in an evolutionary manner. But it appears to have one major, probably crucial, failing. The number of nodes in the total graph must be decided in advance.

A binary N-cube is generated with what is in ways an even simpler rule:
**Let the total structure be one node.
  Make one copy of the total structure.
  Link each node in each copy to the corresponding node
    in the other copy, producing a new total structure; repeat.**

Each repetition of this process creates an N+1-cube, with twice as many nodes. Thus we start with one node (a 0-Cube), then build a 2-node line, a 4-node square, an 8-node cube, and so on.

Although the procedure for generating an N-cube is simple, it is hard to see how N-cubes could possibly evolve. Although a copy might be cloned, the individual nodes would be correctly joined only with great care. And if more than one copy were cloned it would be hard to keep each linking only to the same single other.

Compounds like the N+1 N-node graphs described above are probably easier to construct, and they give denser graphs. But they have many of the same problems.

8.2 *Graphs That Can be Grown Little by Little*

The second type of graph can be exemplified by the following, which will continue to grow a degree 3 tree (often called a "binary" tree):
**Create a one node graph.
  Successively create a new node and link it to any old node that has fewer than 3 links.**

Two slight modifications will give an algorithm that will generate much more regular trees, of any degree:
**Create a one node graph.
  Successively create a new node and link it to the old node that is nearest to the root and has the most links but fewer than d links.**

The following generates augmented-tree graphs, using a rule that first generates a tree, then augments it with links designed to pull distant nodes close together, to give denser graphs:
**Create a one node graph.
  Successively create N new nodes, linking each to the old node that is nearest to the root and has the most links but fewer than d links.
    Next, directly link the pair of nodes that are most distant from one another, until every node has d links.**

This rule is easy to state; but it would be very hard to implement. The system might well have to make an exhaustive search to find the two nodes that are most distant. This is not an especially good algorithm for generating dense graphs. Better, and even more difficult to implement, would be an algorithm that pulled whole sub-graphs of nodes closer together.

The following modifications result in an algorithm that generates functionally augmented tree graphs that reflect and embody information that the larger learning system has chosen to attempt to incorporate in usable form:
**Choose two entities to link together (this choice is an extremely complex matter; the procedures involved are for the moment assumed).
  See whether these two entities are already represented anywhere in the structure:
    a) If both are, and
        i) If they are already linked together (either directly or over a suitably short path), do nothing (or, depending upon the full algorithm, appropriately raise a weight associated with the link).
        ii) If they are not linked, link them (if neither already has too many links - again, this entails a complex evaluation).
    b) If one is, create a new node to represent the new entity, and link it to the already-represented node.
    c) If neither is, create a new node to represent each, and link them together.**

Note that for the first time the algorithm is a function of the information that a node represents. This algorithm will generate several unconnected graphs (but with the possibility of connecting them at a later time). The elimination of step a), which creates two new nodes, would guarantee a single connected graph.

In contrast to De Bruijn graphs and N-cubes, trees and augmented-tree graphs can be generated a little bit at a time, with new nodes linked wherever judged desirable. When real-world constraints (e.g, the maximum possible number of links to a node) impose restrictions, the algorithm can be modified to suit.

The following are two important extensions that will give algorithms that generate a far wider variety of graphs:
. Several nodes can be linked together into a structure that is linked to other nodes. (This might be used to handle n-ary relations, like "a is between b and c" or "a gives b to c"; or to build compounds, like "concatenation('d','o','g') gives 'dog'", "above(back,seat) gives chair", or "touch-bottom-left('|','_') gives L-angle".)
. Rather than add one node and/or one link, the system can add 2, 3, ..., or many. (This might be used to attain a suitable amount of redundancy and variability for robust and flexible behavior.)

Procedures of the sort outlined in this section can be placed under the control of a set of learning rules. They should be designed to build graphs that functionally reflect the relations between pieces of information and between the different structures of information incorporated into different regions of the total system.

9. *Toward Plausible Learning Rules for Choosing to Generate Nodes and the Links Joining Nodes*

This section gives a first rough sketch of a system for learning. Such a system should be capable of building, modifying and un-building a successively larger, more complex and more powerful structure that can handle appropriately a successively greater variety of problems. The building of such a system can be viewed as the construction of a graph - a graph that represents, and completely specifies the behavior of, a multi-computer network and all the processes that make use of it and that it uses.

Since the basic system must build and use graphs, its primitives are nodes, links and symbols. Links join nodes; each node and each link represents one or more symbols that are associated with it. (These symbols are often called "names", "labels" or "attributes".) For example, a node might represent the symbol "d" and have a link representing "concatenation" to the symbol "o" which is similarly concatenated to the symbol "g". This whole 1-dimensional concatenated string must in turn be linked to a node named "dog" which represents the entire structure. (Note that concatenation is a binary relation, and can be handled with a simple link between two nodes. But linking the entire structured string "d--conc--o--conc--g" ["conc" is the symbol for concatenation that labels the link] to a single node needs an n-ary relation, that is, a link with more than two endpoints.) This "dog" node can now be linked to other nodes, e.g., nodes representing "noun", "canine", "mammal", "idea(dog)".

What rules might be used to grow structures of nodes? The following appears to be one of the very simplest:
**When the system judges that two things are related, link the nodes in the hardware/software structure that represent them.
If only one of them is already represented by a node, create a new node to represent the other.
If the newly created node represents a whole structure (e.g., the string "dog" or an image of a dog), build the whole necessary structure (this can be an extremely complex process needing a long sequence of learning experiences; but in very simple cases it can merely mean compounding several contiguous letters into a word, or several spots in an image into a template-like representation).**

There are a great variety of reasons to judge two nodes related. They might be close to one another on some dimensions, for example, time, space, color, texture, or some other extracted or abstracted quality (e.g., jaggedness, made-of-wood, furniture, vehicle, desirable, good, beautiful). Some entity might say words, react, or in some other way indicate (by what is commonly thought of as "feedback"), that they are related. Or the system might infer and generate hypotheses that they are related, and how they might best be represented. (These are major, extremely complex, issues that can only be mentioned here.)

To take a simple example, when two things co-occur close together in space-time, they are more likely to be linked. Thus, when a system is input a picture of a chair [call it p(chair)] and the word "chair" it might do the following:

**Check whether a node already represents p(chair) or "chair".

   a) If both are represented, and

      i) If they are already linked, raise the weight of the link as a function of the value of this instance.

      ii) If they are not reasonably directly linked, create a link joining them (unless no more links are allowed, which may entail additional processing to decide whether to remove old links), with a weight that reflects its value (choosing and using such weights entail still another complex set of problems).

   b) If only one is represented, represent the other using a newly created node and weighted link.

   c) If neither is represented, create the needed nodes and link (this entails a whole complex structure, capable of recognizing the word "chair" or the pictured chair; additional little understood learning is needed to generalize this capability to handle all views of all chairs).

The same algorithm should also be able to handle inputs like a 2-dimensional image of a chair and a couch, or a chair with a person sitting; or a 1-dimensional string where words are juxtaposed, e.g., "chair couch," "a chair is a small couch," "chair for sitting, couch for reclining," or "a chair is a seat." From information of this sort it should be able to build whole complex structures of information that can then be accessed and used to recognize, describe, and discuss these complex concepts (chair, couch, seat, sitting, etc.).

There are a number of possible variations, refinements and improvements to this algorithm. For example, the system might check whether two nodes were already linked through some path (that is, not directly), and if so decide whether to add a link, or even move the old link, to make a shorter path. Nodes and/or links can be erased, or combined. And there are a variety of ways in which nodes might be turned into sets of nodes, representing classes, and structures of nodes, representing generalizations and concepts.

But this is simply a first example, to give some feeling for the ways that learning algorithms might map into very large multi-computer networks. The suggestion is that the massive sizes of multi-computers that are rapidly becoming feasible, and the very large number of possible topologies, should encourage more people to begin to investigate robust, realistic hardware/software structures for learning.

10. *Summary, Comments, Conclusions; and To Be Done*

This paper explores the possibility of developing computer/program hardware/software structures that are, potentially, large and fast enough to be capable of learning to perceive, remember, infer, and think. It examines how VLSI technology makes feasible hardware networks with millions or billions of processors, how the basic entity that is a computer can be used to develop suitable data-flow structures, and how graphs with appropriate topologies can be chosen, or created.

To begin to exhibit any but the simplest manifestations of intelligence, the massively parallel multi-computer networks that are needed must be far bigger and far faster than even the largest "super-computers" of today. And they must be appropriately configured, almost certainly in a structure that closely mirrors the information-flow of intelligent processes. The structure of the hardware multi-computer's processors, the structure of the software and its processes, and the structure of rules whereby systems learn/evolve/grow are all so closely intertwined that they are best examined and designed together.

What this paper treats as a single node might be realized in hardware as a large redundant set of slightly varying nodes, to give robust fault-tolerance, a parallel set of each slightly different experiments, and a certain amount of generalization over minor variations. Or a single node might represent a whole structure of nodes, as in Donald HEBB's [19] cell assemblies, or the more recent connectionist models [20]. That is, these are nodes in conceptual graphs that indicate the design of the flow and transformation of information and also the high-level hardware embodiment in which this can take place. At the next lower hardware level might be the cell assemblies or the connectionist system, just as the detailed structure linking each processor's transistors underlies the structure of processors that combine into a multi-computer network.

Conversely, Hebbian and connectionist models usually assume a complete graph, or a graph where each node has many thousands, or millions, of direct links. But such gigantically high degree graphs cannot be realized using today's VLSI technologies. Multi-computer networks with good fan-out/fan-in characteristics and relatively good density may well be among the best structures through which to realize, with emulations, Hebbian and connectionist models.

Dense fan-out/fan-in graphs appear to be attractive structures for very large hardware/software multi-processor networks within which learning can take place. But the range of possibilities is enormous, and very little work has been done to date on this problem. So there are good reasons to expect that better topologies will be discovered.

223

This paper only begins to sketch out one possible structure of learning rules. It merely mentions the complex set of information that the learner must gather and assess (we human beings do almost all of this unconsciously). It pictures learning as though one tiny node is linked, modified, or created. But slight modifications to the rules will replicate these effects N-fold, multiplying them by hundreds, thousands, or millions as desired (and as hardware can accommodate).

Learning is a function of the system's entire internal structure (pre-built hardware, hardware built and modified through previous learning, and stored information) and of the information from the environment that the system judges (usually unconsciously) to be relevant. The difficult key problems revolve around how the learning system notices and culls the relevant information, solves the puzzles of how to use it, and chooses the relevant structures to modify. Once all that has been accomplished, it is relatively simple to add, modify, and subtract links and nodes. But appropriate hardware/software structure are essential to the development of powerful intelligent systems that learn.

11. Acknowledgement

This research is partially supported by the National Science Foundation under NSF Grant No. DCR-8302397.

12. References

1. L. Uhr: *Algorithm-Structured Computer Arrays and Networks: Architectures and Processes for Images, Percepts and Information.* New York: Academic Press, 1984.
2. E. L. Post: Finite combinatory processes, *J. Symb. Logic*, 1936, *1*, 103-105.
3. A. M. Turing: On computable numbers, with an application to the Entscheidungsproblem, *Proc. London Math. Soc. Ser. 2*, 1936, *42*, 230-265.
4. W. S. McCulloch and W. Pitts: A logical calculus of the ideas immanent in nervous activity, *Bull. Math. Biophysics*, 1943, *5*, 115-133.
5. S. C. Kleene: General recursive functions of natural numbers, *Math. Annalen*, 1936, *12*, 340-353.
6. C. A. Petri: Fundamentals of a theory of asynchronous information flow, *Proc. IFIP Congress 62*, 1962.
7. M. J. B. Duff: CLIP4: a large scale integrated circuit array parallel processor, *Proc. IJCPR-3*, 1976, *4*, 728-733.
8. S. F. Reddaway: DAP - a flexible number cruncher, *Proc. 1978 LANL Workshop on Vector and Parallel Processors*, Los Alamos, 1978, 233-234.
9. K. E. Batcher: Design of a massively parallel processor, *IEEE Trans. Computers*, 1980, *29*, 836-840.
10. L. Uhr: Comparing serial computers, arrays and networks using measures of "active resources," *IEEE Trans. Computers*, 1982, *30*, 1022-1025.
11. L. D. Wittie: MICRONET: A reconfigurable microcomputer network for distributed systems research, *Simulation*, 1978, *31*, 145-153.
12. W. D. Hillis: The connection machine, *A. I. Memo 646, MIT AI Lab.*, Cambridge, 1981.
13. L. Uhr and R. Douglass: *Parallel Multi-Computer Architectures for Artificial Intelligence*, in preparation, 1986.
14. A. J. Hoffman and R. R. Singleton: On Moore graphs with diameter 2 and 3, *IBM J. Res. Devel.*, 1960, 4, 497-504.
15. D. G. De Bruijn: A combinatorial problem, *Koninklijke Nederlandsche Academie van wetenschappen et Amsterdam, Proc. Section of Sciences 49*, 1946, *7*, 758-764.
16. L. Uhr: Compounding denser (d,k) graph architectures for computer networks, *Computer Sciences Dept. Tech. Rept., Univ. of Wisconsin, 1981.*
17. J.-C. Bermond, C. Delorme, and J.-J. Quisquater: Strategies for interconnection networks: some methods from graph theory, *Manuscript M58, Phillips Research Lab.*, Brussels, 1983.
18. A. M. Despain and D. A. Patterson, X-tree: a tree structured multi-processor computer architecture, *Proc. Fifth Ann. Symp. on Computer Arch.*, 1978, 144-151.
19. D. O. Hebb: *The Organization of Behavior.* New York: Wiley, 1949.
20. G. E. Hinton and J. A. Anderson: *Parallel Models of Associative Memory.* Hillsdale, N. J.: Erlbaum, 1981.

Pattern Recognition Based on Holonic Information Dynamics: Towards Synergetic Computers

H. Shimizu, Y. Yamaguchi, I. Tsuda, and M. Yano

Faculty of Pharmaceutical Sciences, University of Tokyo,
Hongo, Bunkyo-ku, Tokyo 113, Japan, and
Bioholonics Research Project, Nissho-Bldg., 14-24 Koishikawa 4-chome,
Bunkyo-ku, Tokyo 112, Japan

1. INTRODUCTION

One of the most striking features of *biosystems*, their ability to percept and generate information is in contrast to the properties of conventional machines which only transform information. Conventional pattern recognition machines identify and classify the input information only. Since their ability is limited to strongly restricted situations, they have only deductive but not inductive capacity. This is not the case for biosystems. Biosystems have a high ability of cognition of unknown events. Animals have survived in the long history of biological evolution under severe circumstances, aquiring the ability to see immediately whether an object is food, enemy or indifferent (neither food nor enemy). This is also true even if the object to be proved is new as is frequently seen in the wild world. WITTGENSTEIN pointed out that in cognitive processes input information is interpreted in terms of stored information, "Vorverstä ndniss" [1]. A similar mechanism has been proposed by MARR for pattern recognition in his book "Vision" [2]. However, no clear explanation has been given to the mechanism of the self-interpretation of input information.

Let us begin by questioning "What is the interpretation?". Here we propose as follows.
(A) *The interpretation of input information is in conformity with creating a relation between the external information and some of the internal represented one by generating holistic information in which these two kinds of information are mutually related.*

A synergetic system can be regarded as a kind of analog computers which are computers composed of cooperative processors with a self-organizing property and applicable to pattern recognition [3]. We will name such processors with a synergetic property *holons*. Holons constitute a parallel processor when they are organized by suitable condition. Such a parallel processor, *a holonic processor*, self-organizes holistic information from various kinds of elementary information in the following way. First, each holon picks up elementary information. This is the coding of elementary information. Then, due to the cooperative behaviour of the holons, the elementary information causes mutual interactions between the holons resulting in a certain state of interaction within the holonic system, that is to be called interpretation of the elementary information. This information processing, the self-organization of the holistic information due to information compression, is done by the holons in a parallel way. At this stage, the coded information in each holon acquires a holistic character and will be called *holonic information*. Finally, the self-organized information is fed out in the decoding of information. According to (A) the external information is recognized when it becomes holonic.

In this paper we will propose the principles of the information processing of the self-interpretation, demonstrating it with a new type of machines for pattern recognition. Our machine should be therefore related in some way to the information processing in brain in pattern recognition.

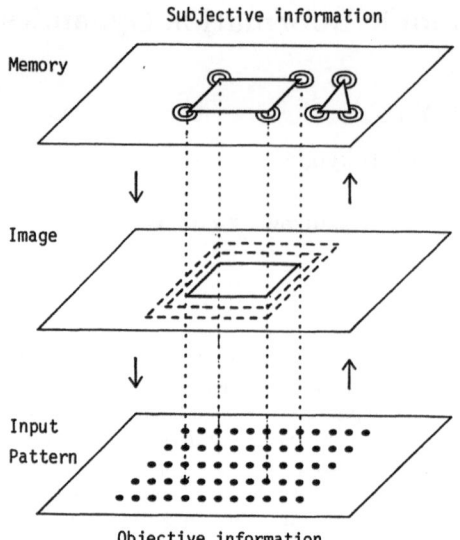

Subjective information

Memory

Image

Input Pattern

Objective information

Fig. 1 Recognition of a square lattice pattern textured on a very wide carpet. On the objective pattern we see various images as parallel lines in various directions or squares of similar shape with various sizes. These images are self-organized from pattern elements of the input pattern under the influence of semantic information stored in long-term memory

2. PATTERN RECOGNITION ON A PLANAR LATTICE

Let us imagine that we are standing on a very wide carpet on which a rather small flower is textured at each lattice point of a regular square lattice as schematically shown in Fig.1. (We will name this pattern the *flower lattice pattern*.) When we look at the flower lattice pattern, we are able to "see" various kinds of virtual pattern such as parallel lines in various directions or squares of various shapes and sizes. Let us discuss the cognition or recognition of these virtual patterns. For instance, having succeeded to find a line on the carpet, other lines are successively found in its neighborhood in the same direction. At this moment we are not aware of lines in other directions. Then, suddenly a line is noticed in another direction followed by successive finding of additional similar lines. Taking no care of the lines   seen before, we only see a group of parallel lines in one direction. When a new line is found before the image of other lines is lost, we shall be aware of the presence of two lines crossing while making a corner, which finally leads us to the image of a square. Then, squares of similar shape with different sizes are successively seen. No parallel lines can be seen when we are seeing squares. Thus, the flower lattice pattern on a carpet could be regarded as a figure with equivocal meanings as figures painted by Escher. The fact that lines in the same direction or squares of the similar shape are successively recognized would be related to the presence of a kind of hysteresis in the recognition of figures with equivocal meanings, as explained by HAKEN in his textbook of synergetics regarding the transitions from a man's face to a girl and from the girl to the face [4]. Thus, from analogy of the Escher's figures, we may regard the lines and squares which are present in our long-term memory as semantic information. The above example of the recognition of virtual patterns on the carpet of a flower lattice pattern clearly indicates that the simplest model of the information dynamics of pattern recognition will have at least three hierarchical levels, i.e. the levels of the objective information, image information and semantic information. Of course we may add subhierarchies to each if we want to discuss the process in more detail.

The above scheme may be explained as follows. First input pattern information is decomposed into a number of pattern elements. This process is necessary in order to enable the self-organization towards an image that is consistent with the already gained semantic information. In other words, this process is crucial in relating the input information to internal stored semantic information. Then, in the next step, pattern elements will begin to form local groups, which grow up to local images. On the other hand, the self-organizing local images recall semantic information, which gives a strategy to complete the organization of local images and to assemble them altogether into an image. Consistency between the self-organized image and the input pattern would be constantly examined during the self-organization of the image. If the strategy is right, a consistent description of the input information in terms of semantic information could be obtained. In each case only one interpretation of the input pattern (governed by semantic information) is stable at the same time. And the same semantic information is used as much as possible for any successive input, therefore it needs a rather long time to modify semantic information. If the actual present semantic information does not enable the self-organization of the objective pattern into a holonic organization of input pattern element consistent to each other because there is too great a difference, one "sees" either a simplified image or even illusions depending on the degree of inconsistency.

4. SELF-ORGANIZATION OF LOCAL IMAGES BY SIMPLE CELLS

Now the problem to be investigated is the principle of information dynamics governing the self-formation of an image from an input pattern signal. What we face is a complex system in which order is self-organized from various kinds of synergetic elements: this order is particularly related to coded information. How can the slaving principle of synergetics [4] be applied to this problem? What are the pattern elements and the order parameters characterizing the image? To add one step more, let us propose the following hypotheses:

(B) *Pattern elements become holonic when they are organized in local images being accompanied with information compression, and the more the pattern elements have to be compressed for the same size of a local image, the easier the formation of a local image. (the principle of the maximum slaving efficiency)*

(C) *A necessary condition to see an input pattern is to produce an image in which all the pattern elements become holonic as a member of local images. (the principle of complete slaving)*

(D) *An image is self-organized from local images by the aid of semantic information which gives a definite relationship to the local images. (the principle of semantic consistency)*

Let us consider the function of simple cells which have been thought of as line detectors in visual cortex [5]. If many dots are aligned with a spacing which is much smaller than the perceptive field of a simple cell, the dots will be detected as a line. Such situation occurs when we see a line in a photograph printed in books or newspapers. However, this is not the case when we see lines on the flower lattice pattern on the carpet because flowers are textured with a wide spacing which would be much larger than the perceptive field. Therefore it would be plausible that simple cells in a hypercolumn are able cooperatively to form a kind of a "simple cell" perceptive field that is much larger than the spacing between the mapping of the neighboring flowers within the visual cortex. The fact that we are much more sensitive to an array of line elements forming a linear line or a line with a constant and rather small curvature than to line elements distributed with random orientations would be related to the known preferable directions in the interactions of simple cells in hypercolumn.

Here, based on a hypothesis of the preference directions of the interaction, we present a holonic image processor "hypercolumn" as illustrated in Fig.2.

(E) *The magnitude of the interaction between neighboring simple cells has the maximum value when their perceptive directions array along a linear line and becomes smaller as they deviate from it.*

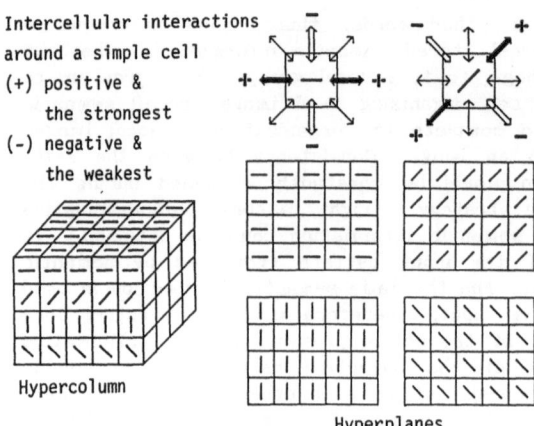

Intercellular interactions
around a simple cell
(+) positive &
   the strongest
(-) negative &
   the weakest

Hypercolumn

Hyperplanes

Fig. 2 The processor of our pattern recognition system, "hypercolumn", which is composed of many "hyperplanes" where a number of "simple cells" with the same orientation are regularly arrayed. These hyperplanes are superposed to obtain a number of "columns" in a perpendicular direction. There are intercelluar interactions among simple cells in a manner to extract lines from distributed pattern elements

More concretely, our image processor is to be explained as follows. It is a parallel processor consisting of a number of vertical columns each of them composed of simple cells with various directions. When we slice our "hypercolumn" horizontally, we have a planar assembly of simple cells with the same orientation. The orientation of simple cells differs from hyperplane to hyperplane, according to the vertical position in the hypercolumn. It is to be noted that our processor is consistent with (E) as well as with (B) and (C) as discussed below.

Imagine the input to our image processor is an assembly of pattern elements forming a dotted line. These elements are projected to the bottom of the hypercolumn and, at that moment, all the simple cells in the columns which receive the projected pattern elements are switched on. We assume that the excitations of these switched-on cells are so weak that they are not perceptible in the absence of interactions among excited simple cells. The presence of the interactions will modify the excitations of the simple cells as follows. According to our hypothesis (E), a group of simple cells of which orientations coincide with the direction of the input line excites most strongly. Owing to the inhibitory mechanism which makes the strongest excitation in each column outstanding by suppressing the activity of the other cells, only the simple cells along the direction of the line "fire" in the next moment. Due to the cooperative excitation of simple cells, excitation could be often induced in those simple cells which were not switched on directly by input signals. Thus we are able to detect a line in the assembly of the pattern elements even when the pattern elements are not placed exactly on the lines or are aligned with wide spacing. This means that we are able to extract pattern order such as lines from distributed pattern elements. Thus our model satisfies (B) and substantially also (C). Furthermore, to detect curved lines, interactions are introduced among simple cells in different hyperplanes.

5. ENTRAINMENT AS A MECHANISM OF SELF-ORGANIZATION OF INFORMATION
The "simple cells" in our processor must have a holonic character. Two kinds of holons will be available, i.e. artificial holons and native holons which are synergetic molecules or organelles isolated from biosystems. Artificial holons easily available at present will be nonlinear oscillators. Therefore, we have

studied how to design a pattern - recognition machine, utilizing nonlinear oscillators as holons. The "firing activity" of each cell is represented by the amplitude of an oscillator, the elementary oscillator. The cells have a resting and an excited state. In the resting state, the damping effect of the oscillation is larger than the activating effect. Therefore, the oscillations have a vanishingly small amplitude. On the other hand, when the oscillators are brought to the excited state due to input signals, the activation exceeds the damping, and a limit cycle with finite, small radius starts in the oscillators. Some of the excited oscillators are more strongly activated, due to positive feedback and gain larger amplitudes, depending on the extent of the activation.

In order to study our problem in more detail. let us start from the following hypothesis:
(F) *A definite relation is established between elementary information coded by two holons when a definite and stable phase relation is established between elementary oscillators which code the information.*

Thus entrainment can produce relations among different pieces of information. Elementary information coded in the elementary oscillators will be compressed by the entrainment of oscillations. What is to be studied here is the dynamics of coded information rather than that of the holons itself. To proceed one step further, we have to give the prescription for the coding/decoding of information to/from holons, elementary oscillators. We propose a three-level system shown in Fig.3. At the lowest level there is a meshed plane "retinal screen", on which input pattern information is projected, being decomposed into pattern elements. The size of each grid of the screen coincides with the size of simple cells in the next upper level, where we have our "hypercolumn", an assembly of parallel and identical columns composed of "simple cells". The columns are perpendicular to the screen. This means that "hyperplanes" are parallel to the screen. Above each grid point of the retinal screen a column is placed. Only those of all the elementary oscillators belonging to columns which receive elementary pattern signals transmitted from the screen are switched on. Then, the hypercolumn has a crucial role in the interpretation of input pattern because it functions to find relations among pattern elements, elementary information, according to semantic information. It is an image synthesizer which is more powerful than a line detector. This is hypothesised to be also the case for related processing levels of the brain (Shimizu & Yamaguchi).

The orientation of a line element can be coded in the phase of one of the van der Pol oscillators that oscillate with the same period (Yamaguchi). Then spontaneous grouping of the phases corresponding to the self-organization of local images occurs in the assembly of the oscillators. Generally an image is organized from more than one local image when the mutual relations are introduced among the local images. The memory part of our model has a structure bearing some similarity to that of our hypercolumn with some additional functions. It is composed of a number of columns with unspecified spatial position. In each column, "memory column", having a structure identical to that of a column in the hypercolumn, an image is stored as an assembly of local images, i.e. lines, each of which is memorized in a memory cell which keeps the line direction specific to the local image. The length of the line as well as the relations among lines corresponding to local images are stored in auxiliary parts of the column which are independent of the dynamics of the cells. In more elaborate treatments, we may introduce "complex cells" [5] for the memory of the relations among the local images. One of the merits of our model is its independence of the translation and size of the input pattern; it is sensitive only to the shape, because only the mutual relationship among pattern elements is processed as the mutual relationship among the phases of elementary oscillators. It would not be a difficult task to make it independent of rotation.

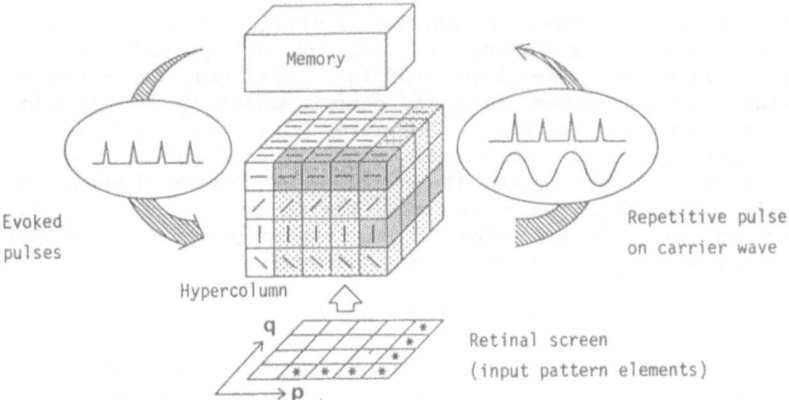

Fig. 3 *A schematic representation of our pattern recognition system. It is composed of three hierarchical levels of information: retinal screen, hypercolumn and long-term memory*

6. EQUATIONS OF MOTIONS OF THE THREE-LEVEL MODEL

Let us explain our model quantitatively. For the sake of later discussion, let us define coordinate systems. The space-fixed coordinate system (p,q) is defined as schematically shown in Fig.3. There is a coordinate system fixed to the (n,i)th simple cell, (r_{ni}, s_{ni}), which s_{ni} axis label the direction of the orientation of the cell. The direction of s_{ni} relative to p is denoted by the angle θ_{ni} in the real space. In the above notation n and i refer to the numbering of columns in the hypercolumn and of simple cells in each column. Besides that coordinate system in the real space there are coordinate systems describing the motion of the (n,i)th oscillator, representing the physiological activity of the (n,i)th cell. Corresponding to (p,q) in the real space, we will define (x,y) in the phase-space. The motion of the (n,i)th oscillator in (x,y) is given by microscopic variables (x_{ni}, y_{ni}). In addition to this there is another physiological coordinate system (u_{ni}, v_{ni}) corresponding to (r_{ni}, s_{ni}) in the real space. The phase difference ϕ_{ni} of the oscillation observed in these two kinds of physiological coordinate system is related to θ_{ni} by $\phi_{ni}=2\theta_{ni}$ since θ_{ni} varies only inbetween 0 and π. All simple cells in the same hyperplane i have the same θ_{ni}. Hence the suffix n in ϕ_{ni} will be omitted. The angle of the direction of the vector from column n to column k relative to p axis is denoted by $\alpha_{nk}/2$.

In our system oscillations are described by the following equations of motion for three kinds of variables: microscopic variables $\{x_{ni}, y_{ni}\}$ for the elementary oscillators, "macroscopic" variables (U, V), image order parameters that denote the extent of the self-organization of image, and "macroscopic" variables $\{X^m_i, Y^m_i\}$ for evoked memories $\{m\}$. Equations of motion for the elementary oscillators will be given by

$$(d/dt)\begin{pmatrix} x_{ni} \\ y_{ni} \end{pmatrix} = F(x_{ni}, y_{ni}; d_{ni}, D_s) + G_{ni}(\{x_{kj}, y_{kj}\}) \tag{1}$$

with

$$F(x_{ni}, y_{ni}; d_{ni}, D_s) = \begin{pmatrix} y_{ni} \\ -x_{ni}+e(d_{ni}-x_{ni}^2)y_{ni} \end{pmatrix} -D_s\begin{pmatrix} x_{ni} \\ y_{ni} \end{pmatrix} \tag{2}$$

230

with a smallness parameter e and a damping constant D_s. The quantity d_{ni} is 1 if the (n,i)th cell is switched on by the input signal and -1 otherwise. Because of the damping term, the second term in RHS of Eq.(2), the amplitude of the oscillator will decay in the absence of cooperative interactions with other oscillators, which are given by G_{ni}. The second term in RHS of Eq.(1) represents the interactions of the (n,i)th oscillator with the order-parameter oscillator, other elementary oscillators and memory oscillators as explained below:

$$G_{ni} = G_{o,ni} + G_{s,ni} + G_{M,ni} \tag{3}$$

with

$$G_{o,ni} = C_r \begin{pmatrix} \cos\phi_i & -\sin\phi_i \\ \sin\phi_i & \cos\phi_i \end{pmatrix} \begin{pmatrix} U \\ V \end{pmatrix} \tag{4}$$

$$G_{s,ni} = (C_s/M_s) \sum_j S_{ni,nj} \begin{pmatrix} X_{nj} \\ Y_{nj} \end{pmatrix} + (C_t/M_t) \sum_k T_{ni,ki} \begin{pmatrix} X_{ki} \\ Y_{ki} \end{pmatrix} \tag{5}$$

$$S_{ni,nj} = [\{\cos(\phi_i-\phi_j)+1\}^2/2] - 1 \tag{6a}$$
$$T_{ni,ki} = [\{\cos(\phi_i-\alpha_{nk})+1\}^2/2] - 1 \tag{6b}$$

$$G_{M,ni} = C_z \begin{pmatrix} 0 \\ z_{ni}(\sum_{mk} b_{mk} Z^m{}_k - \mu) \end{pmatrix} \tag{7}$$

with

$$z_{ni} = [\tanh\{h(y_{ni}-y_o)\}+1]/2 \tag{8a}$$
$$Z^m{}_k = [\tanh\{g(Y^m{}_k-Y_o)\}+1]/2 \tag{8b}$$

with constants h, g, y_o and Y_o and $\mu = 0.1$. In Eq.(7) b_{mk} refers to the amplitude of the kth memory oscillator of the mth memory column normalized as $\sum_{mk} b_{mk} = 1$. Elementary oscillations above "perceptive threshold" y_o are transformed into a kind of repetitive pulse with sharpness h according to Eq.(8a). Similar pulses are generated in the memory according to Eq.(8b). Our hypercolumn processor communicates with the long-term memory in the form of pulse-wave interactions. *Image order parameters U and V* are activated by the elementary oscillators as

$$(d/dt) \begin{pmatrix} U \\ V \end{pmatrix} = F(U, V; 1, D_o) + C_c \sum_{ni} \begin{pmatrix} u_{ni} \\ v_{ni} \end{pmatrix} \tag{9}$$

with

$$\begin{pmatrix} u_{ni} \\ v_{ni} \end{pmatrix} = \begin{pmatrix} \cos\phi_i & \sin\phi_i \\ -\sin\phi_i & \cos\phi_i \end{pmatrix} \begin{pmatrix} x_{ni} \\ y_{ni} \end{pmatrix} . \tag{10}$$

The self-organization of an image proceeds in the assembly of the elementary oscillators under three kinds of forces. As we explained above, local images are synthesized due to cooperative interactions G_s of simple cells in the hypercolumn and then they are combined together to fit to the two kinds of information, input information and stored information. As given by the second term in RHS of Eq.(5) with Eq.(6b), the interactions which are introduced only between neighboring elementary oscillators in each hyperplane have the maximum positive value not only if the orientations of the interacting simple cells are parallel but also if the cells are placed along a linear line. The amplitudes of such oscillators are enhanced above perceptive threshold due to the positive interactions, indicating that the corresponding simple cells are strongly excited. On the other hand, when the orientations of neighboring simple cells are perpendicular, their interactions change to inhibitory interactions. The amplitudes of such oscillators will be damped. In addition, we also assume "vertical interactions", the first term in RHS of Eq.(5), among elementary oscillators in the same column. Therefore, the oscillators with different phases are able to interact and to produce curved lines by the vertical interactions.

Due to the direct interactions G_s local images are self-organized among elementary oscillators. Then an image is self-organized from the local images according to the global feature of the input pattern. (Partial overlapping may occur between these two kinds of processes.) However, the global feature of the input pattern is not reflected in mutual phase relationship of the elementary oscillators by G_s only. Interactions G_o with the order parameters (U, V) is therefore given to result in this relationship. Eq.(10) indicates that (U, V) are related to microscopic variables $\{x_{ni}, y_{ni}\}$ to give proper phase relations $\{\phi_i\}$ to the oscillators. Consequently, the interactions G_o activate the elementary oscillators in a way that proper phase relations are spontaneously introduced among the elementary oscillators through (U, V). The transformation Eq.(10) is necessary to treat holons with particularity as if they are identical ones. We have positive feedback between the microscopic variables and the order parameters. This feedback loop may be compared to LASHLEY's *dynamic engram* [6] carrying short-term memory [7]. When the amplitudes $\{y_{ni}\}$ exceed the perceptive threshold y_o, they are transformed to repetitive pulses $\{z_{ni}\}$ according to Eq.(8a). On the other hand, the order parameters are transmitted as a "carrier wave" [8] to the memory and "evoke" memory oscillators due to the switching effect of damped oscillators by an oscillating field proposed by OHSUGA et al. [9]. Pulse-wave interactions between the processor and the memory are given by G_m in Eq.(7).

The equation of motion for the kth memory oscillator in the mth memory column will be written as

$$(d/dt) \begin{pmatrix} X^m_k \\ Y^m_k \end{pmatrix} = F(X^m_k, Y^m_k; 1, D_m) + G^m_k \tag{11}$$

with

$$G^m_k = G_{r, m k} + G_{h, m k} \tag{12}$$

$$G_{r, m k} = C_p W_m \begin{pmatrix} \cos\phi_k & -\sin\phi_k \\ \sin\phi_k & \cos\phi_k \end{pmatrix} \begin{pmatrix} U \\ V \end{pmatrix} \tag{13}$$

$$G_{h, m k} = C_z \begin{pmatrix} 0 \\ Z^m_k (\Sigma\, n_i z_{ni}/N_e - b_{m k}) \end{pmatrix}, \tag{14}$$

where N_e denotes the total number of excited columns.

The carrier wave which activates memory oscillators is partitioned among the memories with the partition coefficients $\{W_m\}$. The evoked memories have a contest in the fitness of their pulses $\{Z^m_k\}$ to the input pulses $\{z_{ni}\}$ from the hypercolumn. The contest determines the selective value a_m for mth memory by

$$da_m / dt = -C_o (\Sigma\, z_{ni}/N_e - \Sigma\, b_{m k} Z^m_k)^2 a_m + C_a (\Sigma\, n_i z_{ni}/N_e - \mu)(\Sigma\, k b_{m k} Z^m_k - \mu). \tag{15}$$

The partition coefficient W_m obeys to a kind of logistic equation

$$dW_m / dt = (a_m - \Sigma\, n a_n W_n / \Sigma\, n W_n)(1 - W_m^2) W_m, \tag{16}$$

which has a similarity to the equation used by EIGEN and SCHUSTER [10] in their study of hypercycle.

232

The first term in RHS of Eq.(15) lowers a_m gradually while the second term changes a_m with a much more rapid rate. Since pulses $\{z_{ni}\}$ and $\{Z^m_k\}$ are considerably larger than μ in their amplitudes, the second term gives a positive contribution to a_m if both $\sum z_{ni}$ and $\sum Z^m_k$ are nonzero, and a negative contribution if only one of them is nonzero. In other words, if the two kinds of pulses, those from the hypercolumn and the mth memory, match each other, the memory gets award, i.e. increase in a_m, while if they mismatch, it receives penalty, i.e. decrease in a_m. If pulses are absent in the memory, a_m gradually decays to a very small quantity. Therefore, for memories which are only rarely used, a_m becomes negligibly small. Such memories are not so easily evoked because of very small w_m. Eqs. (7) and (14) show that the corresponding elementary and memory oscillators are activated or deactivated according to the matching or mismatching, respectively. It is to be noted that not only the image but also the long-term memory are properly modified due to entrainment caused by the pulse-wave interactions. The stored memory is gradually changed, due to repeating of the matching process, in such a way that a kind of concept from analogous input patterns is organized.

In the absence of input pattern information, all the oscillators in our system are in the resting state, since the damping always exceeds the activation. However, the input signal switches simple cells on, which weakly activates the corresponding elementary oscillators. These elementary oscillators successively turn on various kinds of oscillators to the excited state, due to the switching mechanism of nonlinear oscillators [9]. In particular, we will name the sponta neous *evocation* of relevant memories due to input signals the "active memory mechanism", to distinguish it from the "passive memory mechanism" utilized in conventional computers. The order-parameter oscillator gives the elementary oscillators not only activation through positive feedback but serves also to support coherence proper to the input pattern. Therefore, the pattern fed in our system organizes an image and evokes memories followed by the selection of a relevant one. All these processes automatically proceed in our model regardless of any arbitrary input pattern.

7. SIMPLE APPLICATIONS

For the sake of the examination of our system, we will report some results obtained as the zeroth order approximation. We assumed a retinal screen with 8x8 grid points and a hypercolumn composed of four hyperplanes, each of which is a planar array of 64 simple cells of the same direction. In other words, each column in our hypercolumn is composed of four kinds of simple cells, which are sensitive to lines in four different directions. For the sake of simplicity, there is no cell which is particularly sensitive to an edge where two lines cross one another. This simplification did not produce any serious defect in roughly knowing what was going on in our system. Three kinds of polygons, a right triangle, a regular square and a polygon higher than pentagon, are assumed as semantic information stored in the long-term memory.

In Fig.4a we show an input pattern to our hypercolumn. Each point with bars indicates a column with simple cells, the directions of which are shown by those of the bars crossing at that point. This figure illustrates simple cells initially switched on by the input signal. Fig.4b shows the enhancement in the amplitudes of the elementary oscillators in each hyperplane after a suitable time. The size of small circles indicates the amplitude of elementary oscillators. Due to the perceptive threshold, smallest circles are not detectable, only one group aligned along a line, corresponding to a local image, is detected in hyperplane A, B and D. At the top of Fig.4c a carrier wave from the hypercolumn is shown as a function of time. The amplitude of the carrier wave becomes stronger as local images are gradually self-organized in the hypercolumn. Accompanied with this carrier wave, a pulse pattern shown in the second line is transmitted. The specific feature of

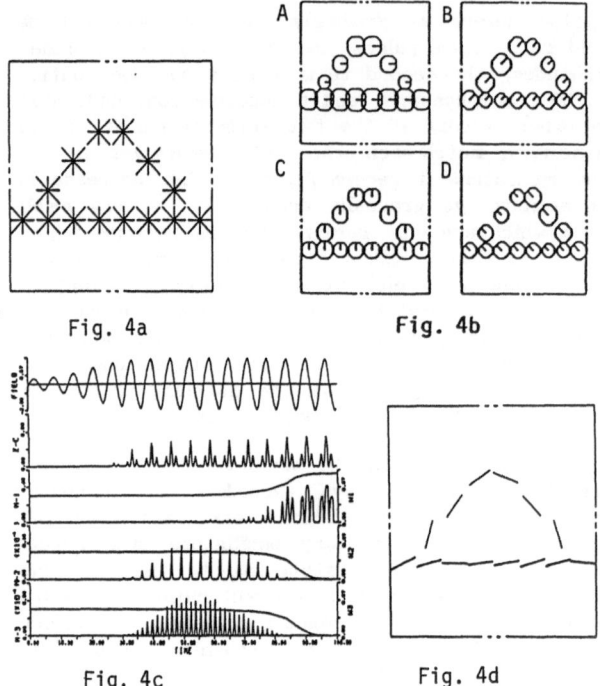

Fig. 4a   Fig. 4b

Fig. 4c   Fig. 4d

Fig. 4 *Recognition of our*
system. Pattern in Fig. 4a was
inputted on the retinal screen.
Fig. 4b shows excitation of
simple cells in each hyper-
plane. Fig. 4c, the evoca-
tion of various memories
followed by contest, which
determined that the self-
organized image.Fig. 4d was
related to a triangle. (The
magnitude of the pulses for
the triangle is denoted by a
scale which is smaller by an
order than those for the other
cases.)

this pattern reflects the self-organized image; the height of each pulse is related to the length of the corresponding line. The carrier wave clamps the repetitive pulse pattern of the hypercolumn. This is necessary to enable the matching between the image and memory. The next three pulse patterns in Fig.4c illustrate the pulses from spontaneously evoked memories of the triangle, square and polygon, respectively. It should be noted that only the signal of the triangle is represented in a scale which is smaller by order one than those for the others. (Similar representations will be used below.) Three kinds of memory compete each other for a while and only the one corresponding to the triangle survives and grows while the others decay soon. This is caused by the change in the partition coefficient of excitation, w_m, which is indicated by a curved line above each pulse sequence. The time-dependence of w_m is governed by the selective value a_m of which time-dependence is determined by the fitness between input pulses $\sum n_i z_{ni}$ and memory pulse $\sum_k Z^m_k$ through Eq.(16). In this example, the image self-organized in the hypercolumn which is recognized as a triangle is shown in Fig.4d. This image is the pattern which is "seen" by our system when the triangle shown in Fig.4a is fed in.

In Fig.5 we illustrate the pattern recognition of our model system when patterns Fig.5a and Fig.5b are the input. In this case, the input patterns are seen as shown in Fig.5c and Fig.5d, and are recognized as a square and a polygon.

Next, we will show results which indicate how our system "interpretes" rather umbiguous patterns. We input square like patterns with defects as shown in Figs.6a and 6b. Our system sees these patterns as shown in Figs.6c and 6d, and it interpretes them as a square. However, the manner of defects is crucial for the pattern recognition as it is shown in the next example. The input pattern is shown in Fig.7a. Our model self-organized an image as shown in Fig.7b, but could not relate it to any one of the memories as is seen in Fig.7c when the magnitude of

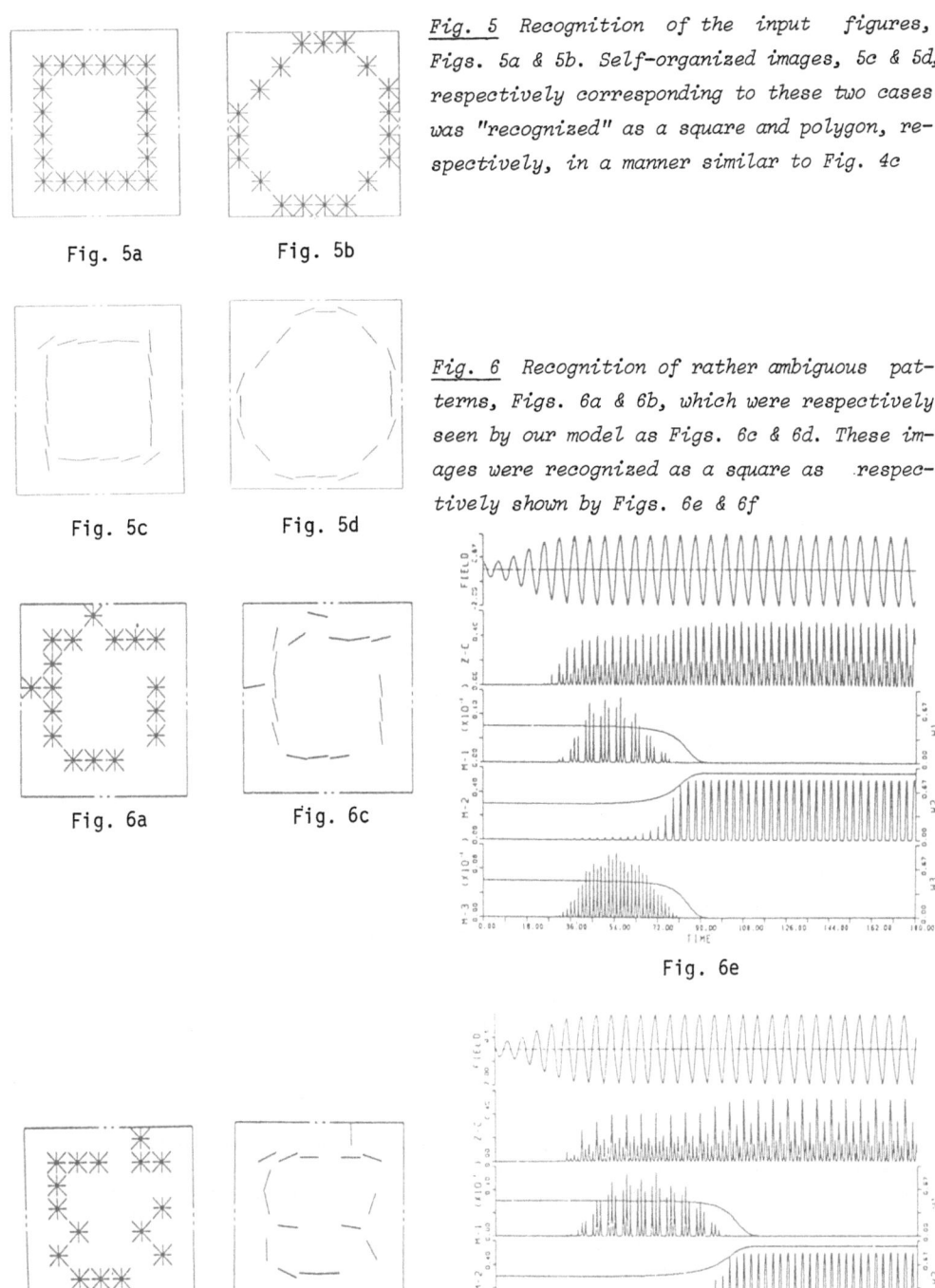

Fig. 5 Recognition of the input figures, Figs. 5a & 5b. Self-organized images, 5c & 5d, respectively corresponding to these two cases was "recognized" as a square and polygon, respectively, in a manner similar to Fig. 4c

Fig. 5a Fig. 5b

Fig. 5c Fig. 5d

Fig. 6a Fig. 6c

Fig. 6 Recognition of rather ambiguous patterns, Figs. 6a & 6b, which were respectively seen by our model as Figs. 6c & 6d. These images were recognized as a square as respectively shown by Figs. 6e & 6f

Fig. 6e

Fig. 6b Fig. 6d

Fig. 6f

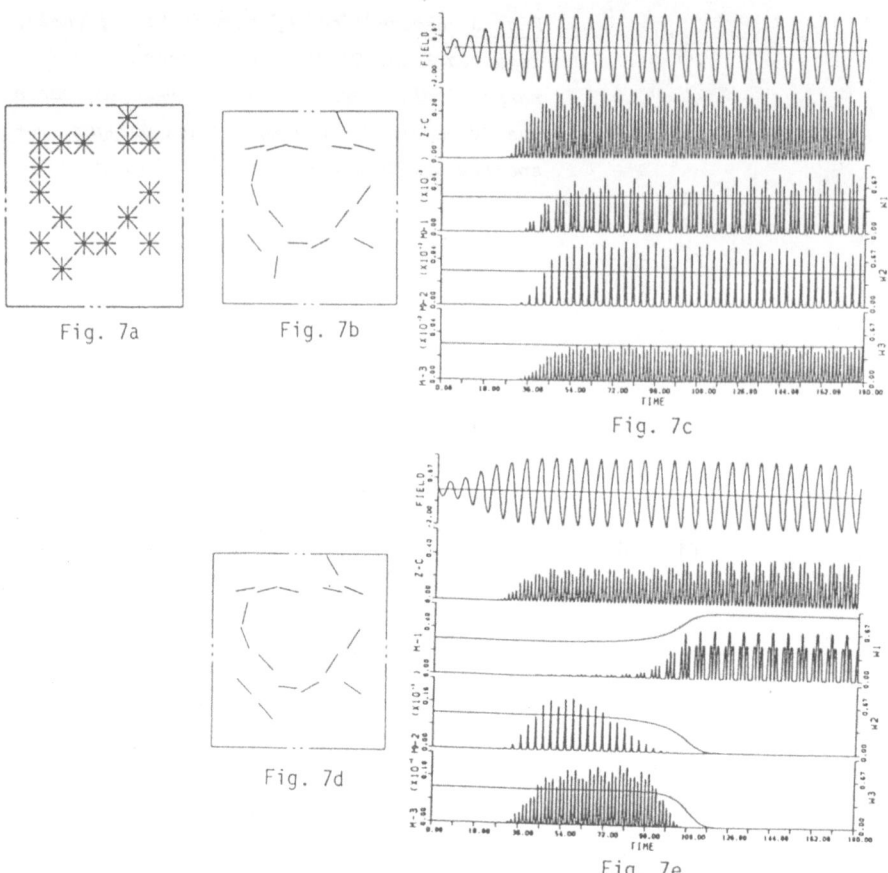

Fig. 7a   Fig. 7b

Fig. 7c

Fig. 7d

Fig. 7e

*Fig. 7   Recognition of an ambiguous pattern, Fig. 7a.   When the interaction between
the hypercolumn and the memory was relatively weak, the pattern was seen as Fig. 7b.
However, our system could not interpret it as shown in Fig. 7c.   When the interaction was increased to the normal level, a small change appeared in the image as
seen in Fig. 7d, which was recognized as a triangle as shown in Fig. 7e.   It is to
be noted that only a small difference is present between Figs. 6b and 7a*

interactions between the image and the memories is relatively small. Then, the
"interpretation" was encouraged by enhancing the interactions.   A slight change
was produced in the image as indicated in Fig.7d, which was recognized as a
triangle as it is shown in Fig.7e. Probably this pattern might be seen as a right
triangle pointing its corner with the right angle in the downward direction.

   Let us demonstrate the ability of our model in the recognition of a square
lattice pattern shown in Fig.8a. We stored a vertical and horizontal line in the
long-term memory of our model. As illustrated in Fig.8b with 8c, our model was
able to find parallel lines in the vertical direction if suitable sets of initial
condition were given. And it also recognized horizontal parallel lines as shown in
Fig.8d with 8e if we started from other sets of initial condition. These two cases
differ in the initial condition of the generation of random numbers used for the
initial distribution of the phases of the elementary oscillators in the resting
state before the pattern was fed in. Thus our model is able to see lines on a

236

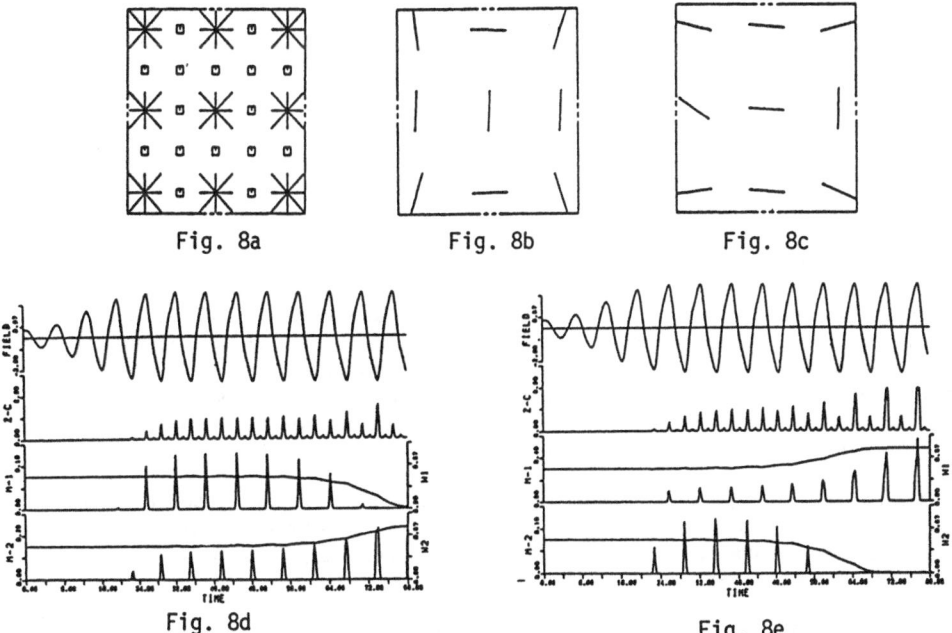

Fig. 8a   Fig. 8b   Fig. 8c

Fig. 8d   Fig. 8e

Fig. 8 Recognition of a square lattice pattern, Fig. 8a. Images, Figs. 8b & 8c, were self-organized in hypercolumn, depending on a small difference in the initial condition. These patterns were recognized as vertical and horizontal parallel lines, respectively, as shown in Figs. 8d & 8e

square lattice pattern only in one direction, as in our pattern recognition of a flower lattice pattern on the carpet.

8. DISCUSSIONS

The model explained above demonstrated one of the possible ways to process information according to our basic understanding of the pattern recognition in the brain, i.e. the interpretation of input pattern in terms of semantic information stored in long-term memory. This interpretation is done automatically by self-organizing an image from the input pattern . such that a matching with any stored pattern is tried. There are many problems left to be studied further.

(a) *Self-Organization of Semantic Information* : In the above examples information on three kinds of patterns was given to the memory at the initial stage. In biosystem semantic information is partly given by teaching and partly learned by the system itself. We have not yet shown the latter process. Success and failure in pattern recognition, i.e. the correct and incorrect interpretation of input patterns, gradually reorganize stored information so that the concept of dog is self-organized from a number of inputs of individual dog patterns with various shapes. Such conceptualization should be introduced to obtain semantic information stored in the memory. Here we will only point out that this conceptualization is brought about by the self-organization of similar images, for instance, through properly utilizing entrainment mechanism. Such entrainment would be accompanied with pattern recognition. Therefore, semantic information on pattern will be properly modified after every experience of pattern recognition. For gaining semantic information the presence of mechanisms to teach the success or failure of the result to the system is crucial, at least at the initial stage of the self-organization of conceptualized information on pattern. Mapping of memories according to the hierarchical order of such conceptualized information should facilitate pattern recognition. There are problems other than this to be studied further on the structure and dynamics of the memory. One of them would be a distributed memory with spatial information on patterns.

(b) *Use of a Global Feature of Pattern for the Recall of Memory* : Although we have a tremendous number of memories on various kinds of pattern, we are able to recognize input pattern information instantaneously. There might be an efficient mechanism to recall memory. One mechanism for this would be the use of suitable mapping based on selective values. Another possibility is the use of information on the global feature of the pattern. This is convenient for recalling a certain category of pattern from the memory (*quick recall*), prior to the self-organization of a detailed image (*detailed recall*). In the above model all the three kinds of memory were recalled instantaneously.

Imagine that a figure of biped, quadruped and bird as illustrated by MARR [2] shown in Fig.9 are stored in another way than above polygons. We like to evoke the latter figures separately from the formers by properly using information on the global feature of these figures.

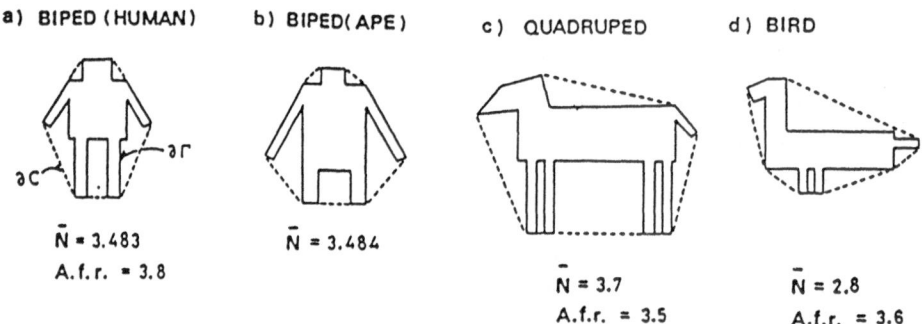

a) BIPED (HUMAN)   b) BIPED (APE)   c) QUADRUPED   d) BIRD

$\bar{N}$ = 3.483 $\bar{N}$ = 3.484

A.f.r. = 3.8

$\bar{N}$ = 3.7 $\bar{N}$ = 2.8

A.f.r. = 3.5   A.f.r. = 3.6

$$\bar{N} = \frac{2|\partial r|}{|\partial C|}$$

A.f.r. = Average firing rate

Fig. 9   A global feature of input pattern, $\bar{N}$, can be obtained by the average firing rate, A.F.R., of excited simple cells. Such a quantity can be used for the selective evokation of memories as rapid recall mechanism

As one of such indices representing the global feature of pattern, we may define a quantity $\bar{N}$, that is the ratio of twice of the total length of the contour of the pattern against its convex envelope. As is given in Fig.9, this quantity is entirely different from 2 that is $\bar{N}$ for triangles, squares and polygons. It is remarkable that $\bar{N}$ becomes almost the same with the average firing rate, A.f.r., of excited simple cells in the hypercolumn when these patterns are percepted, provided that a reasonable function is used for the dependence of the firing rate of a simple cell on the direction of input line (Tsuda & Shimizu). If the frequency of our elementary oscillators is proportional to $\bar{N}$, we are able to have two kinds of carrier waves of different frequencies. The carrier wave with the lower frequency will evoke triangles, squares and polygons while bipeds, quadrupeds and birds will be evoked by the other carrier wave with the higher frequency. More elaborate indices could be found if we want to separate figures more detailed.

(c) *Adaptive Resonance Theory of Grossberg.* Our model has a similarity with the adaptive resonance theory of GROSSBERG [11] in some points. Grossberg presented his model based on extensive considerations on the cognitive process in the brain. In this theory a pattern $x^{(1)}$ caused by the excitation of cells at level $F^{(1)}$ is sent to a higher level $F^{(2)}$ with suitable filtration and evokes a learned pattern $x^{(2)}$, which feedbacks to $F^{(1)}$ as expectation. The both patterns are matched due to "adaptive resonance". Adding the level of input information in the form of elementary information, his model may be regarded as a kind of three-level model. Thus, the presence of three hierarchical levels which are different in the extent of compression of information with an adjustable matching mechanism would be a common feature of elementary models of cognitive information processing, not only in the brain but also in biosystems with cognitive ability. Entrainment of the nonlinear oscillations would be the key mechanisms for the self-organization of images (internal patterns), the evocation of stored information, and the compromised matching between objective (input) and subjective (stored) information.

Acknowledgments ; The authors would express their sincere thanks to Professor Hermann Haken for his kind interest and important discussions from the initial stage of the present work. We thank Dr. E. E. Körner of the bioholonics project group for his kindly correcting the English on our manuscript.

REFERENCES

1. L. Wittgenstein: *Philosophical Investigation* ed. by G. E. M. Anscombe & R. Rhees with an English Translation by G. E. M. Anscombe (Basil Blackwell 1953, 1958, 1967)

2. D. Marr: *Vision — A Computational Investigation into the Human Representation and Processing of Visual Information* (W. H. Freeman & Company 1982)

3. H. Haken: "Pattern Formation and Pattern Recognition — An Attempt at Synthesis" in *Pattern Formation by Dynamic Systems and Pattern Recognition* ed. by H. Haken. (Springer Verlag 1979)

4. H. Haken: *Synergetics, An Introduction*, 2nd ed. (Springer Verlag 1978)

5. D. H. Hubel & T. N. Wiesel: "Functional Architecture of Macaque Monkey Visual Cortex", in Proc. R. Soc. Lond. B., 198, p. 1-59 (1977)

6. K. S. Lashley: "In Search of the Engram" in Symp. Soc. Exp. Biol., 4, p. 454-482 (1950)

7. J. C. Eccles: *The Understanding of the Brain*, 2nd ed. (McGrow-Hill 1973)

8. W. J. Freeman: *Mass Action in the Nervous System — Examination of the Neurophysiological Basis of Adaptive Behavior through the EEG* (Academic Press 1975)

9. M. Ohsuga, Y. Yamaguchi & H. Shimizu: "Entrainment of Two Coupled van der Pol Oscillators by an External Oscillation — As a Base of "Holonic Control" in Biol. Cybern., 51, p. 325-333 (1985)

10. M. Eigen & P. Schuster: *Hypercycle — A Principle of Natural Self-Organization* (Springer Verlag 1979)

11. S. Grossberg: "How does a Brain Build a Cognitive Code?" in Psychol. Rev., 84, p. 1-51 (1980)

Self-Organization of the Dynamical Channel

I. Tsuda

Bioholonics Project, Research Development Corporation of Japan,
Nissho-Bldg. 5F, 14-24, Koishikawa 4-chome, Bunkyo-ku, Tokyo 112, Japan

H. Shimizu

Biophysics Dept., Faculty of Pharmaceutical Sciences,
University of Tokyo, Hongo, 7-3-1, Bunkyo-ku, Tokyo 113, Japan

1. Introduction

There are many important concepts in information sciences. Among others, the concept of information channel is of particularly interest, since the information is selectively transmitted through the channel according to the nature of the channel.

In biological organisms one can observe many examples of this type of channeling. In the nervous system the neuron can be viewed as a dynamic channel, whereby the term "dynamic" labels that information arriving from the environment or from other nerve cells are processed dynamically and transmitted selectively.

It is known that a system is completely controlable only if the system shows both the character of orbital stability and structural stability. However, we are forced to alter the concept "complete control" to "partial control", because of the fact that complicated behaviour like chaos caused by nonlinearity can appear even in simple deterministic dynamical systems [1].

In the course of evolution biological organisms achieved a methodology of partial control. In biological organisms, the partial control is a self-control. This is one of the reasons why we are particularly interested in the investigation of biological systems.

It is hypothesized that biological systems in order to establish a self-control construct dynamic information channels inside and process information, according to the relation of the system's state and the environment as well as the interrelation between innerstates of the system.

Conceiving these topics the paper investigates the character of the dynamic channel consisting of specific nonlinear oscillators and deal not only with the Shannonian information transmission but also with the non-Shannonian information transmission through the channel.

2. Dynamical Systems as Dynamic Information Channel

Let us consider a set of k coupled one-dimensional nonlinear difference equations $\underset{\sim}{X}_{n+1} = \underset{\sim}{F}(\underset{\sim}{X}_n)$. As shown in Fig. 1 a), in the case of one-dimensional mappings (the case k=1) dynamical systems can be viewed as an information channel [2],[3]

a)

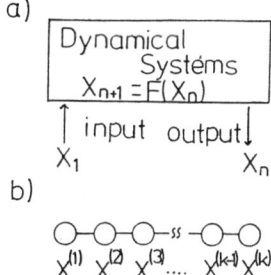

Fig. 1: Dynamical systems viewed as an information channel. a) shows the one-dimensional map case and b) the case of coupled map lattice. In b), the value of the variable $X_1^{(i)}$, or $X_t^{(i)}$ is viewed as the input and the value of the variable $X_t^{(j)}$ as the output.

when we identify the initial value of the variable as the input information and the value of this variable at another time n as the output[3].

In the case of coupled map lattice (see Fig. 1 b)) we can identify the value of the variable $X_1^{(i)}$ or $X_t^{(i)}$ as the input and the value of the variable $X_t^{(j)}$ as the output. Then we can regard the dynamical systems as the dynamic information channel. We take this viewpoint throughout this paper.

3. Coupled Neuron-like Model

3-1 Characteristics of the Local Model as a Basic Element of the Coupled System

We here use the following map as the local model as a basic element of the coupled system.

$$X_{n+1} = F(X_n;b) = f(X_n) + b$$
$$= \text{const} \{\arctan[\beta(X_n-0.2)] + \arctan(0.2\beta)\}/[1+(2X_n)^{19}] + b \qquad (3-1)$$

where b is a bifurcation parameter and const. chosen so as to give max. f(X) =0.8. The steepness parameter β is fixed to 200.0 throughout this paper. The feature of the map is shown in Fig. 2.

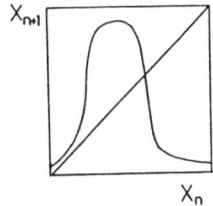

Fig. 2: Schematic drawing of the map treated in this paper as a basic element of the coupled system.

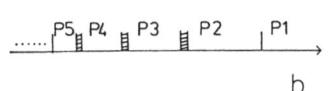

Fig. 3: All maps defined by eqs. (3-1) (3-3) have the same set of solutions. Shaded parts show the parameter regions of more complicated periodic solutions and chaos. Pn denotes period-n oscillation.

241

As indicated in Fig. 3, the map produces periodic solutions with an integer period in a wide parameter region and solutions with more complicated periodicity and with non-periodicity (chaos) in a narrow parameter region[4],[5]. Note that this type of set of solutions can be observed also in the other maps having the similar feature. This feature is that the slope on the left-hand side of the map is rather steep. As examples, also in the following maps (3-2) and (3-3) one can observe the same set of solutions.

$$
\begin{cases}
X_{n+1} = [(X_n - 0.125)^{1/3} + 0.50607357] \exp(-X_n) + b & \text{for } X_n < 0.3 \\
X_{n+1} = 0.121205692[10X_n \exp(-10X_n/3)]^{19} + b & \text{for } X_n > 0.3 \quad (3-2)
\end{cases}
$$

$$
X_{n+1} = 0.8(0.3)^{-\alpha} e^{\alpha} X_n^{\alpha} \exp(-\alpha X_n/0.3) + b, \tag{3-3}
$$

where α is the steepness parameter.

Furthermore, we indicate the similarity between our model and the simple neuron model having the threshold and memory term. Nagumo and Sato[6] reduced the delay-difference equation of Caianiello to the piece-wise linear equation by replacing the memory term with a single variable (the Caianiello equation). They also discussed the devil's staircase-like response of the firing rate of Caianiello's model which was found to be similar to observed data in Harmon's neuron model[7] consisting a mono-stable circuit of transistor. In our case, if we associate "0" (it means "resting state") with the left-hand side of the map and "1" ("firing state") with the right-hand side of the map, then we can calculate the firing rate. The result of the calculation indicates the similarity of our model to the neuron model. Furthermore, the steepness of the left-hand side of the map reminds of the threshold of neuron. In this respect, we call our map neuron-like model.

3-2 Coupled System

We introduce in this subsection the concrete model for information channel. The information channel is constructed by a set of k one-dimensional mappings each of which is a neuron-like model described by eq. (3-1). The connection type is linear and asymmetric. Moreover, each map excites the right-hand nearest neighbour map and inhibits itself (see Fig. 4). So, the overall channel is described as follows.

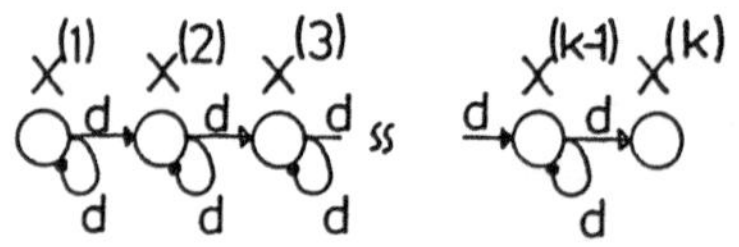

Fig. 4: The neuron-like map is connected linearly and asymmetrically. Arrow indicates the excitatory connection and solid circle the inhibitory connection.

$$\left\{\begin{array}{l} X_{n+1}^{(1)} = F(X_n^{(1)} \; ; \; b^{(1)}) - d \, X_n^{(1)} \\ X_{n+1}^{(2)} = F(X_n^{(2)} \; ; \; b^{(2)}) + d(X_n^{(1)} - X_n^{(2)}) \\ \qquad \cdot \\ \qquad \cdot \\ \qquad \cdot \\ X_{n+1}^{(i)} = F(X_n^{(i)} \; ; \; b^{(i)}) + d(X_n^{(i-1)} - X_n^{(i)}) \\ \qquad \cdot \\ \qquad \cdot \\ \qquad \cdot \\ X_{n+1}^{(k)} = F(X_n^{(k)} \; ; \; b^{(k)}) + d \, X_n^{(k-1)} \; , \end{array}\right. \qquad (3\text{-}4)$$

where d is a coupling constant.

We investigate what kind of information is transmitted through this channel. In the next section, we study the transmission of Shannonian information. In section 5, we study the transmission of non-Shannonian information.

4. Transmission of Shannonian Information

From information theoretical viewpoint [8], chaos can be measured by Shannon's information [9] since chaos has a probabilistic distribution as an invariant density.

To investigate the possibility of transmission of Shannonian information through our channel, we add to the third cell $X^{(3)}$ the further connection of the map Y whose parameter is set to produce chaos. Therefore, the equation of the third cell becomes

$$X_{n+1}^{(3)} = F(X_n^{(3)} \; ; \; b^{(3)}) + d(X_n^{(2)} - X_n^{(3)}) + d'Y_n, \qquad (4\text{-}1)$$

where d' is an additional coupling constant.

As indicated in Fig. 5, the chaotic burst is successfully transmitted. The velocity of the transmission is 0.5 unit space/unit time which is just a half of the light velocity (maximal velocity) 1 unit space/unit time of this system. The velocity depends on the set of the channel parameters $(b^{(1)}, b^{(2)}, b^{(3)}, \ldots, b^{(7)})$.

We calculated the mutual information between Y and $X^{(i)}$ (i=3,4,...,7) in various channel parameters [3]. From this calculation it was seen that Shannon's information was transmitted without decay through our channel. We checked that no matter what value   the channel parameters, Shannonian information was transmitted through the channel. On the other hand, if the logistic map is used as the element of the channel, then the mutual information decays very rapidly, so Shannonian information does not succeed in being transmitted.

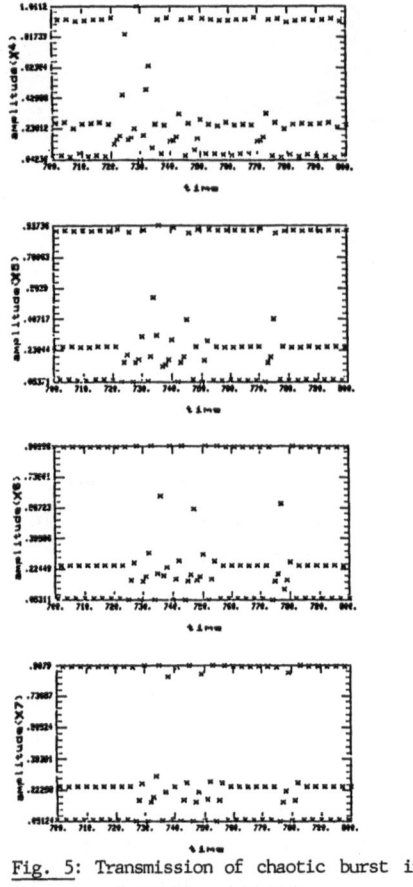

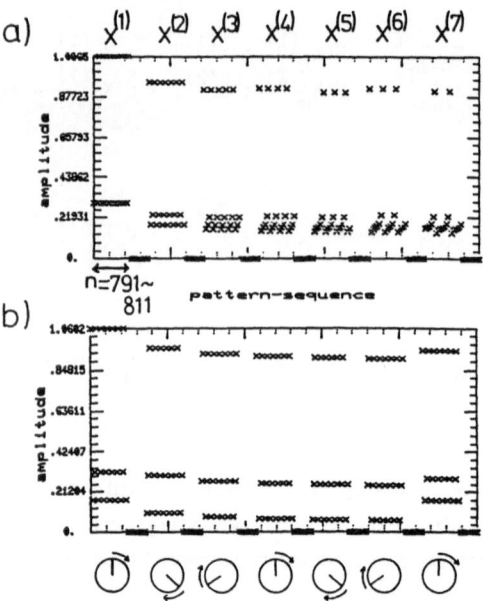

Fig. 6: (a) shows time series (n=791-811) of each basic oscillation. Period of the oscillation $x^{(i)}$ is set to i+1 ($b^{(1)}$=0.3, $b^{(2)}$=0.18, $b^{(3)}$=0.158, $b^{(4)}$=0.15, $b^{(5)}$=0.145, $b^{(6)}$=0.1425 and $b^{(7)}$=0.1405). (b) the case that the coupling constant d=0.12. Every oscillations are entrained to period-3 but their phases shift in metacronal form as shown in the lower part of the figure.

Fig. 5: Transmission of chaotic burst in the case of d=0.12 and d'=0.2.

5. Transmission of Non-Shannonian Information

In this section we investigate the possibility of transmission of non-Shannonian information through the channel. By non-Shannonian information we mean here periodic oscillations. Periodic oscillations cannot be measured by Shannon's information which is based on the probability description. In this respect it seems to be adequate to call periodic oscillations non-Shannonian information.

We observed the transmission of a periodic oscillation only in the case that all oscillations at the right-hand side of the input are entrained to the same period and their phases are shifted in a metacronal form. This condition of entrainment can be easily realized when we choose a period of the local element so as to increase a period by one toward the right (see Fig. 6 a)). In Fig. 6 b), the feature of the time series of each element is depicted and the feature of the phase shift is also indicated. Concerning the phase shift we investigated its dependency on the initial condition. In the case of entrainment to period-3,

244

the phase shift between elements having period-2 and 3 respectively as an original period depends on the initial conditions of these elements. On the other hand, the phase shifts between other elements are fixed even changing their initial conditions. To avoid the dependency on the initial conditions we feed non-Shannonian information into the element after the second element. Fig. 7 shows one example of the transmission of periodic oscillation. Note that not only the period but also the feature of the amplitude are almost completely transmitted. In Fig. 8, the transmission of period-4 is represented by binary coding { 0,1}. If we associate "0" with the left-hand side of each map and "1" with the right-hand side of each map, then we obtain the symbolic representation of the system. In our case, the oscillation with period-4 can be coded by "0001 0001 ....". Fig. 8 shows that the symbol "0001" is eventually transmitted with a constant velocity.

Here, one question arises : How robust is this type of transmission for the external noise? We applied the uniform noise [$-\sigma,\sigma$] to every element of the channel. The transmission of period-5 survives the noise with $\sigma = 0.005$, but it

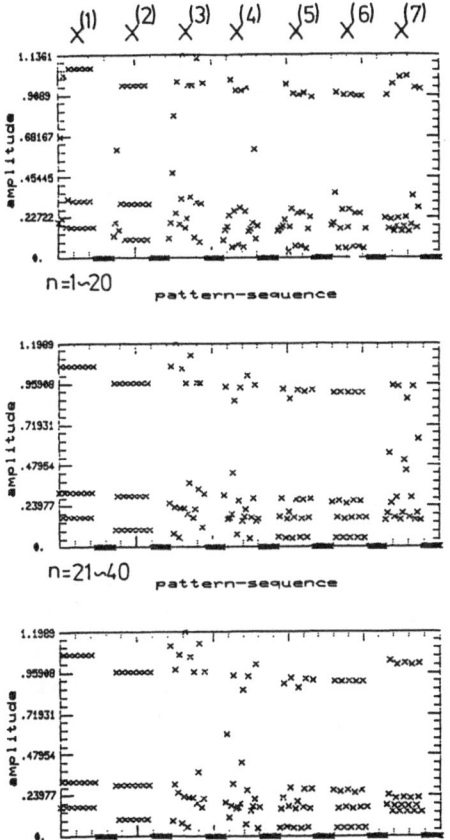

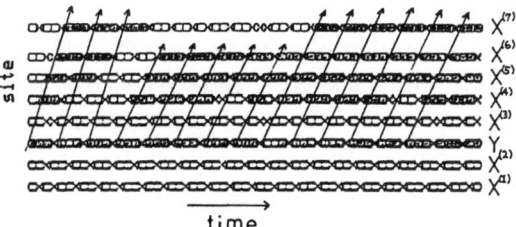

Fig. 7: Transmission of periodic oscillation. Period-4 oscillation is fed to $X^{(3)}$. After n=40, the complete transmission of period-4 oscillation is seen at the position of the seventh cell $X^{(7)}$. In this case, d=0.12, d'=0.2.

Fig. 8: The square denotes "0" and the cross "1". The symbol "0001" is denoted by a set of the subsequent three solid squares and the cross.

is destroyed by the noise with longer amplitude. In the case of period-4, the transmission survives the noise up to $\sigma = 0.015$. So, it can be said that the transmission of the periodic oscillation is comparatively robust for the external noise although the robustness is different between different periods.

We checked that the transmission of one periodic oscillation succeeded also through the channel consisting of eight local elements (k=8). However, the larger value of coupling constant d' between the input and the channel in this case than in the case of seven elements (k=7) is required for the transmission. Thus, for the transmission of non-Shannonian information through the channel one can conceive that the higher the number of local elements of the channel, the stronger the coupling constant, i.e., the higher the energy supply from outside. This characteristics means that the range of transmission of non-Shannonian information is limited if the coupling constant between the input and the channel is fixed. For an application to technology we consider how we can transmit non-Shannonian information to a long distance.

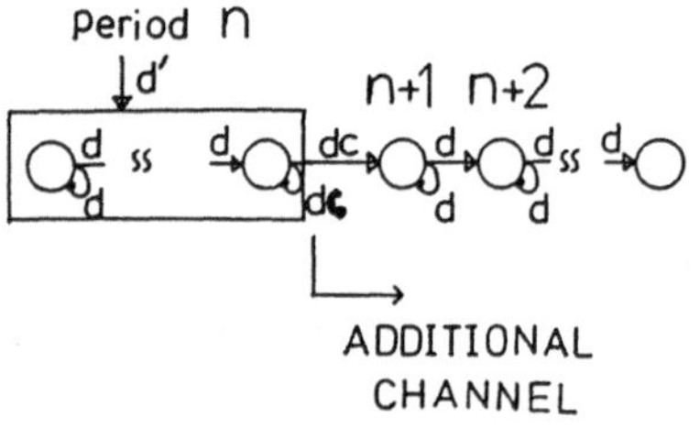

Period n

n+1 n+2

ADDITIONAL
CHANNEL

Period 4

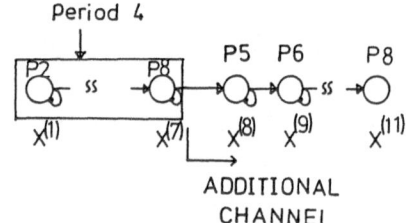

ADDITIONAL
CHANNEL

Fig. 9: For the effective transmission of period-n oscillation, a period of the local element of the additional channel begins at n+1, not at k+1.

Fig. 10

We found a clear method for this purpose. As indicated in Fig. 9, regardless of the periods of the local elements of the channel, a period of the local element of the channel to be added begins at n+1 and increases one by one for the period-n-input. As an example, we show the case of the period-4-input in Fig. 10. A specific feature is obtained : we can control the coupling constant dc between the elements $X^{(7)}$ and $X^{(8)}$. The possible range of dc for the transmission of period-4 is $0.05 \leqq dc \leqq 0.25$ when d and d' are fixed to 0.12 and 0.2 respectively. This gives the maximum number of possible branching (see below). In this case, 5 branchings are possible. For the period-5-input, we obtained $0.04 \leqq dc \leqq 0.18$ as the possible range for period-5-transmission under the condition that d=0.08 and d'=0.2, so the maximum number of branching is 4.

These facts implicate an important aspect concerning a parallel processing of information. As shown in Fig. 11, if the coupling constant dc is fixed to 0.05, the input oscillation with period-4 is copied at every 5 terminals of branches. Additional copies according to additional branchings are possible almost

246

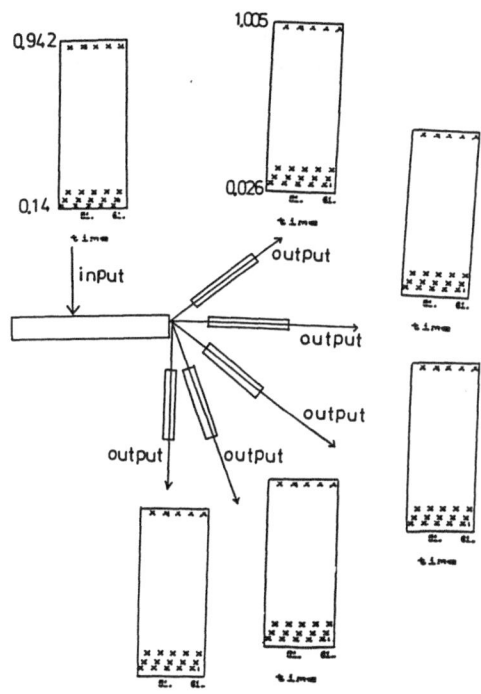

Fig. 11: Period-4 oscillation is copied at every 5 terminals of branches (d=0.12, d'=0.2, dc=0.05).

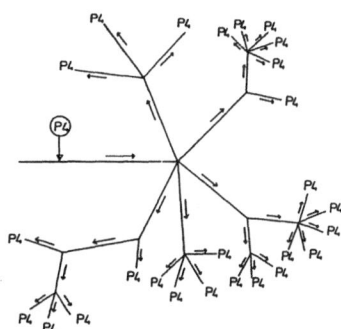

Fig. 12: Schematic drawing of possible additional copies of period-4 oscillation.

unlimitedly as indicated in Fig. 12. These copies can be used as working in parallel at each terminal.

Furthermore, we obtain the additional copy of another input with different period. This is shown in Fig. 13.

Using these branched channels, we can construct local memories. As indicated in Fig. 14 a), let us introduce an additional coupling from $X^{(11)}$ to $X^{(8)}$. Now assume that the information is fed in in the form of periodic oscillation with period-4 only during 100 unit times and is switched off after n=100, and the coupling constant dc is weakened after n=100. Then the input information is stored as the activity of period-4-oscillation into the loop. This type of memory directly stems from the multi-basin of the dynamical systems. In this way, we can select one active state out of many possibilities according to the input information .

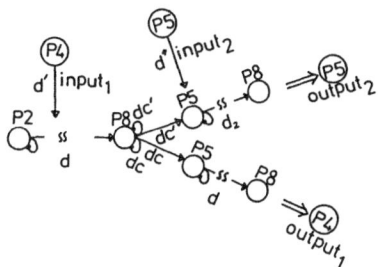

Fig. 13: Period-5 oscillation is also copied. We observed the simaltaneous copy of period-4 and period-5 oscillations at different terminals. In our observation, d=0.12, d_2=0.08, d'=d''=0.2, dc=0.05 and dc'=0.03.

a)

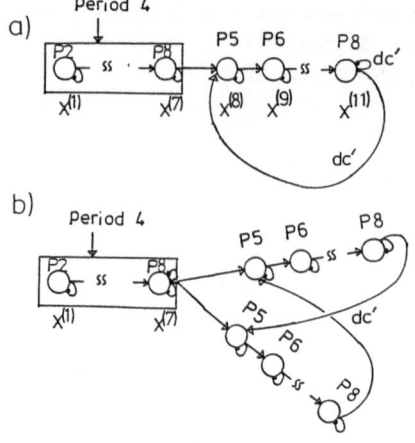

Period 4

b)

Period 4

Fig. 14: a) Local memory of input information is constructed by adding a new connection from $X^{(11)}$ to $X^{(8)}$ and by decreasing the strength of the coupling constant dc. Parameters : d=0.12, d'=0.2 (for n=1~100) and d=d'=0.0(for n>100), dc=0.1 (n=1~100) and 0.01(n>100), and dc'=0.1.

b) One of the other possibilities of the connection for the local memory of period-4 oscillation.

Another construction of the local memory can be realized in the form of intersection of different branches (see Fig. 14 b)).

The construction of this type of memory succeeds also in the case of period-5-input. Therefore, as shown schematically in Fig. 15, we can obtain many memory loops consisting of the activities of period-4 and 5.

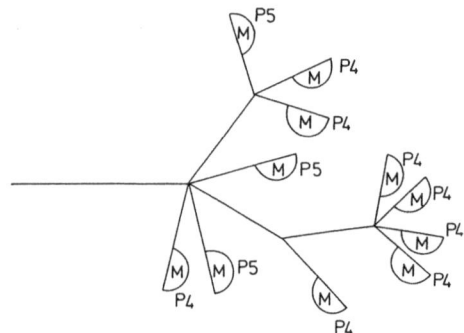

Fig. 15: Schematic drawing of many memory loops consisting of the activities of Period-4 and 5. M denotes memory loop.

We can easily conceive the possibility of the other kinds of memory, for instance, period 6, 7, etc., although we have not yet checked it.

Note that it is possible to destroy the memory by changing dc'. Moreover, once the coupling constant dc is weakened, the local memory survives any external perturbations fed into the channel except to the part of the loop. In this way, we easily construct and destroy the local memory by changing only a few definite parameters, which could be performed by an architecture of computer software.

6. Possible Mechanism of Dendritic and Terminal Branching of Neurons

In this section, we consider the possibility of a new interpretation for the arborization and terminal branching of a neuron in terms of the branching of the channel.

As is well known, a typical nerve cell has dendrites (arborization) and terminal branches of axon. Certainly, a number of branches enables many contacts, which can be used for the processing of information.

According to our results concerning the information transmission, we can imagine the another reason why the nerve cell has dendrites and terminal branches. If the neuron extends according to the original strategy (which is determined by gene), the information would decay and would not be transmitted effectively. So, if the neuron generates the additional parts of the channel according to the input information the information would be effectively transmitted for distances. Furthermore, note that in our channel, the nature of branching depends on the input information. In generating effective branching like the ones indicated in Fig. 11-15, the neuron would possibly transmit information (both Shannonian and non-Shannonian information) precisely and effectively according to the input information.

It seems this idea is supported by the fact that the patterns of arborizations or the patterns of terminal branches correlate with the function of the nerve cells.

For instance, three classes of cells are distinguished according to those patterns of arborization (dendritic branching of ganglion cells in the cat retina [10]) : α, β and γ cells show respectively radiate dendrites, busy dendrites and varied dendritic geometry. These α, β, and γ cells correspond respectively to the classes Y, X, and W. The α and β cells are further classified into off-center cells and on-center cells according to the place in the innerplexiform layer at which the dendrites arborize and from which the number of synaptic contacts from the bipolar cells rapidly increases.

For pattern of terminal branches [11], it is known that various sorts of specialized endings exist and different types of nerve endings correspond respectively to different nature of information transmission from the environment : patterns of nerve endings for pressure, heat, touch, pain and motor function are respectively different.

Furthermore, Sano et al. have observed a reticulated pattern of the branches of the serotonin fibers in various submammalian (the carp, frog, turtle and chicken) and even in mammalian (the rat and cat) central nervous systems[12].

In the submammalian paraventicular organ, proximal processes of the serotonin-containing cerebrospinal fluid-contacting neurons protrude into the venticular lumen with a globular and triangular shape, and distal processes project to the pars distalis and form a fine plexus. In the neurohypophysis of adult cats, serotonin neurons distribute in the internal and external layers of the infundibulum. The branches of the serotonin fibers form a plexus with an irregular pattern. They also investigated the serotonergic innervation of the pia mater of rat. It has been found that the pia mater covering the ventrolateral surface of the medulla oblongata is innervated by the serotonin neurons of the brainstem. In our context, it is of particularly interest that the branches of the serotonin fibers formed a plexus with an irregular pattern. A reticulated pattern of branchings suggests a possibility of local memories and of, perhaps, even some calculation (see Fig. 14).

Our present idea concerning the nerve branching can be one of the mechanisms controling the process of cell differentiation, although one may not overlook the existence of chemical substances in these processes.

7. Summary

We investigated the information transmission through the dynamical channel constituted in the form of coupled neuron-like map.

It was proved that both Shannonian and non-Shannonian information were transmitted through the channel. Concerning the Shannonian information transmission, the functional form of the local element, namely, the neuron-like map, is basically and the order of the local element in space gives only the change of the transmission speed. On the other hand, concerning the non-Shannonian information (periodic oscillation in our case) transmission, the metacronal type of phase shift is also necessary, therefore, the position of the local element in space is of importance.

In the following, we discussed the possibility of further elongation of the channel which is necessary for transmitting information for distances. The maximum number of branching depends on the type of input information. In this context, we successfully made the local memory and the groups of the local memories.

These characteristics of the channel could be useful for implementing it to the parallel processing.

Finally, we discussed a possible relation between the nature of branching of the channel and the dendritic arborization or the terminal branching of physiological nerve cells.

Acknowledgments
We would like to thank Dr. Körner for stimulating discussions concerning dendritic and terminal branchings of nerve cells and also for critical reading of the manuscript.

References

1. Chaos and Statistical Methods ed) Y. Kuramoto (1984) - Springer Series in Synergetics -
    Topics on Nonlinear Dynamics in Dissipative Systems Prog. Theor. Phys. Supplement 79 (1984), particularly, see K. Tomita (pp 1-25).

2. J. S. Nicolis and I. Tsuda, to be published in Bull. Math. Biol. (1985)
   J. S. Nicolis (preprint)
3. K. Matsumoto and I. Tsuda, preprint
4. K. Tomita and I. Tsuda, Prog. Theor. Phys. $\underline{64}$, 1138 (1980)
5. K. Matsumoto and I. Tsuda, J. Stat. Phys. $\underline{31}$, 87 (1983)
6. J. Nagumo and S. Sato, Kybernetik $\underline{10}$, 155 (1972)
7. L. D. Harmon, Kybernetik $\underline{1}$, 89 (1961)
8. R. Show, Zeit. für Naturforschung $\underline{36a}$, 80 (1981)
9. C. E. Shannon and W. Weaver, The Mathematical Theory of Communication (Illinois Press, Urbana, 1949)
10. P. Sterling, in Changing Concepts of the Nervous System (Proceedings of the First Institute of Neurological Sciences, Symposium in Neurobiology) ed) A. R. Morrison and P. L. Strick (Academic Press, 1982), p 281
11. J. G. Taylor, in Lecture Notes on Biomathematics 4 (1974), p 230
12. Y. Sano et al., Histochemistry $\underline{75}$, 293 (1982), ibid. $\underline{76}$, 277 (1982), ibid. $\underline{77}$, 423 (1983)

Note: Concerning the coupled map lattice, see also K. Kaneko (Prog. Theor. Phys. $\underline{72}$, 480 (1984)).

Part V

Theoretical Concepts

Macroscopic Prediction

E.T. Jaynes

Arthur Holly Compton Laboratory for Physics, Washington University,
St. Louis, MO 63130, USA

1.   Introduction

Our topic is the principles for prediction of macroscopic phenomena in general,
and the relation to microphenomena. Although physicists believe that we have
understood the laws of microphysics quite well for fifty years, macrophenomena
are observed to have a rich variety that is very hard to understand. We see not
only lifeless thermal equilibrium and irreversible approaches to it, but lively
behavior such as that of cyclic chemical reactions, lasers, self-organizing
systems, biological systems.

On a different plane, we feel that we understand the general thinking and
economic motivations of the individual people who are the micro-elements of a
society; yet millions of those people combine to make a macroeconomic system whose
oscillations and unstable behavior, in defiance of equilibrium theory, leave us
bewildered. Governments seem helpless to understand what is happening well enough
to predict it, much less control it.

Why is it that knowledge of microphenomena does not seem sufficient to understand macrophenomena? Is there an extra general principle needed for this?

Our message is that such a general principle is indeed needed and already
exists, having been given by J. Willard Gibbs 110 years ago; but it is not fully
recognized in the current thinking of either statistical mechanics or economics.
A macrostate has a crucially important further property (entropy) that is not
determined by the microstate. We hope to discuss the problem in some generality,
give a mathematical scheme embodying that principle, and show a few of its consequences.

Of course, any particular mathematical scheme will apply only in a particular
area, so we do not want to take a provincial viewpoint tied too strongly to the
mathematics. Fortunately, understanding of the problem has advanced to the point
where it is possible to explain the principles in a very simple way with almost no
mathematics. Only from the simplest viewpoint can one see the full range of
application of a principle.

A moment's thought makes it clear how useless for this purpose is our conventional textbook statistical mechanics, where the basic connections between micro
and macro are sought in ergodic theorems. These suppose that the microstate will
eventually pass near every one compatible with the total macroscopic energy; if
so, then the long-time behavior of a system must be determined by its energy.

What we see about us does not suggest this. Solids appear to maintain indefi-
nitely whatever macroform they happen to have, although many others would have the
same energy. In Egypt's Karnak temple there are stone columns that have supported
heavy weights for thousands of years without flowing. In the crystalline texture
of the granite facing stones on the building where I am writing this we see the
record of the pressure and temperature changes when they cooled, preserved
unaltered for over a billion years.

Persistence of macrostructure is not limited to solids. The atoms comprising an elephant could, if reassembled, make instead ten tigers, any twenty humans who have ever lived, or a half-million butterflies. A man who speaks English but not German would, by a convenient rearrangement of some atoms in his brain, be converted into one who speaks both languages.

All of these different configurations of the same atoms, with the same total energy, would preserve their structure and behavior for the life of the organism. Indeed, the macrostructure and behavior persists even when the original atoms are replaced by others.

In short, virtually all the phenomena of Nature are manifestations of the fact that the behavior of macrosystems, over any times of human concern, is *not* determined merely by their energy. Over all the observation time available to us, they remain in extremely small subregions of the energy shell. Surely, no fact is more familiar to us; then we shall not waste any time on ergodicity. Even if ergodic theorems should prove to be true over infinite times, our problem is of a totally different nature.

The problem to which we address ourselves is, then: We have some information about a few macroscopic quantities $(A_1, A_2, \ldots)$, for example, their values in certain space-time regions; call this A for short. Given this and any other relevant prior information I that we may have, what are the best predictions we can make of some other macroscopic quantities $(B_1, B_2, \ldots)$ in some other space-time regions? Call this B for short.

Logically, the problem is one of inference (i.e., plausible conjecture) rather than deduction, since in almost all real problems of physics, biology, and economics our information is far too meager to permit any deductive proof that our predictions must be right. Indeed, it is often too meager to justify any definite predictions at all. But even in such cases (or rather, especially in such cases) a normative theory of inference could be useful if it indicates to us what kind of further information would be needed in order to make good predictions.

We are therefore concerned primarily with optimal information processing, only secondarily with physical law (although we shall of course not ignore any physical law that is known and relevant); so if there are general principles for solving it they ought to apply equally well in or out of physics. More specifically, we are asking whether probability theory can be applied to this problem, in the form: given prior information I and data A, what is the probability $p(B|A,I)$ that any particular prediction B will be right?

Such a question has nothing to do with the "random variables" that are so prominent in conventional probability theory. The reason we need inference is not chance or "randomness", but incomplete information. But if the question makes sense in probability theory, it should make sense also in biology and economics, as well as in physics.

Historically, however, such problems first arose in physics; in the next Section we recall briefly what was found there. Section 3 then explains the principles that we propose to use, in essentially verbal rather than mathematical terms. The reader who wants to get on with the new technical content of this work may turn at once to Section 4.

2. Historical Background

It is now more than 160 years since Carnot started the science of thermodynamics with the first statement of what is perhaps the most fundamental of all principles for predicting macroscopic events. But we have not yet understood and exploited all its implications.

At first, progress came rapidly; from Carnot's principle (no heat engine E can be more efficient than a reversible one R with the same upper and lower temperatures t, t') it follows that all reversible engines must have the same efficiency depending only on t, t'. So, as Kelvin saw, the reversible efficiency e(R) must define a universal temperature scale T(t), independent of the properties of any particular substance:

$$e(R) = 1 - T'/T \quad . \tag{1}$$

From the first law the actual efficiency is e = 1 - Q'/Q; so as Clausius saw, Carnot's inequality e ≤ e(R) implies the existence of a new function of the thermodynamic state

$$S = \int dQ/T \tag{2}$$

which he named "entropy", with the property that in a macroscopic change that begins and ends in thermal equilibrium, the total entropy of all bodies involved cannot decrease:

$$S(\text{final}) \geq S(\text{initial}) \quad , \tag{3}$$

ergo, if it increases, the process is irreversible.

All this was accomplished within thirty years of Carnot's work; but further progress has been painfully slow. The theory has been hampered by a serious restriction; the definition (2) of entropy refers only to equilibrium states (the path of integration being a reversible path; i.e., a locus of equilibrium states). In consequence, the Clausius statement of the second law (3) gives us only minimal information about a change of macroscopic state; it predicts that it will go in the general direction of greater final entropy; but not how fast, or along what path. Nor does it predict what final equilibrium state will be reached. Indeed, as the more careful writers have noted, it does not even predict that the entropy will increase monotonically, since the entropy of an intermediate nonequilibrium state was not defined.

It required twenty more years for the next advance of Gibbs (1875-78) which gave the variational principle for determining the final equilibrium state; but the world was not yet ready for him, and only after another fifty years was the first level of understanding of Gibbs' work reached, by G. N. Lewis. After still another fifty years, rereading Gibbs convinced me that G. N. Lewis extracted only about half of the deep insight in that work, which is by no means of mere historical interest today. After over 100 years, we can still learn important "new" technical facts from Gibbs.

We think that this is due mainly to the fact that Gibbs did not live long enough to complete his work. He was much too far in advance of his time to have left any students capable of carrying it on, and he was not the world's clearest expositor; so his thinking had to be partly re-discovered before we could recognize that Gibbs already had it.

Turning now to macrophenomena, it is clear that far more than (3) can be said about how they proceed. We enunciate a rather basic principle, which might be dismissed as an obvious triviality were it not for the fact that it is not recognized in any of the literature known to this writer:

> If any macrophenomenon is found to be reproducible,
> then it follows that all microscopic details that
> were not under the experimenter's control must be
> irrelevant for understanding and predicting it.

Now in the laboratory we find that irreversible processes follow a definite course; control of a few initial macroscopic quantities is sufficient to yield a reproducible result. We observe this not only close to equilibrium as in heat conduction and viscous laminar flow, but also far from equilibrium as in shock waves, fast chemical reactions, and lasers.

Ergo, information about those initial macroscopic values, together with suitable general prior information I whose nature we have not yet specified, must be sufficient for theoretical prediction of the result. It cannot be necessary to specify the millions of microvariables that were not controlled and would not be the same on successive repetitions of the experiment.

But the Clausius rules (2), (3) presumably defined the initial state by specifying the macrovariables that the experimenter did control or observe; that is his definition of the initial "thermodynamic state", and in his applications it is what we have called the "data" A. Why then could he not predict far more than the increasing tendency of the entropy? He must have ignored a great deal of relevant information; but if he had the same data as we would have today, this could only be the unspecified prior information I.

In biology, it appears that a seemingly small amount of microscopic information (configuration of a few DNA molecules) is sufficient to determine thousands of details of the macroscopic structure of an organism. Biological phenomena are also highly reproducible; *ergo*, given proper theoretical understanding they must be also highly predictable.

One might have thought, then, that progress since Clausius would show us how to recognize and use this missing information I. Indeed, that is just what Gibbs did. But surprisingly, the Clausius statement of the Second law remains the culmination of the subject as taught to physicists. Recent works from which we might have learned something better [1,2] are still ignored in our pedagogy and research literature.

For decades the thermodynamics of physical chemists, trained instead in the Gibbsian principles as expounded by G. N. Lewis, has been ahead of that of physicists. But this theory -- at least, the firmly established "exact" part of it that comes from Gibbs -- is still limited to equilibrium predictions. A few extensions beyond that domain, such as the law of Mass Action, have the nature of useful approximate rules of thumb, rather than parts of an exact theory.

This is not to say that attempts to extend equilibrium theory have not been made. So many different approaches have appeared -- all of the nature of *ad hoc* rules of thumb without foundation in first principles -- that the field is reduced to chaos, even the meanings of such common terms as "entropy" and "reversible" being in contention.

3. The Basic Idea

In the following we conjecture about common features that might be in any theory of macrobehavior; thermodynamic, biological, or economic. But the very simplicity of the idea we want to expound makes it difficult conceptually. Our first thoughts run thus: the macrostate is only a kind of blurred image, or projection, of the microstate, with the fine detail removed. *Ergo*, macrobehavior must be determined by microbehavior; there is no room for other considerations. Extra principles beyond those of microbehavior could not be even relevant, much less needed, to predict macrobehavior. As a physics student in the 1940's, the writer was so strongly indoctrinated with this view that it required decades of hard thinking to escape from it.

Of course, we do not suggest that macrosystems can violate the laws of microphysics, or that those laws can be suspended in the interest of any "final cause" or "vital principle". Our extra-microscopic principle is not of this nature at all; surely, if we knew the exact microstate and could integrate the equations of motion exactly, no additional principle would be needed.

The difficulty of prediction from microstates lies, not in any insufficiency of the laws of microphysics to determine macrophenomena, but in our own lack of the information needed to apply them. We never know the microstate; only a few aspects of the macrostate. Nevertheless, the aforementioned principle of reproducibility convinces us that this should be enough; the relevant information is there, if only we can see how to recognize it and use it.

The problem that Gibbs [3] faced in his *Heterogeneous Equilibrium* of 1875 was: given a few macrovariables defining a nonequilibrium state of a system, predict the final equilibrium macrostate that it will go to. But this problem is ill-posed; not enough information is given to determine any unique solution. Nevertheless, Gibbs gave a solution; he took a hint from the Clausius statement of the second law, but changed its logical status drastically. Where Clausius and Planck had tried to see the second law as an "absolute" law of physics like conservation of energy, Gibbs deprived it of that logical certainty -- but at the same time made it a more powerful tool for prediction.

Instead of the Clausius weak statement that the entropy "tends" to increase, Gibbs made the strong statement that it *will* increase, up to the maximum value permitted by the constraints (conservation laws and experimental conditions).

But Gibbs recognized also that this prediction cannot claim the certainty of deductive proof. There are initial microstates, allowed by the data, for which the system will not go to the macrostate of maximum entropy. There may be further constraints, unknown to us, that make it impossible for the system to go to that state; perhaps new constants of the motion. So Gibbs was only doing *inference* rather than *deduction*.

In spite of its seemingly shaky logical status, Gibbs' principle has proved to be as successful in practice as any of the "certain" laws of physics. Whenever the equilibrium macrostate is reproducible, Gibbs' rule predicts it with quantitative accuracy. Physical chemistry has been based on this variational principle for two generations [4].

Clearly, then, the missing information "I" that was needed to resolve the ambiguity of the ill-posed problem was contained somehow in the entropy function. Gibbs pointed out that entropy is a property only of the macrostate, not (like energy) of the microstate. But, cautious man that he was, Gibbs never undertook to tell us what entropy "really is" and why his principle works.

Today, macroscopic prediction has barely advanced beyond the level reached by Gibbs. Yet the way to generalize it has been staring us in the face for 80 years; for the meaning of entropy was seen already by Boltzmann, Einstein, and Planck before 1906. It is carved on Boltzmann's gravestone in the Zentralfriedhof in Vienna:

$$S = k \log W \quad . \tag{4}$$

The thermodynamic entropy of a macrostate (defined by specifying pressure, volume, energy, etc.) is essentially the logarithm of the classical phase volume (or in quantum theory the number of microscopic quantum states) consistent with it; i.e., the "number of ways" the macrostate can be realized.

Gibbs' variational principle is, therefore, so simple in rationale that one almost hesitates to utter such a triviality; it says "predict that final state that can be realized by Nature in the greatest number of ways, while agreeing with your macroscopic information."

But then the generalization we seek is equally obvious and trivial: to predict the course of a time-dependent macroscopic process, choose that behavior that can happen in the greatest number of ways, while agreeing with whatever information you have -- macroscopic or microscopic, equilibrium or nonequilibrium. The prediction is not required to be right in the sense of deductive proof; but it is the best prediction we can make from the information we have; i.e., it is an inference.

From simplicity comes generality: everything we have said would seem to apply as well to biology and economics as to physics. The basic property of a macrostate that is needed for prediction is its multiplicity W. The second law of thermodynamics was only the first case of this to be discovered.

That is the entire content of our theory; but its simplicity makes it hard for sophisticated scientists to understand it and accept it. We were hoping to find a theory that would be subtle and recondite conceptually, elegant and intricate mathematically; how could one ever trust such a childishly simple rule?

First, we reassure the reader that, in spite of the conceptual simplicity, its full mathematical expression does prove to be elegant and intricate after all. To treat general space-time dependences we need a functional integral formalism very similar to that of modern quantum field theory.

Conceptually, we cannot understand why either Gibbs' rule or our generalization of it works until we recognize that principle of reproducibility. Our initial macroscopic information A confined the possible microstates to some class C of possibilities, containing $W(A)$ microstates. That information corresponds to a generalized entropy $S(A) = k \log W(A)$, equally meaningful for equilibrium and non-equilibrium cases. The subsequent macroscopic behavior could not be reproducible unless it was true that for *each* of the great majority of the microstates in C, we would have the *same* macroscopic behavior. But there remain a small minority of "dissenting" microstates that would lead to a different result, and therefore deny our rule the status of logical deduction.

Nevertheless, if our information determining C includes all the conditions that are needed in the laboratory to yield a reproducible result, our rule must predict that result; for it in effect takes a majority vote over class C, thereby suppressing that small minority. The smaller the minority, the more reliable we expect our predictions to be.

As is well known, the minorities are extremely small in the usual thermodynamic situations. For two macrostates A and B, if there is a tiny entropy difference $D = S(B) - S(A)$ corresponding to one microcalorie at room temperature, the ratio of multiplicities is about

$$W(B)/W(A) = \exp(D/k) = \exp(10^{15}) \quad .\tag{5}$$

The macrostate of higher entropy can be realized in overwhelmingly more ways; this is the basic reason for the high reliability of the Gibbs equilibrium predictions.

In other applications we cannot expect the ratios to be so large; but there is a long way to go. If the ratio were only $\exp(10)$, as it might be in a problem of economics, we would still expect the predictions to be reliable enough for most purposes (in any event, to make any better ones would require more information).

Clearly, new information which does not cause an appreciable contraction of the class C cannot appreciably affect our predictions. Thus the cogency of new information -- the degree to which it could help to improve our predictions -- is indicated in a general way by how much further reduction in entropy it would achieve.

Now the equations of probability theory are usually presented as rules for calculating frequencies of "random variables" in "random experiments." But we are not forced to think of them in that way; abstract mathematics may be interpreted in whatever way serves our purpose. Probability theory is also the mathematical tool *par excellence* for locating the class C and taking that majority vote.

Indeed, probability theory was originally conceived, by James Bernoulli and Laplace, as a tool for conducting inference. A probability distribution may be used to describe a state of incomplete knowledge, where there is no "random experiment" involved. Its power for this purpose was demonstrated in massive detail by Sir Harold Jeffreys [5], and its uniqueness as the only consistent such tool was proved by R. T. Cox [6]. In recent years there have been exciting new applications of this viewpoint in many areas of science, such as spectrum analysis and image reconstruction; and indeed, any situation where we are obliged to draw the best conclusions we can from incomplete information [7].

4. Mathematical Formalism

In setting up our generalization of Gibbs' variational principle we shall use probability distributions over microstates. It almost never makes sense -- and it can weaken our results -- to think of the probability of a microstate as a real frequency in any "random experiment". In thermodynamics the imaginary experiment would have to be repeated for perhaps $\exp(10^{24})$ times before there was much chance of that particular microstate appearing even once. In geophysics or economics it is seldom possible to repeat an experiment at all. Therefore we should think of our probability distributions in the original Bernoulli-Laplace sense, as representing simply our state of knowledge when we have the incomplete information A and some prior information I consisting of whatever we know about the laws of microbehavior and any other information that seems relevant. This gives us the freedom to take advantage of whatever information we have.

For our purposes the quantity W is not yet well defined. Specifying a macrostate is never so precise that it makes one microstate easily possible and an adjacent one impossible. Rather the probabilities on microstates that describe real macroscopic information must shade off to zero in some smooth way, and we need a refined definition of W that takes this into account.

We have noted before [8] that the asymptotic equipartition theorem of information theory has an application to this problem, relating Boltzmann's W to the Gibbs H function in classical theory, and to the von Neumann-Shannon information entropy $H = -\mathrm{Tr}(\rho\ln\rho)$ in quantum theory. Consider the latter case; almost everything we say holds *mutatis mutandis* in the former. Given any density matrix ρ with eigenvalues $r_1 \geq r_2 \geq \ldots$ etc., let $W(\varepsilon)$ be the smallest integer for which

$$\sum_{i=1}^{W(\varepsilon)} r_i \geq 1 - \varepsilon , \quad 0 \leq \varepsilon \leq 1 . \tag{6}$$

Intuitively, $W(\varepsilon)$ is the number of reasonably probable microstates, "reasonable" being defined by ε. Now we can associate $\log W(\varepsilon)$ with H in various ways. The strongest supposes that we have N particles, and correlations fall off to zero at some distance. Then as $N \to \infty$ with the intensive parameters held constant, $N^{-1}[\log W(\varepsilon) - H] \to 0$ provided ε is not 0 or 1. Remarkably, in the limit it does not matter what we mean by "reasonably probable". This theorem is discussed more fully by Feinstein [9].

Numerical experimentation shows that a similar result holds under wider conditions than have been proved. But rather than going into all the minute details of more rigorous theorems about this connection -- which would be in the end irrelevant to our problem -- we now recognize that the definition $W \equiv \exp(H)$ is itself the appropriate refinement of the notion of "number of ways" that takes into account the smooth variation of probability. This is obviously correct in the case were W was exactly defined (i.e., when the probabilities r_i are uniform on a subset, zero elsewhere); and any other choice would lead us into conflict with masses of known results in equilibrium statistical mechanics, where the H of the canonical density matrix gives the experimental entropy to quantitative accuracy, and therefore $S = k \log W$ will remain valid.

This gives us a mathematically well posed, and formally elegant, variational principle; given incomplete information A, the best predictions we can make of other quantities are those obtained from the "ensemble" ρ that has maximum information entropy H while agreeing with A. By "agreeing" with A we mean, of course, that the information A can be extracted back from ρ by the usual rule of prediction: $<A> = Tr(\rho A)$. This represents the taking of that majority vote.

If A consists of values of constants of the motion (energy, mole numbers, angular momentum, etc.) this prescription leads, as is well known, back to the Gibbs canonical, grand canonical, and rotational ensembles. Thus conventional equilibrium statistical mechanics, with the mathematical apparatus of partition functions, etc., is contained in our proposed rules as a special case.

Extensions to other kinds of information A are straightforward mathematical generalizations of that standard apparatus. We indicate briefly in two stages what is described more fully elsewhere [10]. Let A stand for a set of m real quantities $\{A_1, \ldots A_m\}$ such as energy, pressure, magnetization, concentration gradient, etc. and denote their observed values at time $t = 0$ (the data) by A'_k ($1 \le k \le m$). We define an m-component vector $\lambda = \{\lambda_1, \ldots \lambda_m\}$ of Lagrange multipliers, the scalar product $\lambda \cdot A$, and the partition function

$$Z(\lambda) = Tr \exp(-\lambda \cdot A) \quad . \tag{7}$$

Then the density matrix that agrees with the data A' while assuming nothing beyond -- i.e., which spreads the probability as uniformly as possible over all microstates subject to the constraint $Tr(\rho A) = A'$, is

$$\rho = Z^{-1} \exp(-\lambda \cdot A) \quad . \tag{8}$$

The Lagrange multipliers are found from the m conditions

$$A'_k = -\partial/\partial \lambda_k \log Z \quad , \quad 1 \le k \le m \tag{9}$$

and from this information the "best" (in the sense that it minimizes the expected square of the error) prediction of any other quantity B is

$$<B> = Tr(\rho B) \tag{10}$$

which we may think of as the majority concensus of the likely microstates. If some of the A_k are not constants of the motion, specifying their values at only one time $t = 0$ would not in general lead to equilibrium predictions of other quantities. The density matrix, no longer a function of constants of the motion, would be time-dependent in the Schrödinger representation, and the above algorithm would give time-dependent predictions.

The situation is not essentially different if some of the $A_k = A_k(x)$ are functions of position, and their observed values $A'_k(x)$ specified in certain regions. The Lagrange multipliers then get promoted to functions $\lambda(x)$, the partition function to a partition functional $Z[\lambda_k(x)]$, all generalizing the above relations in the most obvious way. This is also a rather common algorithm, often miscalled the "local equilibrium" theory on the grounds of a loose analogy with the grand canonical ensemble if the space-varying A's are particle densities. It has a puzzling "induction time" phenomenon; irreversible fluxes do not seem to be going at the initial time $t = 0$, but require a short transient startup period, after which they often settle down to "plateau" values.

Of course, if data referring to time $t = 0$ are all we have, the "local equilibrium" algorithm will represent the best predictions we are able to make. But now knowing the values of the A's at other times becomes relevant and can improve our predictions. In fact, real macroscopic information can hardly refer to only one instant of time; we really know that the values A' have persisted for a short time in the past. Taking this seemingly unimportant information into account removes the induction time phenomenon, the corrected algorithm yielding irreversible fluxes by direct quadratures over the initial ensemble. An example [11] shows calculation of the diffusion coefficient; others are entirely analogous.

This illustrates why we lay such stress on the interpretation of probabilities. If one believes that a probability distribution describes a real physical situation, then it would seem wrong to modify it merely because we have additional information. Indeed, the very idea that a probability distribution describes our state of information is foreign to almost all recent expositions of statistical mechanics; but this precludes any possibility of a full prediction of irreversible processes, as may be seen as follows.

The local equilibrium density matrix, that has maximum information entropy subject to given macroscopic values A' at only one instant of time, assigns equal probability to every possible microstate   compatible with A', regardless of its past history. Thus it represents a mixture of every possible history by which the system could have reached the macrostate A'. But the characteristic feature of an irreversible process, which one would think it the main purpose of theory to predict, is the appearance of fading memory effects; the behavior of a system depends on its past history. Since in the local equilibrium distribution all memory of the past has been thrown away, it is in principle incapable of predicting those parts of future behavior that depend on the past.

Indeed, the local equilibrium distribution cannot even distinguish the past from the future; from symmetry, every possible micromotion and its reversed motion are present with equal weight, so the expectation of every flux is zero. The induction time phenomenon is thus only an artifact of the incomplete information being used; but it cannot be corrected until one recognizes that our probability distributions describe information.

The extension to take into account time-dependent information is straightforward. If the given information consists of the values of

$$A'_k(x,t) \quad , \quad 1 \le k \le m \tag{11}$$

in the space-time region R_k, then the corresponding Lagrange multiplier functions are defined in the same regions, so that

$$\lambda \cdot A = \sum_{k=1}^{m} \int_{R_k} d^3x \, dt \, \lambda_k(x,t) A_k(x,t) \tag{12}$$

in which $A_k(x,t)$ is the Heisenberg representation operator, and the rest of the formalism is extended in the obvious way.

However, when time-dependent information is used, a new terminology will help to avoid confusion. When information entropy is maximized subject to constraints A', the maximum attained is of course a function of the constraints:

$$S(A') = \log Z + \lambda \cdot A' \quad . \tag{13}$$

If the A_k are ordinary thermodynamic parameters, this is the Clausius experimental entropy of thermodynamics. For the "local equilibrium" case, it becomes a functional

$$S_o = S[A'_k(x)]$$

still conventionally called "entropy". The same functional, with the $A'_k(x)$ taken at any time t, then defines a time-dependent entropy S_t which is a property of the macrostate at time t and for which various inequalities can be proved [10].

The notion of entropy varying with time is so ingrained in our thinking that the next stage of generalization, in which the maximum information entropy becomes a functional

$$\sigma_A = \sigma[A'_k(x,t)] \tag{14}$$

262

of the entire space-time history of the macroscopic process (over the regions R_k where we have information), calls for a new name. We have ventured [10] to call it the *caliber* of the process, since it measures the cross-section of a tube, a bundle of world-lines in "phase space-time", each line representing the time development of a possible microstate consistent with all the given information.

We have shown [11] that this space-time generalization of the canonical ensemble, stated in terms of the partition functional

$$Z[\lambda_1(x,t) \ldots \lambda_m(x,t)] \tag{15}$$

leads automatically to such known results as the Wiener prediction theory and the Kubo formulas for transport coefficients, but in a more general form free of re-striction to quasi-stationary or near-equilibrium conditions. For example, a single formula for predicted particle density encompasses both static diffusion and ultrasonic dispersion and attenuation.

Indeed, near equilibrium the second functional derivatives of log Z

$$K_{ij}(x,t;x't') = \frac{\delta^2}{\delta\lambda_i(x,t)\delta\lambda_j(x't')} \log Z$$

$$= \int_0^\beta du <A_i(x,t-ihu)A_j(x',t')> - <A_i(x)><A_j(x')> \tag{16}$$

are the Kubo-type space-time covariance functions of transport theory.

A different -- although equivalent -- mathematical form of the theory, based on the caliber instead of the partition functional, makes the relation to the work of Einstein, Fokker-Planck, and Onsager clearer. The first functional derivatives of the caliber generate the Lagrange multipliers, or "potentials":

$$\lambda_k(x,t) = \frac{\delta\sigma}{\delta A'_k(x,t)} \tag{17}$$

and at a point where σ_A is locally convex, by which we mean that under a slight change in the problem

$$A'_k(x,t) \rightarrow \delta A'_k(x,t) + \delta A'_k(x,t) \tag{18}$$

with $\delta A'_k$ not identically zero, we have

$$\delta\lambda \cdot \delta A = \sum_{k=1}^m \int_{R_k} d^3x \, dt \, \delta\lambda_k(x,t)\delta A'_k(x,t) < 0 \quad , \tag{19}$$

the second functional derivatives generate a new set of space-time functions

$$G_{ij}(xt;x't') \equiv \frac{\delta^2\sigma}{\delta A'_i(x,t)\delta A'_j(x',t')} \tag{20}$$

which are the functional inverses of the covariance functions $K_{ij}(xt;x't')$.

If the A_k are ordinary thermodynamic variables, our convexity condition becomes positive definiteness of the matrix G, and they reduce to inverse matrices, $G = K^{-1}$. If in addition we are at an equilibrium point, the matrix G defines the quadratic form in Onsager's expansion of the entropy

$$S = S_o - \frac{1}{2} \sum_{ij} G_{ij} \, \delta A_i \, \delta A_j + \ldots \tag{21}$$

about that point. Onsager thought of the entropy gradient

$$X_i = \frac{\partial S}{\partial A_i} = - \sum_j G_{ij} \, \delta A_j \tag{22}$$

as the "force" that drives a system to equilibrium according to his phenomenological equations

$$\dot{A}_j = \sum_j L_{jk} \, X_k \tag{23}$$

and argued for the reciprocal relations $L = L^T$.

In Onsager's work there was no apparent connection with the earlier work of Einstein on diffusion in ordinary space or of Fokker-Planck on diffusion in momentum space. But now, if we restate our general prediction algorithm in terms of the caliber, a relation appears.

5. The Maximum Caliber Principle

Although the mathematical details needed to carry it out can become almost infinitely complicated, the principle itself remains almost trivially simple in content. We shall describe the principle in generality, then give a simple application of it that combines the Fokker-Planck and Onsager ideas.

We are given macroscopic information A which might consist of values of several quantities $\{A_1(x,t) \ldots A_m(x,t)\}$ such as distribution of stress, magnetization, concentration of various chemical components, etc. in various space-time regions $\{R_1 \ldots R_m\}$. This defines a caliber $\sigma_A = \log W_A$, which measures the number of time-dependent microstates consistent with the information A. What predictions should we make of the quantities $\{B_1(x,t) \ldots B_n(x,t)\}$ (some of which might be the same as some of the A's) in space-time regions $\{U_1 \ldots U_n\}$ (some of which might be the same as some of the R's)?

Suppose we just make any guess we please about B, and consider the caliber σ_{AB} subject to both the A and B constraints. Of course, since we have imposed a new constraint, we shall have $\sigma_{AB} \leq \sigma_A$. If $\sigma_{AB} > 0$, then our guess was a possible one; there exist microstates whose time development would reproduce both the A and B macrobehaviors. But the choice of B which renders σ_{AB} a maximum is the "majority vote" macrobehavior that could be realized by Nature in more ways than could any other consistent with A. So it is an obvious generalization of the principle given by Gibbs in 1875, that this guess is our optimal prediction -- optimal in the sense that it takes into account all our knowledge of the microphysics and all our macroscopic data; and makes no arbitrary assumptions beyond that. Whether the prediction is correct or not, to make any better one would require more information than we had.

The caliber of a space-time process thus appears as the fundamental quantity that "presides over" the theory of irreversible processes in much the same way that the Lagrangian presides over mechanics. That is, in ordinary mechanics we learn first that a variational principle (minimum potential energy) determines the conditions of stable equilibrium; then we learn how to generalize this to the Lagrangian, whose variational properties generate the equations of motion. In close analogy, Gibbs showed that a variational principle (maximum entropy) determines the states of stable thermal equilibrium; now we have learned how to generalize this to the caliber, whose variational properties generate the "equations of motion" of irreversible processes.

But these equations of motion are not, except in a certain approximation, differential equations. In general they turn out to be nonlinear integral equations. Close to equilibrium they become linear integral equations, which contain the conventional Kubo relations for linear transport phenomena as special cases. Or, in a short-memory approximation, they reduce to generalized Fokker-Planck-Onsager equations showing how the approach to equilibrium is both steered and stabilized by the entropy function.

Thus define $\{\delta A_k, \delta B_k\}$ as the departures from the equilibrium values. Then close to equilibrium we may use the expansion (21); in a shorthand notation

$$\sigma_A = S_0 - \frac{1}{2} \delta A \cdot G_{AA} \cdot \delta A + \ldots \tag{24}$$

and

$$\sigma_{AB} = S_0 - \frac{1}{2} [\delta A \cdot G_{AA} \cdot \delta A + \delta B \cdot G_{BA} \cdot \delta A + \delta A \cdot G_{AB} \cdot \delta B + \delta B \cdot G_{BB} \cdot \delta B] \quad . \tag{25}$$

Now the reciprocities

$$G_{AB} = G_{BA} \tag{26}$$

turn out to hold trivially in this theory. Therefore σ_{AB} is maximized with respect to B for fixed A, if

$$G_{BB} \cdot \delta B = - G_{BA} \cdot \delta A \tag{27}$$

which is a set of simultaneous linear integral equations determining $\{\delta B_1(x,t) \ldots \delta B_n(x,t)\}$. If the caliber is convex, the kernel G_{BB} is positive definite and there is a formal inversion

$$\delta B = -K_{BB} \cdot G_{BA} \cdot \delta A \quad . \tag{28}$$

Of course, this is extremely compact notation, to demonstrate how simple the underlying ideas are. But it is probably beyond our mathematical ability to do the indicated calculations explicitly for any really nontrivial problem; that is perhaps a task for the computers of the next Century. Nevertheless, many formal relations can be extracted from (27), which are capable of being tested experimentally even if we are unable to calculate the covariance functions from first principles. There is no reason to be surprised by this, since it is still true of the equilibrium canonical ensemble that it predicts testable relations like $(\partial P/\partial T)_V = (\partial S/\partial V)_T$ but we can seldom calculate $(\partial P/\partial T)_V$ exactly from first principles.

6. Bubble Dynamics

Finally, we sketch hurriedly the short-memory approximation to a prediction problem. Our given information consists of the past behavior of a few macroscopic variables:

$$A_k(t) \quad , \quad 1 \leq k \leq m \quad , \quad -\infty < t < 0 \quad . \tag{29}$$

The quantities B_k that we want to predict are their future values. Our full prediction equations would in principle make use of the values of the $A_k(t)$ arbitrarily far into the past. But we may be able to make a "short memory" approximation, in which only information for a short period, perhaps of the order of the mean time between collisions -- or in some cases even the duration of a collision -- needs to be considered.

We must warn against a common fallacy here. When we say "short memory", we do not mean to imply that the physical system itself can "forget" where it has been. Rather, beyond a certain time further information -- unless it is very unexpected -- does not help us to improve our predictions. Mathematically, this is because it does not reduce σ appreciably. Conceptually, when we know the macrohistory A over a certain time in the past, information about still earlier times often becomes nearly redundant, in the sense that it is nearly what we would have predicted -- or rather retrodicted -- from A. We should really say "short relevance".

But about the only thing that can happen to the $A_k(t)$ in a very short time is a kind of diffusion. A given macrostate $(A_1 \ldots A_m)$ can be realized by an enormous number $W(A_1 \ldots A_m)$ of microstates, which would lead to slightly different macrostates a short time later. Recall that the ordinary diffusion equation for particle density $n(x,t)$

$$\dot{n} = -\nabla \cdot J = D\nabla^2 n \tag{30}$$

expresses a short memory approximation to the correct constitutive equation for particle flux

$$J(x,t) = \int d^3x' \int_{-\infty}^{t} dt' \ K(x-x';t-t')n(x',t') \tag{31}$$

which recognizes that the particles that are here now came on the average from a mean free path away a mean free time ago. The phenomenological diffusion coefficient D is really a space-time integral over the memory function K. In a similar way, we may introduce a time-dependent probability density $P(A_1 \ldots A_m;t)$ which represents a "bubble" of probability in the macroscopic state space, which expands due to diffusion and is found after some long arguments to satisfy a phenomenological equation of "bubble dynamics"; slightly oversimplified, it is

$$\frac{\partial P}{\partial t} + (D/k)\nabla \cdot (P\nabla S) = D\nabla^2 P \tag{32}$$

where $S(A) = k \log W(A)$ is the ordinary entropy of thermodynamics. Generally, the diffusion coefficient D should be written as a diffusion tensor.

Equation (32) is like a Fokker-Planck equation in that it has a diffusion effect given by the right-hand side. But it is also like an Onsager equation in that the local entropy gradient is present. It turns out to have beautiful solutions that exhibit both the Fokker-Planck and Onsager behavior. When the entropy is a quadratic function of the A_i, the Green's function is an expansion in Hermite polynomials that can be summed exactly in closed form, so all questions about the solution can be answered. An arbitrary initial distribution $P(A_1 \ldots A_m;0)$ relaxes quickly to a gaussian shape, which is retained as it moves to the final equilibrium state:

$$P(A_1 \ldots A_m;\infty) \propto \exp\{k^{-1} S\} = W(A_1 \ldots A_m) \tag{33}$$

which is also the final solution of (32) for any entropy function and is of course just the supposition that Einstein made in his discussion of fluctuations; the size of the bubble (33) is the range of fluctuations to be expected.

But because this size is so small, almost any entropy function is accurately a quadratic function in the region of appreciable probability. The result is that, after an initial transient period in which a nongaussian bubble becomes gaussian, to all the accuracy one could want the solution is a gauss bubble moving along a trajectory in A-space and readjusting its size in accordance with the local entropy curvature. In the one-dimensional case, putting $A_1 = x$, the very accurate solution is

$$P(x,t) = \frac{1}{\sqrt{2\pi R(t)}} \exp\left\{-\frac{[x - q(t)]^2}{2R(t)}\right\} \tag{34}$$

where

$$\dot{q} = k^{-1} D S'(q) = L S'(q) \tag{35}$$

266

$$\dot{R} = 2D[1 + k^{-1} S''(q)R]$$

$$= -\alpha [R(t) - R_\infty] \tag{36}$$

in which $L \equiv k^{-1} D$, $R_\infty = k|S''(q)|^{-1}$. Equation (35) is of the Onsager form, the system moving along a trajectory defined by the local entropy gradient. In general the Onsager phenomenological coefficients are related to our diffusion tensor by $L_{ij} = k^{-1} D_{ij}$ and the Onsager symmetries $L = L^T$ reduce to the triviality

$$<\delta A_i \, \delta A_j> = <\delta A_j \, \delta A_i>.$$

But (36) shows a welcome new feature; the curvature of the entropy function stabilizes the bubble if it is convex ($S'' < 0$). If $S''(q) = 0$, and we start from a delta function $R(o) = 0$, (36) gives for the variance

$$<\delta x^2> = R(t) = 2Dt \quad, \tag{37}$$

the Einstein Brownian motion spreading law. With a linear entropy function, bubble dynamics reduces to Einsteinian Brownian motion with superposed steady Onsager drift. But if $S''(q) < 0$, the spreading (37) does not continue forever; the bubble constantly tries to adjust its size to conform to the local entropy curvature.

Thus we can get a very good intuitive understanding of an irreversible process simply from (35), (36) without further mathematics. In fact, the theory leads us to the following picture of macrostate change, which seems general enough to apply its macroeconomics as well (even though we still lack an "economic Liouville theorem" to help us define the right microvariables)

In physics or economics, even though a neighboring macrostate of higher entropy is available, the system does not necessarily move to it. A pile of sand could convert some of its potential energy into heat, thus increasing its entropy, by levelling itself; but it does not actually do this unless there is an earthquake. Any system might just stagnate where it is, unless it is shaken up a little by what an Englishman might call a "dither" of some sort.

An Einstein saw in physics and as we conjecture to be the case also in economics, the dither that prevents stagnation and drives us up the entropy hill is a kind of turbulence injected into the macrovariables by fluctuations in the underlying micro-variables. By this means, the macrostate is constantly "exploring the possibilities" of neighboring states. But in this exploration the system is always more likely to move to one of higher than lower entropy, simply because there are more of them (greater multiplicity).

In thermodynamics, the dither is provided by what are usually called "thermal fluctuations". In the present view of the problem we see these fluctuations, not merely as a small diffusion manifestation of the microphysics, but as the active driving force that makes an irreversible process go. Thus "fluctuation-dissipation theorems" and their Hilbert transforms, "fluctuation-response theorems" are not merely curious accidental relations; they express directly the physical mechanism of the irreversible process.

Generally, the dither not only puts random uncertainty into macrovariables, it drives a systematic movement of the macrostate. at a drift velocity proportional to:

(entropy gradient) x (mean-square fluctuations)   .

Thus we see that stagnation can have two quite different causes: loss of entropy gradient, and loss of dither. Without an entropy gradient, the sense of direction is lost and the system drifts aimlessly in a way not determined by any macrovariables. In thermodynamic systems at very low temperatures, the dither is nearly lost, and systems may approach equilibrium so slowly that they appear to be frozen in nonequilibrium states.

There is a close analogy in the mechanism of biological evolution. There the dither that drives the process is spontaneous random mutations, as a result of which every species is constantly exploring the possibilities of a slightly different design. Darwin hypothesized that a "good" mutation has a better chance than a "bad" one of surviving, and thus introducing a new, slightly altered form of the species. Of course, this is necessarily true to the extent that a sufficiently bad mutation cannot survive at all. However, species might also diverge merely as a matter of chance, the resulting different subspecies having no particular survival advantage. Ecology may represent a case where the steering from the entropy gradient is very weak, but the dither still operates.

In economics, perceiving the dither but not the entropy factor, Keynes did not find the phenomenon that our model considers one of the fundamental causes of macroeconomic change. This is conceivably the reason why microeconomic theories do not account well for the facts of macroeconomic change.

We find that deterministic, random, and unstable behavior are all exhibited by this model, as follows. A strongly convex (i.e., downward curving, $S'' < 0$) entropy function is strongly stabilizing, leading to a small bubble; i.e., nearly deterministic behavior. When the curvature of the entropy function decreases, stabilizing forces are weaker and the bubble enlarges, representing more "random" behavior (by which we really mean "less predictable from macrovariables").

When the entropy curvature is zero (i.e., entropy is a locally linear function), the restoring forces are zero, and the bubble spreads following Einstein's law of Brownian motion (dimensions growing as the square root of the time). As noted, it seems plausible that this may be the case for biological evolution.

At a point where convexity of the entropy fails altogether (i.e., it becomes upward curving) we have instability, the bubble stretching out and usually splitting into two smaller bubbles that go their separate ways. In thermodynamics, such a bifurcation signifies a phase transition, the two bubbles representing the development and eventual equilibrium of the two phases. In economics it signifies perhaps a political revolution, the two bubbles representing the different possible societies growing out of the revolution. So, if these economic conjectures prove to be right, enlightened Governments of the distant future will keep an eye not only on the local tilt of the entropy function, but also on its local curvature.

Gibbs explained the phase transitions that occur in multidimensional thermodynamic spaces as "catastrophes" arising from the various kinds of local loss of convexity that can occur in an entropy function in a multidimensional space. We do not know whether all the types classified by René Thom are realizable in thermodynamic systems.

However, as the name implies, bubble dynamics has more content than catastrophe theory, which describes only the equilibrium states and not the dynamics telling us along what path and at what velocity the system gets to those states. Bubble dynamics can, for example, describe the fine details of time development at a bifurcation point, determining what fraction of the bubble will move to the left or right (i.e., given the size, shape, and position of a bubble approaching a bifurcation point, what is the probability that the system will move ultimately to the left or right?).

7. Conclusion

We should correct a possible misconception that the reader may have gained. Most recent discussions of macrophenomena outside of physical chemistry concentrate entirely on the dynamics -- microscopic equations of motion or an assumed dynamical model at a higher level, deterministic or stochastic -- and have ignored the entropy factors of macrostates altogether. Indeed, we expect that such efforts will succeed fairly well if the macrostates of interest do not differ greatly in entropy. But there are puzzling cases, as noted in the Introduction, where macrobehavior seems hard to understand in terms of any reasonable dynamics alone. In these cases, the entropy factors may be the missing ingredient; as we learned from Gibbs, prediction of chemical equilibrium could not have succeeded at all until the macroscopic entropy was recognized.

It may have appeared from the above that we have gone to the opposite extreme, and ignored the dynamics. But this was only for expository purposes, to emphasize what is new in our approach -- the effect of entropy on macroscopic predictions. In realistic problems the dynamics will reappear automatically in our general equations, since the Heisenberg operators needed to define the covariance functions K and inverse covariance functions G contain the full dynamics. In a mechanical problem, the bubble dynamics equations will acquire new terms representing inertial effects that we have left out here.

References

1. H. C. Callen, Thermodynamics, J. Wiley & Sons, Inc. New York (1960).
2. C. Truesdell, Rational Thermodynamics, McGraw-Hill Book Co., New York (1969);
   Second enlarged edition, 1985.
3. J. Willard Gibbs, Heterogeneous Equilibrium, Trans. Conn. Acad. Sci. (1875-1878).
   Reprinted in The Scientific Papers of J. Willard Gibbs, Longmans, Green
   & Co., New York (1906) and by Dover Publications, Inc., New York 1961).
4. G. N. Lewis and M. Randall, Thermodynamics and The Free Energy of Chemical
   Substances, McGraw-Hill Book Co., New York (1923).
5. H. Jeffreys, Theory of Probability, Oxford University Press (1939); also
   numerous later editions.
6. R. T. Cox, The Algebra of Probable Inference, Johns Hopkins University Press
   (1961).
7. E. T. Jaynes, "Predictive Statistical Mechanics", in Proceedings, NATO Advanced
   Study Institute, Santa Fe, New Mexico, June 3-16, 1984 (in press).
8. E. T. Jaynes, "Gibbs vs. Boltzmann Entropies", Am. Jour. Phys. 33, 391-398
   (1965).
9. A. Feinstein, Foundations of Information Theory, McGraw-Hill Book Co., New York
   (1958); Chap. 6.
10. E. T. Jaynes, "The Minimum Entropy Production Principle", in Annual Review
    of Physical Chemistry, S. Rabinovitch, Editor, Annual Reviews, Inc.,
    Palo Alto, California (1980).
11. E. T. Jaynes, "Where do we Stand on Maximum Entropy?", in The Maximum Entropy
    Formalism, R. D. Levine and M. Tribus, Editors, MIT Press, Cambridge, Mass.
    (1978).

Entropy-Like Potentials in Non-Equilibrium Systems with Coexisting Attractors

R. Graham

Universität Essen GHS, Fachbereich Physik,
D-4300 Essen, Fed. Rep. of Germany

Recent work is reviewed investigating the possibility to describe non-thermo-
dynamical systems in terms of an entropy-like potential. First a general introduction is presented formulating, for the non-specialized reader, the main general
questions which have been investigated and the answers which have so far been obtained. In a second somewhat more technical section the essential points are presented for a particular dynamical system where a limit cycle appears via a discontinuous transition with hysteresis and bistability.

1. Introduction

In order to successfully deal with very complex dynamical systems,it is necessary
to develop concepts which permit us to concentrate on a few relevant aspects of a
system only. To begin with, the concept of the "relevant variables" of the system
itself must be properly defined, the success of the adopted definition being determined by the extent to which it captures the "purpose" of the system and by
the degree of simplification to which it gives rise. Very often in physics and
other fields complex dynamical systems have the property of establishing a well-
defined structure on macroscopic scales in space and time much larger than the
typical scale of the original complex system. The relevant variables of the system
may then be usefully identified with the degrees of freedom of this macroscopic
structure, behind which the complexity of the underlying dynamics on smaller
scales disappears.

The most important example in physics, where a simplified reduced description
in terms of a few relevant macroscopic variables is obtained,is the statistical
mechanics of thermodynamic equilibrium. Indeed, a large part of theoretical
physics is the art of constructing or explaining superstructures in terms of completely different underlying more microscopic structures. Naturally, the method
of equilibrium statistical mechanics is at the heart of these attempts. Often only
the zero-temperature limit of equilibrium statistical mechanics is required.

The most basic feature of the method of equilibrium statistical mechanics is
simplification,achieved by discarding all irrelevant information. The amount of
discarded information, called entropy, depends on the macroscopic state of the
system, and influences that state by acting as an effective potential in which the
system moves, macroscopically, under certain constraints. The driving forces behind this motion are of two kinds. The first kind is due to the impossibility to
retrieve information once it has been discarded, leading to a monotonous increase
of entropy in the simplified description. The rate of that increase is determined
by a matrix of transport-parameters which are of fundamental importance for the
simplified description. The second kind of forces is due to an intrinsic inertia
of the system against further loss of information, leading to conservative motion
in the effective potential provided by the entropy. The former forces give rise
to motion which is irreversible in time, the latter forces, if they acted alone,
would give rise to reversible motion. Now and then the complexity of the underlying microscopic dynamics makes itself felt again, even on the macroscopic level,
by giving rise to a fluctuation. Suddenly information which had been discarded as
irrelevant is fed back into the system and entropy decreases (more precisely the
"coarse grained" or mean-field-type entropy we shall always employ,and which for

its usefulness requires that fluctuations are very weak, i.e. very rare events only). The stiffness of the system against such a fluctuation, i.e. the probability of its occurrence within a given size, is also determined by the effective potential provided by the entropy.

These very attractive but special simple features of dynamical systems emerging from equilibrium statistical mechanics have often led to the conclusion that this kind of a description must be confined to systems in thermodynamic equilibrium. In particular, an effective potential similar to entropy is usually not used or even thought to be available in systems far from thermodynamic equilibrium. Entropy production has long been proposed as an effective alternative potential for non-equilibrium systems, but it has been recognized to serve this purpose in the vicinity of thermodynamic equilibrium only, and the efforts based on the generalization of the concept of entropy production have only met with limited success. On the other hand, it was found some time ago [1 - 3] that simple systems with one relevant variable only, or systems reducible to one variable due to some symmetry, or linear systems in fact have a very similar mathematical structure as systems in thermodynamic equilibrium, and can be usefully described by an effective potential with the same properties as entropy.

Recently, in collaboration with A. Schenzle [4] and T. Tél [5 - 7] we have examined the extent to which the general structure of the macroscopic description of systems in thermodynamic equilibrium is compatible with much more general dynamical systems. The basic questions we asked and partially answered are:

- Is it possible for a given dynamical system, which need not even have any thermodynamical notions attached to it, to define an entropy-like potential, with properties like those described above for systems in thermodynamic equilibrium? (The answer is yes for many systems, but the degree of generality, in particular for systems with more than two macroscopic degrees of freedom, has not yet been determined.)

- Which of the properties known for equilibrium systems can be maintained and which of them have to be compromised? (All properties can be maintained except for the identification of the two kinds of forces as irreversible and reversible. Also the effective potential, in general, has discontinuities in its first derivatives.)

- Which quantities must be known to define the potential uniquely? (The macroscopic equations of motion and a transport matrix. An unambiguous and physically meaningful way to determine the transport matrix is the analysis of the (by assumption) small and short-lived fluctuations of the system. The transport matrix is determined from the second-order correlation coefficients of the fluctuations away from the deterministic state for correlation times short compared to the macroscopic scale.)

- What is the statistical relevance of the potential? (The same as in thermodynamic equilibrium, provided the transport matrix is properly related to the fluctuations.)

- How can the potential be determined in specific examples and what are methods to determine it in some systematic approximation? (By solving a problem in the theory of (generally non-integrable) Hamiltonian systems by the numerical or approximate methods developed in that field.)

Our results are based on a number of explicit examples of systems with two macroscopic variables and on some general arguments drawn from the general theory of non-integrable Hamiltonian systems, on which our problem can be mapped, mathematically. Our results indicate that an entropy-like potential can be defined for a much larger class of dynamical systems than those in thermodynamic equilibrium. (Mathematically closely related questions appearing in the asymptotic theory of stochastic processes have been analyzed by Freidlin and Ventsel (for a review

cf. [8]). Similar mathematical problems appear also in the quasi-classical limit
of quantum theory which have been analyzed by Maslov and Fedoriuk [9].)

For a detailed exposition of the above-mentioned results I refer to the original papers. In the following some detail is given by considering a special example.

2. Example

Let us consider the following dynamical system [7]

$$\dot{x} = v$$
$$\dot{v} = -\gamma v - \sin x + F \tag{1}$$

and let us assume that the transport matrix is given by $Q^{xx} = Q^{xv} = Q^{vx} = 0$,
$\gamma Q^{vv} = 2\gamma\eta$. The choice of this matrix corresponds to fluctuations in the dynamical system which can be described by adding a Gaussian white-noise Langevin
force $F(t)$ on the right-hand side of the second of eqs. (1) with $\langle F(t) F(0)\rangle =$
$= 2\gamma\gamma \delta(t)$. Let us note that with this addition eq. (1) describes
Brownian motion in the potential $V(x) = -\cos x - Fx$ which ramps downward on the
average towards positive x and has no bottom. This problem of Brownian motion
has a number of different very interesting physical applications which have been reviewed - together with some approximate analytical and numerical solutions - in
[10]. The problem was also studied analytically in ref. [11] using different
methods. For a comparison see [7].

The configuration space of (1) is the surface of a cylinder

$$X_{s_1} \leq X \leq X_s = X_{s_1} + 2\pi \quad ; \quad -\infty < V < +\infty \tag{2}$$

and therefore not simply connected. If the "external foce" F in eq. (1) is
sufficiently small $(0 \leq F < F_c(\gamma))$ it is easy to show that eq. (1) has a single
attracting fixed point P_0 and a saddle point S on the cylinder surface. As P_0 for
F=0 corresponds to thermodynamic equilibrium one may call this the thermodynamic
branch of states. For $F_c(\gamma) < F < 1$ the attracting fixed point P_0 coexists with
an attracting limit cycle and the saddle point S. For $F > 1$, P_0 and S disappear and
only the limit cycle remains.

For F=0, i.e. thermodynamic equilibrium, eq. (1) may be written as

$$\dot{x} = v \quad ; \qquad \dot{v} = -\gamma \frac{\partial \phi_0}{\partial v} - \sin x \tag{3}$$

where $\phi_0(x,v) = 1/2 \, v^2 - \cos x$ is an entropy-like potential (more precisely, it is
the free energy if γ is identified with temperature in energy units.) We notice
the irreversible ϕ_0-decreasing "force" and the reversible "forces"
$\dot{x}=v$, $\dot{v}=-\sin x$ which conserve the potential, $(\partial\phi_0/\partial x)v - (\partial\phi_0/\partial v)\sin x = 0$.

We now ask, whether also for $F \neq 0$ a potential $\phi(x,v)$ can be defined uniquely
for each value (x,v) such that eq. (1) reduces to the form

$$\dot{x} = r_x = v$$
$$\dot{v} = -\gamma \frac{\partial \phi}{\partial v} + r_v \tag{4}$$

with $r_x \partial\phi/\partial x + r_v \partial\phi/\partial v = 0$. An obvious choice may seem to be $\phi_0 = v^2/2 - \cos x - Fx$,
but on the cylinder surface this is a multivalued function without lower bound and
therefore not a solution to our question. The latter may be reformulated by eliminating r_x, r_v as

$$\gamma\left(\frac{\partial\phi}{\partial v}\right)^2 + (-\gamma v - \sin x + F)\frac{\partial\phi}{\partial v} + v\frac{\partial\phi}{\partial x} = 0 \tag{5}$$

This equation, purely formally, may be interpreted as a Hamilton-Jacobi equation where ϕ is the unknown minimizing action and x,v are generalized coordinates with canonically conjugate momenta $P_x = \partial\phi/\partial x$, $P_v = \partial\phi/\partial v$. The Hamiltonian is then

$$H = \gamma P_v^2 + (-\gamma v - \sin x + F) P_v + v P_x = 0 \qquad (6)$$

The characterization of ϕ as a solution of the Hamiltonian-Jacobi equation is only local and the question arises, in which of its solutions are we really interested. The monotonous change of ϕ under the deterministic dynamics (1) leads to the requirement that the first derivatives p of ϕ all vanish in the attractors, repellors and saddles of eq. (1). It is easy to convince oneself that these are not only singular sets of eq. (1) but also of the Hamiltonian (6). As shown in [5] this means that $P_x = \partial\phi/\partial x$, $P_v = \partial\phi/\partial v$ determines a separatrix in the phase-space of the Hamiltonian system connecting these singular sets. For a general dynamical system, this separatrix, in general, does not possess a single valued projection on the hyperplane p = 0 but rather has infinitely many wild oscillations leading to an infinitely multivalued function $\phi(x,v)$. For the present example, though, this particular problem does not appear, as the Hamiltonian (6) for H = 0 admits a second conserved (albeit explicitly time-dependent) phase-space function

$$A(v, P_v, t) \equiv \left(\frac{v}{P_v} - 1\right) e^{\gamma t} \qquad (7)$$

which prohibits the appearance of a wild separatrix. Nevertheless it is necessary, in principle, to formulate a unique global characterization of the potential ϕ which then serves to select among the local solutions of the Hamilton-Jacobi equation. This global formulation is achieved by making use of the stochastic process we associated with eq. (1) and the transport matrix and the requirement that the description (4) in terms of ϕ arises as the weak-noise limit of this stochastic process. It leads to a characterization of ϕ by a minimum principle, which, in our present example, takes the form

$$\phi(x,v) = \min_{(A_i)} \left\{ \int_{\substack{(\tau = -\infty, \\ x, \dot{x} \in A_i)}}^{\substack{(\tau = 0, \\ x, \dot{x} = v)}} \frac{d\tau}{4\gamma} (\ddot{x} + \gamma\dot{x} + \sin x - F)^2 + C_i \right\} \qquad (8)$$

Here the symbol A_i denotes the attractors of the system. If several attractors coexist (as in our case for $F_c(\gamma) < F < 1$) the absolute minimum has to be taken separately for each attractor A_i over all trajectories $x(\tau)$ starting in A_i at $\tau = -\infty$ and ending in the desired point (x,v) at time $\tau = 0$, and then also over the two attractors A_i (if $F_c(\gamma) < F < 1$). The coefficients C_i are fixed by the requirement that the minima for the fixed point and the limit cycle give the same results if $(x,v) = S$.

The minimum principle (8) in the general case, eliminates the multivaluedness of ϕ, in the case of a "wild separatrix", in favour of the lowest lying branch, however, only at the cost of introducing discontinuities of the first derivatives of ϕ on certain surfaces in configuration space. In the present example, where a wild separatrix does not appear, the minimum principle still serves to eliminate a different kind of multivaluedness of ϕ (again at the cost of discontinuous first derivatives) which appears because of the existence of several attractors and (or) a multiply connected configuration space. Both features imply that the minimizing trajectory of eq. (8) can change discontinuously as a function of (x,v), leading to discontinuities of the first derivatives of ϕ.

The potential ϕ of eq. (1) has been determined from these principles in ref. [7] to which we refer for further detail. The following 3 figures give the equipotential lines of ϕ for the three qualitatively different cases $0 < F < F_c(\gamma)$; $F_c(\gamma) < F < 1$; $1 < F$. The first of these figures has been drawn using an analytical approximate but very accurate solution for ϕ, while the other two figures were constructed from numerical results obtained by numerically solving the Hamiltonian (6).

273

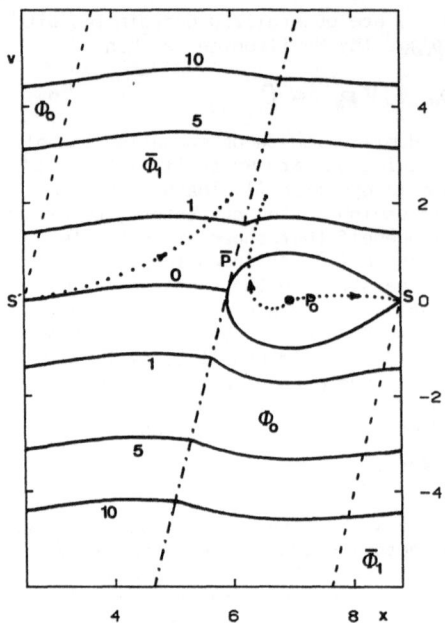

Fig. 1 :
Equipotential lines for $F < F_c$ (γ) (cf. text)

Fig. 1 shows equi-potential lines of $\phi(x,v)$ in the (x,v)-plane for the case $0 < F < F_c$ (γ). The parameter values are $\gamma = 5$, $F = 0.6$. P_0 is the fixed point, S and S' are (equivalent) saddle points. The potential ϕ has been normalized to zero on the equi-potential line through S. The numbers give the respective value of ϕ . Inside the closed contour $\phi = 0$, the potential is given by $\phi = \phi_0 = 1/2\ v^2$ -sin x · - Fx + const. The curve connecting S' with $\bar{P}$ on the curve $\phi_0 = 0$ is given by the instable manifold of S' in the deterministic system (1). Along this curve $\phi = 0$ must also hold, because ϕ (S') = 0 and ϕ ($\bar{P}$) = 0 and ϕ cannot increase along any deterministic trajectory. In $\bar{P}$ and along the dash-dotted curve through $\bar{P}$ the potential ϕ is continuous but has a discontinuous first derivative. This happens, because the minimizing trajectory in eq. (8) from P_0 to a given endpoint P changes discontinuously as P crosses this curve. If P is to the right of this curve the path proceeds directly from P_0 to P as indicated. If P is to the left of this curve the minimizing path from P_0 to P winds around the cylinder in a different way, as indicated. ϕ to the left of this dash-dotted curve is represented by a local function ϕ_1 which differs from ϕ_0 . It can be calculated at least approximately, by solving the Hamilton-Jacobi equation in a local expansion around the manifold $\phi = 0$ connecting S' with $\bar{P}$. The function $\bar{\phi}_1$ joins ϕ_0 continuously with (as one can show [7]) continuous first derivatives along the dashed lines emanating from S and S'. These dashed lines can be shown [7] to be mirror images of the stable manifold of S and S' in the deterministic system, and are obtained from this manifold by the reflection $v \rightarrow -v$, $x \rightarrow x$ on the x-axis. $\bar{\phi}_1$ is restricted to the stripe bounded by the dashed dotted curve through $\bar{P}$ to the right and by the dashed curve to the left and its extension to the rest of the cylinder surface.

Fig. 2 shows the equi-potential lines for the case F_c (γ) $< F < 1$ ($\gamma = 0.13$, $F = 0.83$). A limit cycle C and a fixed point P_0 coexist in this case and ϕ has therefore two local minima. The minimum at P_0 and the potential $\phi = \phi_0$ in its neighbourhood are as before, except that the potential well at P_0 is now more shallow and the domain where $\phi = \phi_0$ holds has shrunk. The remainder of the dashed dotted curve and the dashed curve of fig. 1 are, respectively, the dashed dotted line and the dashed line emanating from S and intersecting mutually in Q. Again the potential ϕ has discontinuous and continuous first derivatives along these curves, respectively. (In fig. 1, as $F \rightarrow F_c$ (γ) the point $\bar{P}$ moves towards S and coincides with S for $F = F_c$ (γ) when the unstable manifold of S' passes through S,

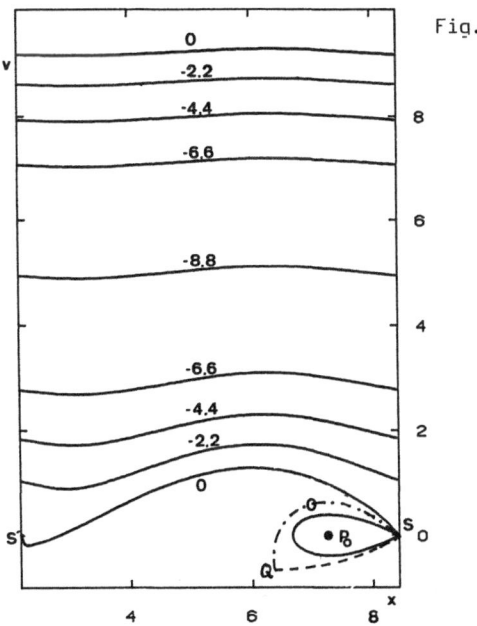

Fig. 2: Equipotential lines for $F_c(\gamma) < F < 1$ (cf. text)

i.e. when S becomes homoclinic. At this moment the limit cycle is born.) Inside the loop formed by these two curves (the dashed curve still being formed by the mirror image of a piece of the stable manifold through S) the potential ϕ is represented by the function ϕ_0. Outside that loop ϕ is represented by a different function, which has been obtained [5] by solving the Hamilton Jacobi equation numerically with the boundary condition ϕ = const. on the limit cycle. In S this solution must be normalized to $\phi_0(S)$ which determines the relative depth of the two minima of ϕ. Again we normalized $\phi_0(S) = 0$. Only equi-potential lines for $\phi \lesssim \phi_0(S)$ are shown.

The relative depth of the potential wells gives a very useful global measure of the relative stability of the two attractors, which fully takes into account the nonlinearities of the system and therefore goes well beyond a linear stability analysis. The steepness of the potential wells also gives a useful measure of the stiffness of the system against the fluctuations by which it is perturbed. If F increases towards 1 the potential well around P_0 shrinks in size and depth and disappears (together with the dash-dotted and the dashed curve) at F = 1. From then on only the limit cycle exists as an attractor.

Fig. 3 shows the equi-potential lines for $F > 1$ (γ = 0.16, F = 1.25) around the minimum of ϕ on the limit cycle, which is now normalized to zero. We see that the limit cycle, like a river in its valley, flows through the landscape formed by the potential.

In conclusion, this example demonstrates that entropy-like (or free-energy-like) potentials can be constructed for systems driven in non-equilibrium states. The main compromise which allows their construction in these more general circumstances is the allowance for discontinuities in their first derivatives. In particular, such discontinuities must be expected for systems with topologically non-trivial configuration spaces and coexistence of several attractors.

The potential is useful because it provides immediate intuitive insight and captures the essence of the nonlinear stability properties of the attractors and their susceptibility against the small fluctuations perturbing the system. In

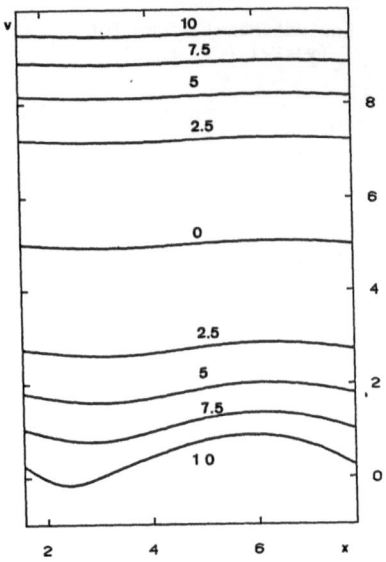

Fig. 3 : Equipotential lines for F > 1 (cf. text)

systems more complicated than the example we have analyzed here, the potential will, of course, be difficult to determine explicitly. Nevertheless its mere existence and the general properties which it is known to have may already be useful in providing a simple qualitative picture of the dynamics. It is also conceivable that the basic dynamical description of the system may not be given in the direct form of eq. (1) (or some generalization) but instead in the corresponding implicit form of eq. (4) with some given form of ϕ . In this case, it is important to realize that, according to our results [3], dynamical systems generated by continuously differentiable ϕ are special and structurally unstable against small arbitrary changes of the right-hand side of eq. (1).

References
 1 H. Haken, "Laser Theory", Encyclopedia of Physics vol. XXV/2c, ed. S. Flügge, (Springer, Berlin 1970)
    R. Graham, in "Coherence and Quantum Optics" ed. L. Mandel and E. Wolf (Plenum, New York 1973)
 2 R. Kubo, K. Matsuo, K. Kitahara, J. Stat. Phys. 9,51 (1973)
 3 K. Tomita, H. Tomita, Progr. Theor. Phys. 51,1731 (1974)
 4 R. Graham, A.Schenzle, Phys. Rev. A23,1302 (1981); Z. Physik B52,61 (1982)
 5 R. Graham, T. Tél, Phys. Rev. Lett. 52,9 (1984); J. Stat. Phys. 35,729 (1984) (Addendum J. Stat. Phys. 37,709 (1984)
 6 R. Graham, T. Tél, Phys. Rev. A31,1109 (1984)
 7 R. Graham, T. Tél, Phys. Rev. A, to appear
 8 M.I. Freidlin, A.D. Ventsel, "Random Perturbations of Dynamical Systems" (Springer, New York (1984))
 9 V.P. Maslov, M.V. Fedoriuk, "Semiclassical Approximation in Quantum Mechanics", (Reidel, Dordrecht 1981)
10 H. Risken, "The Fokker Planck equation" (Springer, New York 1984)
11 E. Ben-Jacob, D.J. Bergman, B.J. Matkowsky, Z. Schuss, Phys. Rev. A26,2805 (1982)

Part VI

Physical Systems; Order and Chaos

Bifurcations in Particle Physics and in Crystal Growth

Ch. Geiger, W. Güttinger, and P. Haug

Institute for Information Sciences, University of Tübingen
D-7400 Tübingen, Fed. Rep. of Germany

I. Introduction

In the course of structure formation, periods of slow and steady development are invariably separated by sudden and discontinuous changes in which new spatio-temporal patterns are created. The latter are triggered when, in the nonlinear domain, competing but continuously driving forces enter conflicting regimes. Such spontaneous changes can be described geometrically in terms of bifurcations. In the vicinity of a bifurcation point a system becomes extremely sensitive to small ambient factors like imperfections, external fields or fluctuations. This enhances the system's ability to perceive its environment and to adapt to it by forming preferred patterns or modes of behavior. Many of the bifurcation processes observed in different systems are qualitatively similar and universal in the sense of being largely independent of system details. This calls for a topological description of the phenomena under consideration which defines, by the notion of equivalence, what it means for two bifurcation processes to be qualitatively similar. The same notion also allows us to account for the basic, though often forgotten fact that physical systems are structurally stable, i.e., that they preserve their quality under small perturbations. This persistence property guarantees that today's experiment reproduces yesterday's result. We do not know how it got that way. But, surprisingly enough, the fundamental topological invariance principle provided by this requirement of structural stability enables us to describe and to classify the bifurcation processes that underly the spontaneous formation of structure on geometrical grounds alone, irrespective of their particular physical origin.

Among the theoretical programs venturing into the area of structure formation, notably dissipative structures [1] and synergetics [2], singularity theory [3] makes explicit use of the concept of structural stability. We outline in Section II an extension of this theory to general bifurcation problems. In Section III the formalism is applied to the formation of patterns in a solidifying liquid, i.e., to crystal growth, and, in Section IV, to spontaneous symmetry breaking in elementary particle physics. While space-time symmetries provide us with conservation laws and equations of motion, the invariance provided by the requirement of a structurally stable geometry relates to the material sector of the former, and so complements the known physical laws. In the case of crystal growth, structural stability implies that the observed solidification processes are insensitive to small material imperfections. This is the source of a definite sequence of solidification patterns. In the case of elementary particle physics, structural stability ensures that basic interactions are insensitive to a fine-tuning of, e.g., coupling constants and that massive particles are generic. This permits us to classify the ways in which symmetries are broken. In both physical situations, the requirement that structurally stable solutions exist determines those bifurcation paths along which new structures must evolve.

II. Bifurcation Geometry

Bifurcation occurs in a nonlinear evolution equation when a variation of a parameter through a critical value causes a qualitative change in the behavior of solutions, e.g., when an equilibrium splits into two. The main result of bifurcation theory is that, in the course of possible bifurcations, a system does not become unstable in an arbitrary way but that the bifurcations occur in certain definite and classifiable ways. The key to this is the concept of structural stability.

1. Structurally Stable Bifurcations

A local bifurcation problem consists of the solution $x=x(\lambda)$ of a system of equations

$$G(x,\lambda) = 0 \qquad\qquad (2.1)$$

with $G(0,0)=0$, $G_x(0,0)=0$, where $G=(G_1,\ldots,G_n)$ is a smooth (C^∞) function of $R^n \times R$ into R^n, $x \in R^n$ are state variables and $\lambda \in R$ is a distinguished, externally controllable bifurcation parameter -- distinct from others describing perturbations -- which represents the driving force in a physical process. We may, for example, assume that (2.1) is the equation for the amplitudes x of the solutions of a system of nonlinear evolution equations obtained by a Lyapunov-Schmidt reduction. The point $0=(x=0,\ \lambda=0)$ is a bifurcation point or a singularity of G and the solution set of (2.1) constitutes the bifurcation diagram $x(\lambda)$.

The pitchfork bifurcation provides a familiar example: The stationary states $\dot{x}=0$ of $\dot{x}=G=x^3-\lambda x$ can be represented by the diagram $x(\lambda)$ of Fig. 1. When λ varies through zero from negative values, the trivial solution of $G=0$ bifurcates into two branches. They may, e.g., be interpreted as the two magnetization directions in a ferromagnet below critical temperature. Bifurcation is an idealized, nongeneric phenomenon because diagrams such as the pitchfork in Fig. 1 are structurally unstable, and small perturbations (e.g., the earth's magnetic field acting on a ferromagnet) will break the diagram. Since material imperfections are present in any real physical system we should require that a description of bifurcations include the effects of variations in the problem other than those of the distinguished bifurcation parameter λ. If, for example, the idealized bifurcation problem is determined by the pitchfork equation $G=0$, one wishes to know the most general form of the solution structure when G is slightly perturbed. In [4] a technique has been developed which allows to determine all possible qualitatively different bifurcation diagrams which arise when the physical system described by G, (2.1), is subjected to small perturbations.

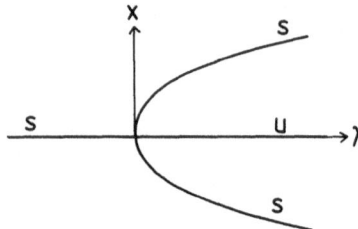

Fig. 1: Pitchfork bifurcation

The notion of equivalence defines what it means for two bifurcation problems G and G', and their solution sets, to be qualitatively similar. G and G' are called contact equivalent ($G \approx G'$) if there exists a local diffeomorphism $x \to X(x,\lambda)$, $\lambda \to \Lambda(\lambda)$, with $\det X_x(0,0) \neq 0$ and $\Lambda_\lambda(0) > 0$, and an $n \times n$ matrix $T(x,\lambda)$ with $\det T(0,0) > 0$, such that

$$G'(x,\lambda) = T(x,\lambda) \cdot G(X(x,\lambda),\Lambda(\lambda)). \qquad\qquad (2.2)$$

The distinguished parameter λ is not allowed to mix with the state variables x in the coordinate transformation $x \to X$, $\lambda \to \Lambda$: λ influences x but not conversely. Since T is invertible, the qualitative properties of the bifurcation diagram $B(G) = \{(x,\lambda)|G=0\}$ are unchanged by equivalence.

A bifurcation problem G is structurally stable or persistent if for any small perturbation term $\varepsilon P(x,\lambda)$ the perturbed problem $G+\varepsilon P$ is contact equivalent to G. The central issue of bifurcation theory is to characterize all stable perturbations of G which respect the special role of λ and to enumerate all qualitatively different bifurcation diagrams. This problem is solved by constructing a universal unfolding of G. A k-parameter family of functions $H(x,\lambda,\alpha)$ is called an unfolding of $G(x,\lambda)$ if $H(x,\lambda,0)=G(x,\lambda)$. The α_i in $\alpha=(\alpha_1,...,\alpha_k)\in R^k$ are called unfolding or imperfection parameters. An unfolding H of G is structurally stable or versal if (up to equivalence) H contains all small perturbations of G, i.e., if any sufficiently small perturbation $G+\varepsilon P$ of G is equivalent to H for some α near 0. H is a universal unfolding of G if it is stable and depends on the minimum number of unfolding parameters needed for stability. This minimum number, k, is called the codimension of G. It is a rough measure of the complexity of a bifurcation phenomenon, bifurcations with lower codimensions being more likely. The unfolding parameters fall into two classes, modal and non-modal. The modal parameters parametrize the largest family of perturbations of G such that no two perturbations in this family are equivalent. The classification of all qualitatively different bifurcation diagrams that may occur rests on the observation that, under certain defining and nondegeneracy conditions on the Taylor coefficients of G at the origin, G is contact equivalent to a polynomial $G'=N(x,\lambda)$. Namely, higher-order terms in G may be absorbed by the nonlinear change of coordinates in (2.2) so that they have no effect on the qualitative behavior of the bifurcation diagram in the small. We have the following theorem: A bifurcation problem $G(x,\lambda)$ of finite codimension is contact equivalent to one described by polynomial normal forms $N(x,\lambda)$ whose universal unfoldings $F(x,\lambda,\alpha)$ are also polynomials.

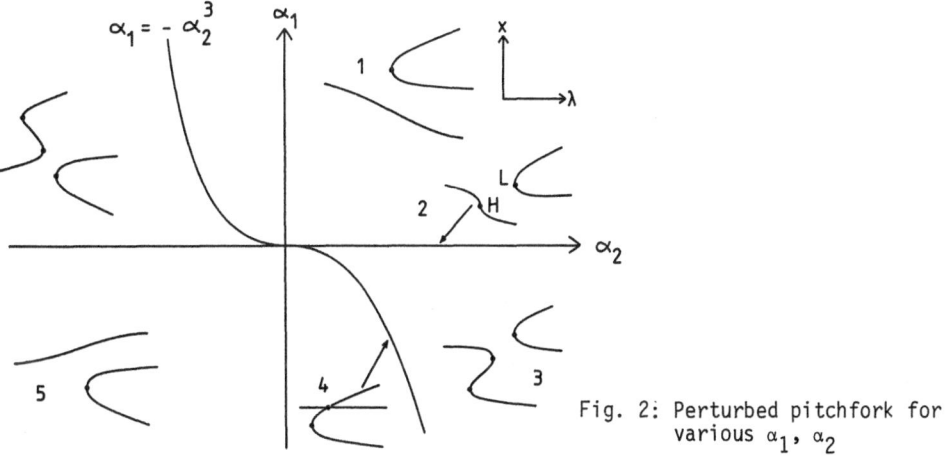

Fig. 2: Perturbed pitchfork for various α_1, α_2

A representative example is this: Let $G(x,\lambda)$ with $x\in R$ be a bifurcation problem such that at $x=\lambda=0$ we have the defining (or degeneracy) conditions $G=G_x=G_{xx}=G_\lambda=0$ and the nondegeneracy condition $G_{xxx}G_{x\lambda}<0$. Then G is contact equivalent to the codimension-2 pitchfork $x^3-\lambda x$ with universal unfolding $x^3-\lambda x+\alpha_1 x^2+\alpha_2$ or, equivalently, $F=x^3-\lambda x+\alpha_2\lambda+\alpha_1$. The bifurcation diagrams of the perturbed pitchfork $F=0$ are shown in Fig. 2. Another example is provided by the unfolding $F=x^3+\lambda^2+\alpha_2 x +\alpha_3 x\lambda+\alpha_1=0$ of the winged cusp $x^3+\lambda^2$; typical bifurcation diagrams are shown in Fig. 3. In the case of one state variable, $x\in R$, the theorem can be proved by relating the bifurcation problem $G(x,\lambda)=0$ to paths through the universal unfolding of a Thom cuspoid [3]. For example, setting $\lambda=-u$ and $\alpha_1+\alpha_2\lambda=v$, the pitchfork's $F=0$ is just the overhanging cliff S of Fig. 4. Fixing (α_1,α_2) defines a straight line in the (u,v)-plane as λ varies, and the bifurcation diagrams of the perturbed pitchfork are the intersection curves of S with the vertical planes having these lines as basis. Crossing a fold branch by a line through the cusp corresponds to a limit point L in the bifurcation diagram of F while a contact of the line with the cusp point produces a hysteresis point H in $x(\lambda)$. This is shown in Fig. 5. Paths generating bifurcation diagrams of the winged cusp are shown in Fig. 6.

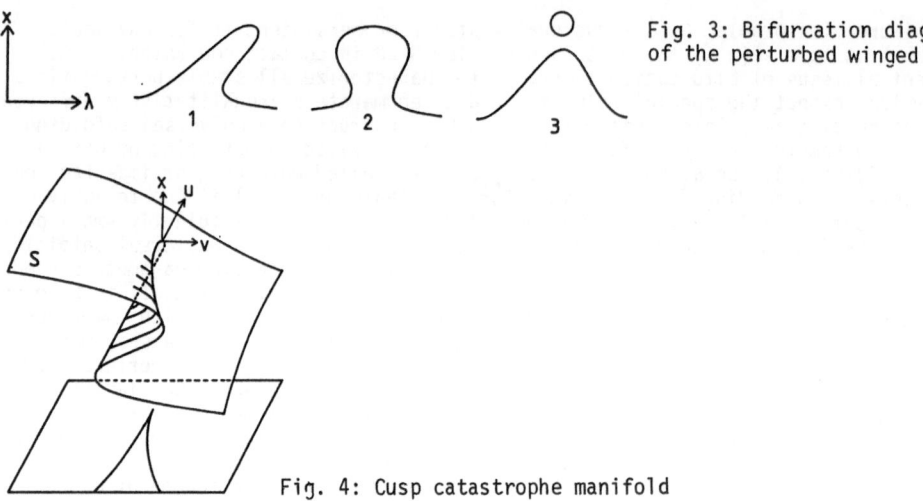

Fig. 3: Bifurcation diagrams of the perturbed winged cusp

Fiy. 4: Cusp catastrophe manifold

The changes in the stability properties of a solution $x=x(\lambda)$ of $F=0$ follow by considering x as equilibrium solution of the system $\dot{x}=F$ and determining the signs of the eigenvalues of the Jacobian J of F evaluated at x. A stability exchange occurs for those values of λ for which either $\mathrm{Tr}J=0$ and $\det J>0$, or $\det J=0$. Given a universal unfolding F of N, the classification of the qualitatively different perturbed bifurcation diagrams proceeds by dividing the unfolding parameter space R^k into finitely many regions such that any two choices of parameter from within one such region give equivalent bifurcation diagrams. These are the stable diagrams that remain unchanged (in the sense of equivalence) if subjected to an additional small perturbation. Structural stability breaks down if the parameters cross the hypersurfaces in R^k that separate the stable domains. For example, instability occurs along the two curves $\alpha_1=0$ and $\alpha_1=-\alpha_2^3$ in Fig. 2.

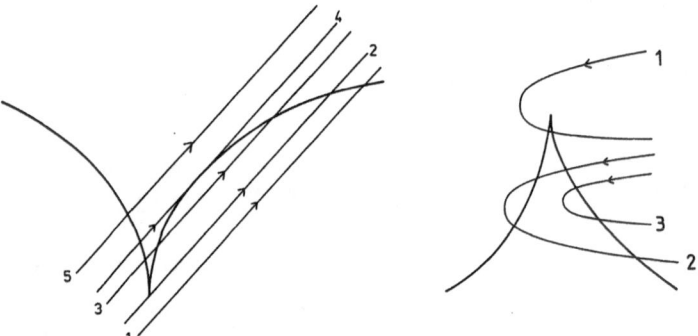

Fig. 5: Straight paths through the cusp producing the perturbed pitchfork

Fig. 6: Parabolic paths through the cusp producing Fig. 3

The bifurcation theory outlined above can be generalized to problems with symmetry [5], [6]. Let G be a bifurcation problem and Γ a compact Lie group acting linearly on R^n. Then G is a bifurcation problem with symmetry group Γ if G is Γ-equivariant, i.e., if $G(\gamma x,\lambda)=\gamma G(x,\lambda)$ for all $\gamma\in\Gamma$. G and G' are Γ-equivalent if (2.2) holds with the symmetry conditions $X(\gamma x,\lambda)=\gamma X(x,\lambda)$ and $\gamma^{-1}T(\gamma x,\lambda)\gamma=T(x,\lambda)$. Γ-equivariant unfoldings, Γ-codimensions etc. are defined in an analogous way. Co-dimensions are changed by symmetry. For example, the Z_2-codimension of the pitchfork is zero. This explains why the pitchfork appears in many idealized bifurcation phe-

nomena whereas the fact that its contact codimension is two explains why it is difficult to observe experimentally. If G=0 is an Γ-equivariant bifurcation problem, a
bifurcating solution will have an isotropy subgroup which is not Γ, i.e., the new
solution will have less symmetry than the old: The symmetry has broken spontaneously.

2. Nonlinear Evolution Equations

The bifurcation equations (2.1) may be viewed, e.g., as the result of a Lyapunov-
Schmidt reduction of a system of evolution equations $F(u,\lambda)=0$, $F(0,0)=0$, where u is
an element of a Banach space and F a nonlinear operator. Assuming that
$dim(ker F_u(0,0))=n$, $N=ker F_u(0,0)=\{\phi_1,\ldots,\phi_n\}$, and denoting by P the projection onto N
one obtains the bifurcation equations (2.1) with $G=PF(v+w(\lambda,v),\lambda)$ where $v=Pu=\Sigma x_i\phi_i$
and $w=(I-P)u$. The x_i are the amplitudes of the bifurcating solutions. The defining
and nondegeneracy conditions of the normal forms into which G is transformed depend
on the parameters in F. A complete list of perturbed bifurcation diagrams with codimension ≤ 3 is available for several bifurcation problems of physical interest. We
give some representative examples for later use.

(i) Hopf Bifurcation [7]. The amplitude x of a periodic oscillation from a steady
state x=0, and that of an odd stationary mode, satisfy an equation with Z_2 symmetry
$G=xa(x^2,\lambda)=0$ with $a(0,0)=0$. Representative bifurcation diagrams, i.e., solutions $x(\lambda)$
of $F(x,\lambda,\alpha)=0$ for various values of α are shown in Fig. 7 where s and u indicate stable
and unstable branches. The formation of hysteresis, instability gaps, mushrooms and
selfoscillating islands is obvious from the diagrams.

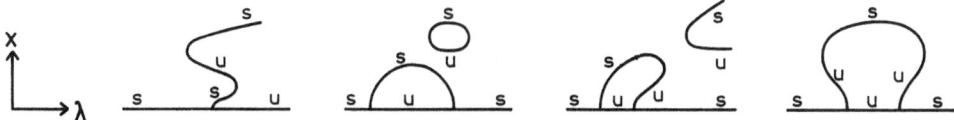

Fig. 7: Typical Hopf bifurcation diagrams for selected values of α_i

(ii) Coupled Hopf and Steady-State Bifurcations [8]. The amplitudes of coupled
Hopf H:y and steady-state S:x modes obey the corank-2, Z_2-covariant bifurcation equations

$$G(x,y,\lambda) = \begin{pmatrix} a(x,\lambda,y^2) \\ yb(x,\lambda,y^2) \end{pmatrix} = 0 \qquad (2.3)$$

with $a(0)=b(0)=a_x(0)=0$. The perturbed diagrams in (x,λ,y)-space shown in Fig. 8 are
representative. The first exhibits Hopf-steady-state hysteresis, the second a tertiary bifurcation point.

Applications of bifurcation geometry are presently springing up in many fields.
Topological problems of bifurcations in optical bistability were discussed in [9]
where, among other phenomena, the appearance of stable islands representing self-

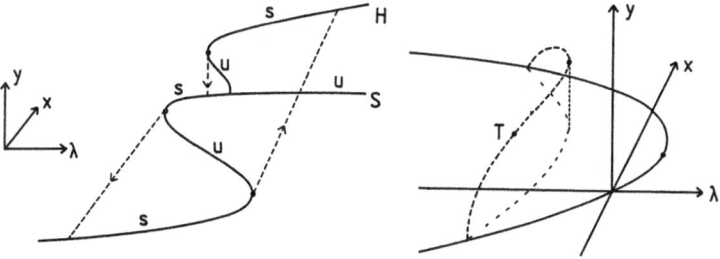

Fig. 8a: Hopf-steady-state-hysteresis   Fig. 8b: Tertiary bifurcation point

confined light-pulsing was predicted. This is shown in the bifurcation diagrams of Fig. 9 in which x denotes the transmitted light intensity of an optical resonator, λ is the incident light intensity, and the amplitude y of the time-periodic oscillation bifurcation from the S-shaped steady-state is projected onto the (x,λ)-plane as a dotted loop. The same island formation is also present in chemical reactions [10].

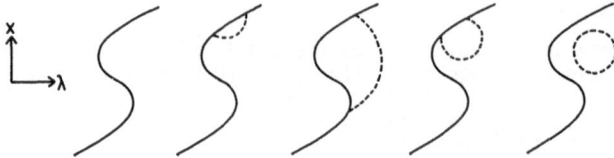

Fig. 9: Island formation in coupled Hopf-steady-state bifurcations

Degenerate Hopf bifurcations leading to hysteresis, instability gaps and jumps to an oscillatory state of increasing amplitude were shown to occur in the Hodgkin-Huxley neuron [11]. As a further example we mention the classification of symmetry-breaking bifurcations in the Taylor fluid [12] which bear some resemblance to those described in Section IV. The fact that many of these bifurcation phenomena, which were predicted on geometrical grounds alone, still await experimental confirmation, demonstrates both the predictive power of this geometric-physical reasoning and the limits of present experimental possibilities. Major problems are, of course, posed by a full dynamical description of the phenomena [13].

III. Solidification as a Bifurcation Problem

Some of the most beautiful examples of spontaneous pattern formation can be found in the solidification of liquids, i.e., in the growth of crystals, and in the reverse process of melting. The morphological transitions in spatial pattern of a solidifying liquid is a nonequilibrium phenomenon of self-organization, whose study is not only of interest to geo- and space-physicists but, in particular, to metallurgists who must deal with them in the design of materials processes. Despite the fact that solidifying systems are accessible in both nature and laboratories, the underlying mechanisms are not well understood, mainly because methods for a full nonlinear analysis were hitherto not available [14]. The linear theory [15] and subsequent asymptotic nonlinear analyses [16], [17] have established the initial stages of the evolution of a planar solidification front separating a binary melt from its solid to develop undulations that lead to a cellular interface and eventually to dendritic crystal growth. In this section we demonstrate that the bifurcation theory outlined in Sec. II provides us with a comprehensive framework to describe analytically the transitions to and between cellular interfaces arising during unidirectional solidification of a binary alloy.

1. The Standard Model

Consider a long thin sample of a binary mixture lying in the (y,z)-plane which is pulled, at constant velocity V, through a fixed temperature gradient established by stationary hot and cold contacts at A and B (Fig. 10). The respective temperatures are chosen so that the sample is molten at A and solid at B, with the solid-liquid interface visible in between. It is assumed [14], [18], [19] that diffusion of the impurity or solute is negligible on the solid side of the interface (one-sided model), and that the latent heat of solidification, which is released at the interface, is small. This implies that the solidification dynamics are primarily governed by solute transport towards or away from the interface which is located at z=s(y,t).

At low pulling velocity, the external thermal conditions impose a planar interface. As V increases, solute concentration builds up in front of the advancing interface until we encounter constitutional supercooling since the concentration profile

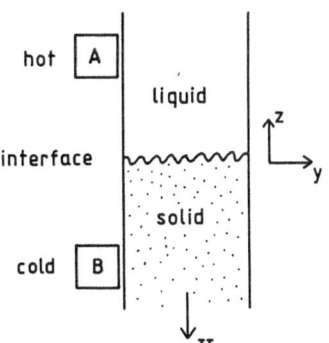

Fig. 10: The directional solidification system

enters a two-phase region. Thus, with increasing V, solute diffusion tends to desta-
bilize the planar front: if the solid bulges, the concentration gradient is increased
in front of the bulge and the solute current is enhanced so that the growth of the
bulge is favored and the interface becomes unstable. This destabilizing effect of the
concentration gradient is counterbalanced by the stabilizing effects of the surface
tension. The resulting interplay between kinetic and capillary effects produces the
complex cellular patterns that we see in the advancing front. In experiments there is
an accumulation of the solute in the grooves trailing along parallel lines of impu-
rity-rich material through the sample (Fig. 11).

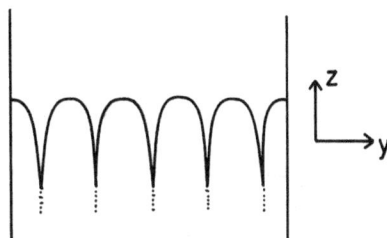

Fig. 11: Shape of the interface observed
in experiment

In the frame of reference moving in the z-direction at the interface velocity V,
the "quasistationary" diffusion of the suitably normalized solute concentration
(chemical potential) W(y,z,t) is determined by the equation

$$D(W_{yy} + W_{zz}) + V \cdot W_z = 0 \tag{3.1}$$

in the region $0 \leq y \leq 1$ (width of the sample), $-\infty < z < \infty$, with $W(y,\infty,t)=-1$, and the Neumann
boundary conditions $W_y(0,z,t)=W_y(1,z,t)=0$. It is assumed that the diffusion of W is
much faster than the displacement of the interface and thus $W_t=0$ during the process.
The solute equation (3.1) is to be solved inside the liquid with the boundary conditions on the interface $z=s(y,t)$

$$W = -d \cdot \gamma(s) - \alpha \cdot s \tag{3.2}$$

and the Rankine-Hugoniot condition which relates the speed of diffusion to that of
the interface, viz.,

$$s_t + V - D(s_y \cdot W_y - W_z) = 0 . \tag{3.3}$$

Here

$$\gamma(s) = -\frac{s_{yy}}{(1+s_y^2)^{3/2}} \tag{3.4}$$

285

is the mean curvature of the interface at $z=s(y,t)$, and d and α are combinations of the various physical constants involved [14]. In virtue of (3.4), the system (3.1)-(3.3) to be solved is strongly nonlinear.

Our objective is to derive an equation of motion for the interfacial displacement $z=s(y,t)$ and then look for stationary solutions of this equation. We represent s by the Fourier series

$$s(y,t) = \sum_{m=-\infty}^{\infty} \varepsilon_m(t) \cdot \exp(im\tilde{k}_o y) . \tag{3.5}$$

In virtue of the boundary conditions at $y=0$ and $y=1$ we have $\tilde{k}_o = \pi/1$ for the fundamental wave number and $\varepsilon_m = \varepsilon_{-m}$ so that s is a cosine series. Since the ε_m change with t, so does s and, therefore, W. Consequently, W is a functional of the ε_m, $W=W(y,z,t,\{\varepsilon_m\})$, and can be expanded into a Taylor series around the flat interface $\varepsilon_m=0$,

$$W = \sum_{m=0}^{\infty} \sum_{|\mu|=m} D^\mu W|_0 \cdot \varepsilon^\mu \frac{1}{\mu!} \tag{3.6}$$

with multi-indices notation understood. Substituting (3.5) and (3.6) into (3.1) and (3.2) yields the coefficients $D^\mu W|_0$ and substituting the resulting W, (3.6), into (3.3) gives the desired equation of motion for the interface s in the form of the following system of equations for the $\varepsilon_m(t)$,

$$\dot{\varepsilon}_m = \sum_{r=1}^{\infty} \sum_{\substack{n_1,\ldots,n_r \\ \Sigma n_j = m}} c^{(m)}_{n_1,\ldots n_r} \cdot \varepsilon_{n_1} \cdots \varepsilon_{n_r} . \tag{3.7}$$

The $c^{(m)}_{n_1,n_2,\ldots}$ can be calculated explicitly [18], [20] but only the lowest ones will be needed to determine the degeneracy of the bifurcation problem to which (3.7) will be reduced below.

2. Stability Analysis

Writing (3.7) in the form

$$\dot{\varepsilon}_m = c^{(m)}_m \cdot \varepsilon_m + M_m(\{\varepsilon_j\}) , \quad m = 0,1,2,\ldots \tag{3.8}$$

and rescaling $V \to v$, $\tilde{k}_o \to k_o$ one finds

$$c^{(m)}_m(k_o,v,u) = (v-1-m^2 k_o^2)(v/2 + \sqrt{v^2/4 + u\, m^2 k_o^2}) - v^2 \tag{3.9}$$

where $u=1/(\alpha d)>0$ is a combination of the physical constants involved. The planar interface ($\{\varepsilon_j\}=0$) loses stability as v increases, if $c^{(m)}_m$ (i.e., the eigenvalues of the linearized problem) go through zero from negative values for at least one m. Thus, considering mk_o as a continuous variable k for a moment, instability sets in if the parametrized surface

$$c = c(k,v;u) = (v-1-k^2)(v/2 + \sqrt{v^2/4 + u\, k^2}) - v^2 \tag{3.10}$$

in (k,v,c)-space intersects the (k,v)-plane from below. Such an intersection in a convex contour $\Gamma(u)$ occurs for $u \geq 27$ (Fig. 12).

Returning to discrete wave numbers mk_o, we observe that the n-th mode becomes unstable if $k_1 \leq nk_o \leq k_r$ (Fig. 12). Since we are interested in the transition from a stable planar interface to a cellular one, we must determine that mode nk_o which becomes unstable first when v increases through v_c, i.e., the unstable mode with the

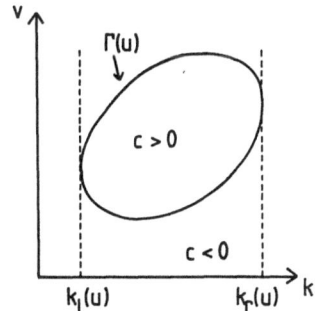

Fig. 12: Critical values of v as a
function of rescaled wave-
number k

lowest critical value v_c. Of course, this mode depends on the values of k_0 and u. A
detailed analysis yields a division of the (k_0,u)-plane into different regions such
that when (k_0,u) lies in region "n", the n-th mode is the first one that becomes unstable (Fig. 13). If for some (k_0,u) the n-th mode is the first that becomes unstable
and if the point (k_0,u) varies in the (k_0,u)-plane, then this mode ceases to be the
first one that becomes unstable when (k_0,u) crosses the boundary of the region "n".

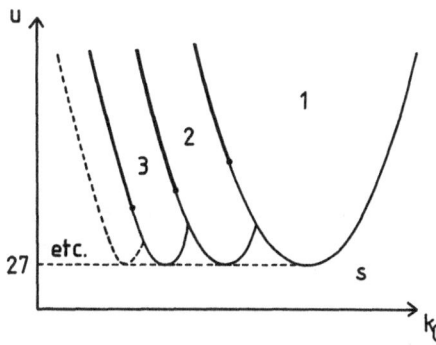

Fig. 13: Mode selection

There are essentially two ways for this to happen.
(1) Let $nk_0=k_1(u)$ and $(n+1)k_0<k_r(u)$, cf. Fig. 14b. If k_0 decreases, then the mode nk_0
does not become unstable at all since the line $k=nk_0$ does no longer intersect $\Gamma(u)$

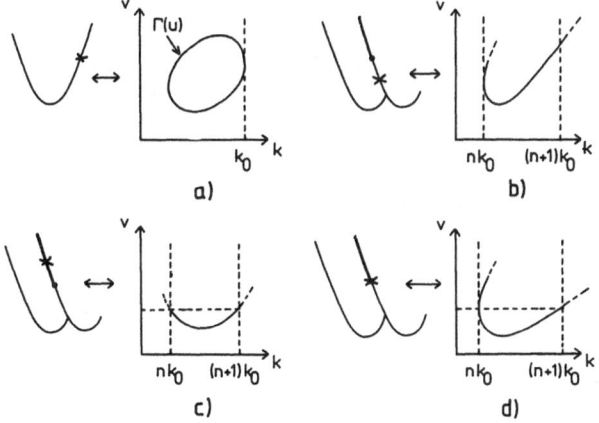

Fig. 14: Correspondence between points (k_0,u),✕, and the onset of instability

and the mode $(n+1)k_0$ will therefore be the first one that becomes unstable.
(2) Let $k_1(u)<nk_0<(n+1)k_0<k_r(u)$. Then decreasing k_0 has the effect of eventually
increasing the critical $v=v_c$ for the mode nk_0 and of decreasing the critical v for
the $(n+1)k_0$-th mode. Thus there exists a point (u,k_0) where these two critical v are
identical, i.e., where the n-th and the $(n+1)$-th modes become simultaneously unstable
as the first ones. This is illustrated in Fig. 14c.
Fig. 14a describes a special case of (1); in Fig. 14d the two situations (1) and (2)
occur simultaneously.

3. Bifurcation Geometry

In looking for stationary solutions of (3.8), $\dot{\varepsilon}_m=0$, we assume that, as v goes through
a critical value v_c, the n-th mode becomes unstable as the first one, i.e., $c_n^{(n)}=0$
and $c_m^{(m)}<0$ for $m\neq n$. Then, substituting the solution $\varepsilon_m=\varepsilon_m(\varepsilon_n)$ of the nondegenerate
$(m\neq n)$ system (3.8) into the remaining equation $G=M_n=0$, we obtain a single algebraic
bifurcation equation

$$G(x,\lambda) = 0 \tag{3.11}$$

for the amplitude $x=\varepsilon_n(0)$ and the distinguished parameter $\lambda=v-v_c$. This procedure corresponds to a Lyapunov-Schmidt reduction. Since the system (3.1)-(3.3) is invariant
under reflection at the axis $y=1/2$ it follows that G is odd in x [20] whence (3.11)
takes the form

$$G(x,\lambda) = xa(x^2,\lambda) = 0 \tag{3.12}$$

with $a(0,0)=0$. The Taylor coefficients of $a(x^2,\lambda)$ can be expressed in terms of the
$c_{n_1,n_2,\ldots}^{(m)}$ of (3.7). The corank-1, $Z(2)$-covariant bifurcation problem (3.12) for the
evolution of the cellular solidifying interface is identical in form with that of the
degenerate Hopf bifurcation discussed in Sec. II so that the machinery developed
there can be immediately applied. It is, of course, perfectly in the spirit of synergetics that the same bifurcation geometry governs such disparate processes as spatial pattern formation in crystal growth and temporal pattern formation in neuronal
activity. Assuming that the solidification process is structurally stable, the results of Sec. II provide us with a complete classification of the bifurcations which
the planar interface undergoes when the velocity is increased beyond v_c and the
effects of system imperfections, e.g., perturbations of u or k_0, are taken into
account in the unfoldings of the normal forms into which a in (3.12) can be transformed. We assume that $a(x^2,\lambda)$ has (via contact equivalence) been transformed into
polynomial normal form and denote the new coordinates X,Λ again by x,λ. The normal
forms and their unfoldings $F(x,\lambda;\alpha,\beta,\gamma)$ up to codimension three, where (α,β,γ) are
imperfection (unfolding) parameters, and the possible bifurcation diagrams are listed
in [7] and [20].

Bifurcation from planar to cellular interfaces occurs in (3.12) if degeneracy conditions such as $a_{xx}=0$, $a_\lambda=0$ etc. are satisfied at the bifurcation point $(x=0,\lambda=0)=0$.
These conditions imply, and are implied by, conditions satisfied by the $c_{n_1,n_2,\ldots}^{(m)}$ in
(3.7). Since the latter are functions of k_0,u and the bifurcation parameter $\lambda=v-v_c$,
degeneracy of G occurs for certain values of k_0 and u or on curves $u(k_0)$ in the
(k_0,u)-plane. For example, from $G_x(0)=c_n^{(n)}$ and $G_{x\lambda}(0)=\partial c_n^{(n)}/\partial v$ one infers from Fig. 14
that in each region "n" there are points (k_0,u) where $a_\lambda(0)=0$ and the corresponding
nondegeneracy conditions are satisfied. In this case from the list of normal forms
[7] one has the unfolding

$$F = x^3 + (\lambda^2 + \alpha)x = 0 \tag{3.13}$$

of N with unfolding parameter α. The corresponding bifurcation diagram is shown in
Fig. 15 with the stability assignments (s=stable, u=unstable). This result is physically plausible because $a_\lambda(0)=0$ implies that $k=nk_0$ equals $k_1(u)$ (or $k_r(u)$) in Fig. 12.
Since the effect of the imperfection α is equivalent to slightly shifting k, e.g.,
to the right of k_1 (or changing u), the n-th mode becomes unstable at a critical

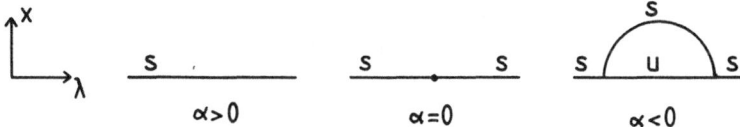

Fig. 15: Bifurcation diagrams for $a_\lambda=0$

$\lambda=\lambda_c$ and restabilizes again at $\lambda=\lambda'>\lambda_c$. This is revealed in the perturbed diagram for $\alpha<0$ of Fig. 15.

The problem of degeneracy and wave number selection has been analyzed in detail by one of us [20]. While the degeneracy conditions leading to mushroom and isola bifurcations cannot be satisfied in the system modelled by (3.1)-(3.3), the following diagrams are of direct physical significance:

(i) $a_{xx}(0)=0$. In this case the unfolding of N is $F=x^5+2\alpha x^3+\varepsilon\lambda x=0$ with imperfection parameter α and $\varepsilon=\text{sign}(a_{xxxx}(0))$. In this case the degenerate points in the (k_0,u)-plane lie in the appropriate region "n" only for $n\leq3$, and for $n\geq4$ this type of bifurcation does not occur. The bifurcation diagram is shown in Fig. 16 for $\varepsilon=-1$.

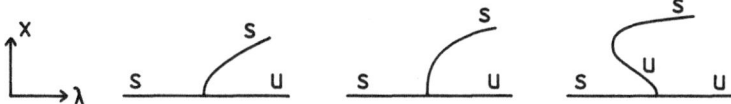

Fig. 16: Diagrams for $a_{xx}=0$

(ii) $a_{xx}(0)=a_{xxxx}(0)=0$. The unfolding of N is now $F=x^7+\beta x^5+\alpha x^3+\varepsilon\lambda x=0$ with $\varepsilon=\pm1$ depending on the sign of $a_{xxxxxx}(0)$. The bifurcation diagrams are shown in Fig. 17 for various values of (α,β). The last two diagrams demonstrate that even in the relatively simple solidification model described by (3.1)-(3.3) there exists hysteretic behavior.

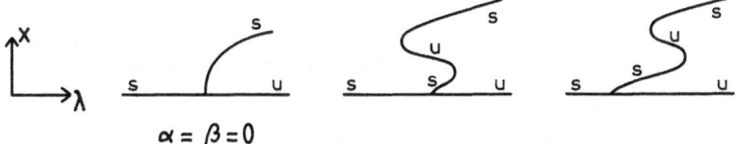

$\alpha = \beta = 0$

Fig. 17: Some diagrams for $a_{xx}=a_{xxxx}=0$

(iii) $a_\lambda(0)=a_{xx}(0)=0$. In this case, only the regions "1" and "2" contain (k_0,u)-values for which the various degeneracy and nondegeneracy conditions of a are satisfied so that only in the cases where the first or second mode is the first that becomes unstable at $v=v_c$ can this type of bifurcation occur. The unfolding is now a polynomial of fifth order,

$$F = x^5 + 2b\lambda x^3 - (\lambda^2-\beta\lambda+\alpha)x = 0,$$

containing a modal unfolding parameter b. Some relevant diagrams are plotted in Fig. 18. It is remarkable that there are parameter λ regions for which there are no stable solutions at all.

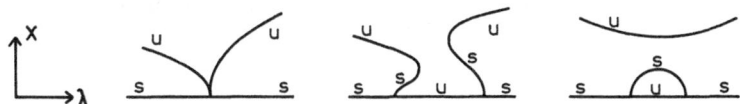

Fig. 18: Some diagrams for $a_\lambda=a_{xx}=0$

We have seen that for certain values of k_0 and u two modes ε_n and ε_{n+1} can become simultaneously unstable when the velocity of the sample goes through its first critical value. In this case a double eigenvalue goes through zero, $c(n)=c\binom{n+1}{n+1}=0$, and instead of (3.12) we obtain now a coupled system of two bifurcation equations for the stationary amplitudes $x=\varepsilon_n(0)$ and $y=\varepsilon_{n+1}(0)$. If $n=1$, the bifurcation equations have the form

$$G_1 = x(a(x^2,y^2,\lambda) + yb(x^2,y^2,\lambda)) = 0$$
$$G_2 = x^2 c(x^2,y^2,\lambda) + yd(x^2,y^2,\lambda) = 0. \qquad (3.14)$$

This corank-2 bifurcation problem arises because the system (3.1)-(3.3) is subject to Neumann boundary conditions. The bifurcation diagrams can be classified [21]. One degeneracy that occurs is $G_{1\,x\lambda}=0$ in which case the associated normal form is

$$F_1 = x(-y-\lambda^2+\alpha\lambda+\beta) = 0$$
$$F_2 = x^2 + y(y^2-\lambda) = 0 \qquad (3.15)$$

with unfolding parameters α, β. The corresponding bifurcation diagrams are discussed in [21]. They describe secondary bifurcation in the evolution of solidifying cellular interfaces, i.e., the evolution of families of interfaces with varying wavelengths.

In this section we have presented a classification of the bifurcation processes which govern the "standard model" of solidification and predicted a number of phenomena such as hysteresis and gap structures which still await experimental verification. The analysis can be extended to the three-dimensional case. The next steps ought to deal with the time-dependence of the solidification patterns, with dendritic growth processes and with the problem of fractal or chaotic behavior of solidifying liquids.

IV. Bifurcation and Symmetry Breaking in Particle Physics

In elementary particle physics symmetry plays the role of a "first principle". Since our present universe is less symmetric than our fundamental equations, the consideration of symmetry goes hand in hand with that of its breakdown. Philosophically speaking, symmetry already contains its own negation; mathematically speaking, symmetry may be spontaneously broken. Since symmetry breaking is a common phenomenon when bifurcation takes place, it is natural to try to apply concepts of bifurcation theory to problems in particle physics. Here the goal is to unify all the known interactions by deriving them from one interaction scheme based on a single symmetry group and one expects to be able to arrive at the empirical symmetries through spontaneous symmetry breaking processes.

The most successful method to achieve this is provided by the Higgs mechanism [22], viz., by an ad hoc introduction of a number of massive scalar fields ϕ incorporated into a potential term $V(\phi)$ in the fundamental Lagrangian. Consider, e.g., a Lagrangian L which is invariant under $SO(n)$ [23]

$$L = -1/4\, F_{\mu\nu}^a\, F_a^{\mu\nu} + |D_\mu \phi|^2 - V(\phi)$$

where $F_{\mu\nu}^a = \partial_\mu A_\nu^a - \partial_\nu A_\mu^a + gf^{abc}A_{\mu b}A_{\nu c}$ and $D_\mu = \partial_\mu - igL_a A_\mu^a$, where f^{abc} are the structure constants and L_a are the $n(n-1)/2$ generators of $SO(n)$. The gauge fields A^a transform under the adjoint action of $SO(n)$. The scalar potential is given by

$$V(\phi) = (\alpha/4)(\phi^i \phi_i)^2 - (\varepsilon\mu^2/2)(\phi_i \phi^i) + \varepsilon^2 \mu^4/4\alpha$$

with n scalar fields ϕ_i. We can expand around the minimum $<\phi>$ of the potential by choosing $\Phi_i = \exp(-iT^b\xi_b(\Phi)/<\Phi>)\cdot(0,0,\ldots,<\Phi>+\eta(\Phi))^T$. T^b are the generators of the algebra of an isotropy subgroup $H=SO(n-1)$ which leaves the minimum invariant. $\xi^b(\Phi)$ describing Goldstone bosons, and $\eta(\Phi)$ describing the Higgs bosons, are new fields in that expansion. One uses the gauge freedom to eliminate the massless Goldstone bosons and arrives at a new Lagrangian containing a mass term $M^2_{ab}=g^2<\Phi>^2T_aT_b$ for some of the gauge bosons. Dim(H) gauge bosons lie in the directions of the generators T^a and remain massless. Dim (G/H) gauge bosons are massive vector fields. In this new gauge the original symmetry is hidden in the sense that for small energies interactions can be mediated only by the $n(n-1)/2$ massless gauge bosons.

In the Weinberg-Glashow-Salam (WGS) theory there are three massive gauge bosons W^+, W^-, Z^0 and a massive Higgs field. The massless fields are the photon of the electromagnetic interaction and 8 gluons mediating the strong interactions. The WGS theory is based on a SU(3)xSU(2)xU(1) symmetry group and the question arises if all gauge interactions can be derived from a single gauge group. The most promising candidate is the SU(5)-GUT of Georgi and Glashow [24]. This "SU(5)-GUT-world" would be present at energies above 10^{15}GeV. The gauge interactions are described by 24 gauge bosons. After spontaneous symmetry breaking 12 of these bosons would acquire mass through a Higgs-mechanism and beyond 10^{15}GeV one is left with the 12 gauge bosons of the WGS-theory. Another group of symmetries arises if one considers the fermionic components of the theory. Here one has to explain why the 12 fundamental fermions possess so different masses. To deal with this problem one considers symmetry groups like SU(3), SU(4), SU(6), chiral symmetries, and so forth. In what follows we discuss spontaneous SU(n) symmetry breaking in the context of bifurcation theory. We consider symmetries which appear to be important in particle theory but for simplicity confine the analysis to the adjoint representation of SU(n). This allows a direct geometric visualization of observable bifurcation phenomena.

1. SU(n) Invariant Algebra

To formulate bifurcation problems with SU(n) symmetry we need SU(n) invariant algebra [25]. SU(n) is the group of nxn unitary, unimodular matrices g. Every g can be written in the form $g = \exp(ix)$. The traceless Hermitian nxn matrices x form an (n^2-1)-dimensional vector space $E=R^{n^2-1}$. The Euclidean scalar product of two vectors x and $y\in E$ is defined by $(x,y)=(y,x)=\mathrm{tr}xy/2$ and the adjoint representation of SU(n) is defined by $x\to gxg^*$. The symmetrical Lie algebra multiplication law is $x_\lor y=\sqrt{n}(xy+yx)/2-\mathrm{tr}xy/\sqrt{n}$. Every x satisfies the characteristic equation

$P_n(x) = x^n - \sum\limits_{k=1}^{n}\Theta_k x^{n-k} = 0$ whose coefficients $\Theta_k(x)$ are (k+1)-th degree homogeneous invariants, $\Theta_{k-1}(x) = k^{-1}\mathrm{tr}(x^k - \sum\limits_{1=2}^{k-2}\Theta_1 x^{k-1})$, which form a Hilbert basis in R^{n-1} and generate the ring E^{n^2-1}. Thus, every smooth SU(n)-invariant function $\Theta(x)$, $(\Theta(U(g)x)=\Theta(x)$ where U(g) is a representation of g), is a polynomial in $\{\Theta_1,\ldots,\Theta_{n-1}\}$. Let $G(x):E\to E$ be a smooth SU(n)-equivariant map, $G(U(g)(x))=U(g)G(x)$. Then there exists an equivariant (i.e., covariant) basis $\{G_1(x),\ldots,G_{n-1}(x)\}=\{x,x_\lor x,(x_\lor x)_\lor x,\ldots\}$ which generates a module M^{n^2-1} over the ring E^{n^2-1} so that every G can be represented in the form

$$G(x) = \sum\limits_{i=1}^{n-1} f_i(\Theta_1,\ldots,\Theta_{n-1})\ G_i(x) \tag{4.1}$$

where f_i are smooth invariant functions. To establish contact SU(n) equivalence we define $T(x):E\to E$ as an equivariant map (written as an operator) such that $G'=TG$ where G, $G'\in M$. Then T can be represented as a linear combination

$$T(x) = \sum\limits_{i=0}^{n-1} g_i(\Theta_1,\ldots,\Theta_{n-1})\ T_i(x) \tag{4.2}$$

of independent equivariant $(U(g)T(x)U^{-1}(g)=T(U(g)x)$ elements T_i which generate a module $\tilde{M}^{n^2-1}$ over the ring of invariant functions E^{n^2-1}. A bifurcation problem with SU(n) symmetry consists of the solution $x(\lambda)$ of the system of equations $G(x,\lambda)=0$ with

291

$G(0,0)=d_xG(0,0)=0$ where G is given by (4.1) with $f_i=f_i(\Theta,\lambda)$ depending on a distinguished bifurcation parameter λ (e.g., temperature) and $\Theta=(\Theta_1,..,\Theta_{n-1})$.

2. Bifurcation Problems with SU(3) Symmetry

The vector space of the adjoint representation of the Lie algebra is R^8, whose elements x are 3x3 Hermitian traceless matrices. They satisfy the equation $p_3(x)=x^3-\bar\Theta_1(x)x-\bar\Theta_2(x)=0$ and the ring of invariant functions is generated by

$\Theta_1=(x,x)=trx^2/2=\bar\Theta_1$, $\Theta_2=(x_vx,x)=\sqrt3\ trx^3/2=3\sqrt3\ \bar\Theta_2/2$. The module M^8 is generated by $G_1=x, G_2=x_vx$. The module $\tilde M^8$ is generated by $T_0=1$, $T_1=x_v$, $T_2=(x_vx)_v$, $T_3=((x_vx)_vx)_v$ viewed as operators to be applied to $G\in M^8$.

A bifurcation problem with SU(3) symmetry consists of the solution of a system of equations $G(x,\lambda)=0$ where $G:R^8xR\rightarrow R^8$ is SU(3) equivariant, i.e.,

$$G(x,\lambda) = a(\Theta_1,\Theta_2,\lambda)x + b(\Theta_1,\Theta_2,\lambda)x_vx = 0 \tag{4.3}$$

where a and b are functions of Θ_i and λ. Consider (4.3) with $a(0)=b(0)=0$, $a_\lambda(0)\neq0$ at the bifurcation point $0=(\Theta_1=0,\ \Theta_2=0,\ \lambda=0)$ and with linear term

$$G_1 = \sum_{p=1}^{2}\sum_{q=0}^{2} u_{pq}\Theta_q G_p$$

with $\Theta_0=\lambda$ and the nondegeneracy conditions $u_{11}\neq0$, $u_{10}\neq0$ and $r=u_{11}u_{22}-u_{21}u_{12}\neq0$, $s=u_{10}u_{21}-u_{20}u_{12}\neq0$. Then G is SU(3) equivalent to the following normal form $N=TG$ in new coordinates x and λ,

$$N(x,\lambda) = (\Theta_1\pm\lambda)x + (\pm\Theta_1 + \bar u\Theta_2)x_vx \tag{4.4}$$

where $\bar u=(u_{10})^2r/s$. The universal unfolding of N with SU(3)-codimension 2 is

$$F(x,\lambda,\alpha,\beta) = (\Theta_1\pm\lambda)x + (\pm\Theta_1 + \alpha\Theta_2 - \beta)x_vx \tag{4.5}$$

where α is a modal unfolding parameter. The situation is analogous to that of the Bénard problem with O(3)-symmetry [26].

3. Bifurcation Problems with SU(4) Symmetry

In the case of SU(4), the bifurcation problem $G:R^{15}xR\rightarrow R^{15}$ is given by

$$G(x,\lambda) = a(\Theta,\lambda)x + b(\Theta,\lambda)x_vx + c(\Theta,\lambda)x_vx_vx = 0 \tag{4.6}$$

where $\Theta=(\Theta_1,\Theta_2,\Theta_3)$. In this case, $E^{15}=R\cdot\{\Theta_1=trx^2,\ \Theta_2=\sqrt3\ trx^3,\ \Theta_3=(7\Theta_1^2-12trx^4)/4\}$, $M^{15}=E^{15}\{G_1=x,\ G_2=x_vx,\ G_3=x_vx_vx=(x_vx)_vx\}$ and $\tilde M^{15}=E^{15}\{T^0=1,\ T^1=x_v,\ T^2=(x_vx)_v$, $T^3=((x_vx)_vx)_v$, $T^4=x_vP_x$, $T^5=(x_vx)_vP_x\}$ where P_x is an operator acting on $y\in E$, $P_{xy}=x(x,y)$. The bifurcation point is $0=(\Theta_0=0,\ \lambda=0)$ and we assume that $a(0)=0$, $a_\lambda(0)\neq0$. The linearized G is

$$G_1 = \sum_{p=1}^{3}\sum_{q=0}^{3} u_{pq}\Theta_q G_p$$

with $\Theta_0=\lambda$ so that the linearized coefficients a, b, c are

$$a_1 = \sum_{q=0}^{3} u_{1q}\Theta_q, \quad b_1 = \sum_{q=0}^{3} u_{2q}\Theta_q, \quad c_1 = \sum_{q=0}^{3} u_{3q}\Theta_q\ .$$

292

The nondegeneracy requirements are $u_{11} \neq 0$, $u_{10} \neq 0$, $u_{21}u_{10} - u_{20}u_{11} \neq 0$, $\bar{u}_{ij} = \bar{u}_{ij}(u_{k1}) \neq 0$ and $\bar{u}_c = \bar{u}_c(u_{k1}) \neq 0$ (cf. (4.8)-(4.11)). Then the following theorem can be proved [27]: G is $SU(4)$-equivalent to a normal form $N=TG$ with universal unfolding F. In particular we have for

(i) $b(0) \neq 0$, $c(0) \neq 0$:

$$N(x,\lambda) = \lambda x + x_v x + x_v x_v x \tag{4.7}$$

whose codimension is zero so that $F = N$,

(ii) $b(0) = 0$, $c(0) \neq 0$:

$$N(x,\lambda) = (\Theta_1 \pm \lambda)x + (\pm \Theta_1 + \bar{u}_{22}\Theta_2 + \bar{u}_{23}\Theta_3)x_v x + \bar{u}_c x_v x_v x \tag{4.8}$$

with a codimension-3 unfolding

$$F(x,\lambda,\alpha,\beta,\gamma) = (\Theta_1 \pm \lambda)x + (\pm \Theta_1 + \alpha \Theta_2 + \beta \Theta_3 - \gamma)x_v x + \bar{u}_c x_v x_v x \tag{4.9}$$

and modal unfolding parameters α, β,

(iii) $b(0) = c(0) = 0$:

$$N(x,\lambda) = (\Theta_1 \pm \lambda)x + (\pm \Theta_1 + \bar{u}_{22}\Theta_2 + \bar{u}_{23}\Theta_3)x_v x + (\bar{u}_{31}\Theta_1 + \bar{u}_{32}\Theta_2 + \bar{u}_{33}\Theta_3)x_v x_v x \tag{4.10}$$

with a codimension-6 unfolding

$$\begin{aligned} F(x,\lambda,\alpha,\beta,\gamma,\delta,\epsilon,\rho) = &(\Theta_1 \pm \lambda)x + (\pm \Theta_1 + \alpha \Theta_2 + \beta \Theta_3 - \gamma)x_v x + \\ &+ (\bar{u}_{31}\Theta_1 + \delta \Theta_2 + \epsilon \Theta_3 - \rho)x_v x_v x \end{aligned} \tag{4.11}$$

and modal unfolding parameters α, β, δ, ϵ.

4. Spontaneous Symmetry-Breaking in SU(n)

Because of spontaneous symmetry-breaking, different solution branches of a bifurcation problem may possess different symmetries and one can classify these solutions by their symmetry. Let the bifurcation problem $G(x,\lambda)=0$ have symmetry group Γ, $U(g)G(x,\lambda)=G(U(g)x,\lambda)$, $g \in \Gamma$, with representation $U(g)$. Let $H \subset \Gamma$ be a subgroup of Γ, $h \in H$ with representation $U(h)$, and consider a bifurcating solution $x=x_0(\lambda)$ of $G=0$. The symmetry Γ is spontaneously broken to the isotropy subgroup H, if x_0 is no longer invariant under the full group Γ but under the group H, i.e., $U(h)x_0=x_0$. The set of all $U(g)x_0$ is the orbit of x_0 under the action of Γ and the set of all orbits with the same isotropy subgroup forms a stratum Σ through x_0. Thus, a stratum is the union of all orbits of a given type. From the topology of the different strata of a specific group representation one can infer typical properties of bifurcating solutions without explicitly solving the bifurcation problem. Here we are interested in the strata of $SU(n)$. One can calculate the equations $Q_i^k(\Theta)=0$ of a stratum Σ_k [28] for a given representation of a symmetry group Γ.

For the adjoint representation of $SU(3)$ three different strata exist (Fig. 19): (i) the origin Σ_0, $\Theta_1=\Theta_2=0$, whose isotropy group is $SU(3)$ itself, (ii) a one dimensional stratum Σ_1 represented through the curve $Q^1(\Theta_1,\Theta_2)=\Theta_1^3(x)-\Theta_2^2(x)=0$ with exclusion of the point Σ_0, (iii) a two dimensional stratum Σ_g, called the generic stratum, consisting of the open dense domain $Q^g(\Theta_1,\Theta_2)=\Theta_1^3(x)-\Theta_2^2(x)>0$.

Figure 20 shows the orbits and strata of the adjoint representation of $SU(4)$. Near the origin only the strata with maximal isotropy subgroup survive. Consequently, if the trivial solution of an $SU(4)$ equivariant bifurcation equation $G=0$ loses stability at the origin, the bifurcating symmetry breaking solutions jump into the maximal isotropy subgroups, i.e., $S(U(2) \times U(2))$ and $U(3)$: There is no bifurcation to a nonmaximal isotropy subgroup from the trivial solution. This fact persists for

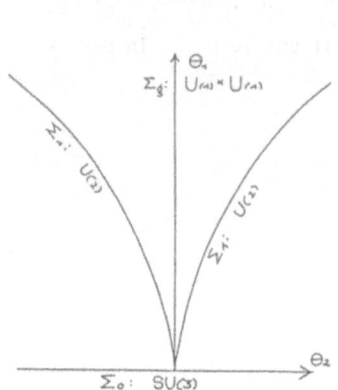
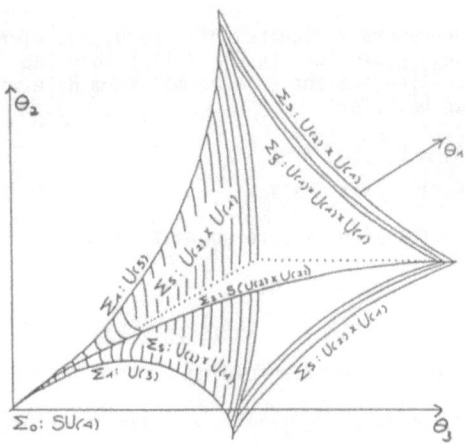

Fig. 19: Strata of SU(3)　　　　　　Fig. 20: Strata of SU(4)
　　　　(adjoint representation)　　　　　　　(adjoint representation)

all adjoint representations of SU(n). For example, in the case of SU(5) the equations
for the strata with maximal isotropy subgroup are: $Q_1=\Theta_3-a_k\Theta_1^2=0$, $Q_2=\Theta_2^2-b_k\Theta_1^3=0$,
$Q_3=\Theta_4-c_k\Theta_1\Theta_2=0$ with $\Theta_k=Trx^{k+1}$. We have two one-dimensional strata: (i) Σ_1: k=1,(a_1,
b_1,c_1)=(13, 9, 17)/20, the isotropy subgroup of the orbits is SU(4)xU(1), (ii) Σ_2:
k=2,(a_2, b_2, c_2)=(7, 1, 13)/20, and isotropy subgroup SU(3)xSU(2)xU(1). Thus, near
the origin Σ_1 and Σ_2 are tangent to the Θ_1-axis.

5. Bifurcation Diagrams

Consider the bifurcation problem (4.4) with SU(3) symmetry, viz.,

$$N = (\Theta_1-\lambda)x + (\Theta_1+\bar{u}\Theta_2)x_v x = 0. \tag{4.12}$$

The only non-trivial solutions near $\lambda=0$ are those with maximal isotropy subgroup
H_1=U(2). Combining the equations $Q_1=\Theta_1^3-\Theta_2^2=0$ for the 1-dimensional stratum Σ_1 with
(4.4), we obtain the bifurcation diagram shown in Fig. 21. Fig. 22 illustrates why no
nonmaximal isotropy subgroup solutions exist: If the symmetry group of a solution is
H_2=U(1)xU(1) then x and x_vx are linearly independent, thus a solution of N=0 is given
by the simultaneous solution of $\Theta_1-\lambda=0$ and $\Theta_1+\bar{u}\Theta_2=0$ under the constraint $\Theta_1^3>\Theta_2^2$.
Fig. 22 shows that all bifurcations from solutions with maximal isotropy subgroup
occur far away from the origin at point P. Unfolding the above normal form, i.e.,
solving F=0, yields the bifurcation diagram shown in Fig. 23. The secondary bifurcation points A and B are intersection points of solutions with maximal isotropy subgroup (on $\Theta_1^3=\Theta_2^2$) and solutions with generic isotropy subgroup represented by the
line L:$\Theta_1+\alpha\Theta_2-\beta=0$ (Fig. 24). In Fig. 23 symmetry breaking occurs through hysteresis.

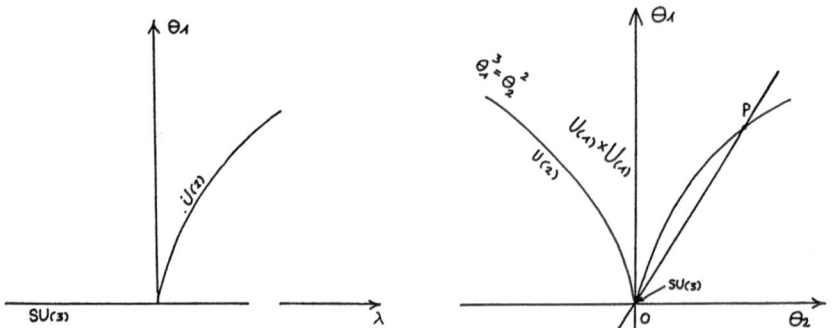

Fig. 21: Bifurcation diagram of (4.12)　Fig. 22: Generic subgroup solutions of (4.12)

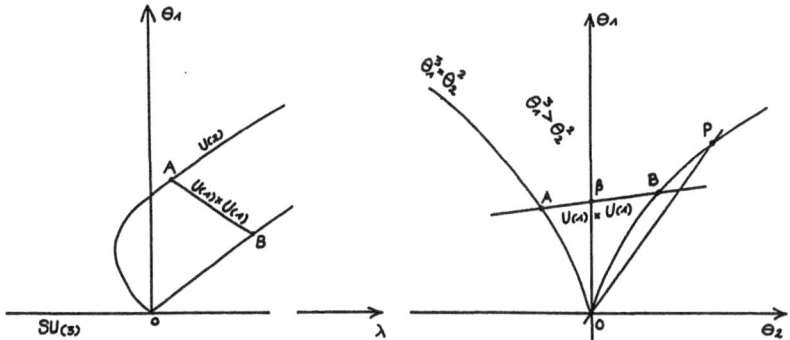

Fig. 23: Bifurcation diagram of (4.5)

Fig. 24: Solutions of (4.5) on the generic stratum of SU(3)

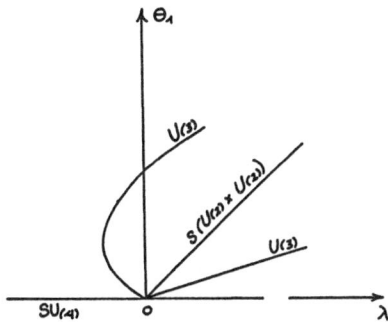

Fig. 25: Bifurcation diagram for the maximal isotropy subgroup solutions of (4.9) and (4.11)

In the case of SU(4) the results are these: The codimension-0 problem has no solution except the one with maximal isotropy subgroup $H_1=U(3)$ or $H_2=S(U(2)\times U(2))$ (and the trivial one). The codimension-3 problem (4.9) exhibits solutions to the nonmaximal isotropy subgroup $H_S=U(2)\times U(1)$. For these solutions $x_V x_V x$ is colinear to x and $x_V x$. The bifurcation point of the secondary bifurcating solutions shrinks toward the origin if γ approaches zero. Fig. 25 shows the bifurcation diagram for the different maximal isotropy subgroup solutions in the cases of codimension 3 and codimension 6. In the codimension 6 case, F=0 (4.11), one finds a very rich bifurcation structure. Varying the nonmodal bifurcation parameter γ away from 0 yields solutions of F=0 on

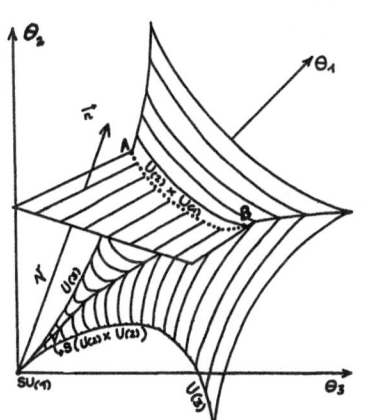

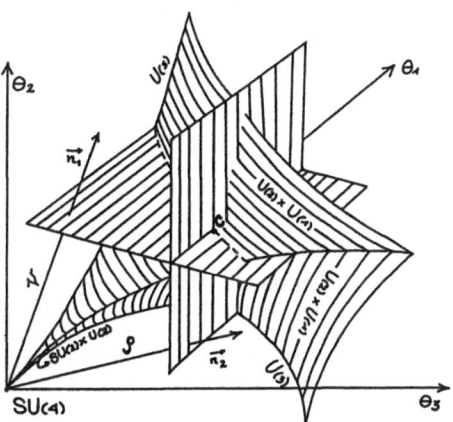

Fig. 26: U(2)xU(1) isotropy subgroup solutions of (4.11) on the two-dimensional stratum of SU(4)

Fig. 27: U(1)xU(1)xU(1) isotropy subgroup solutions on the generic stratum of SU(4)

the two-dimensional stratum Σ_s of SU(4) and in Fig. 26 a secondary bifurcation branch appears between A and B causing a symmetry exchange, viz., U(3)→U(2)xU(1) → → S(U(2)xU(2)). From Fig. 27 one can infer solutions with the generic subgroup by varying the second nonmodal bifurcation parameter ρ from zero to positive values. Then one arrives at an additional (tertiary) bifurcation point C leading to branches in the 3-dimensional stratum Σ_g. On Σ_g, x, x_yx, and $x_v x_v$x are linearly independent and solutions of F=0 are given by the intersection line of the two planes $\theta_1+\alpha\theta_2+\beta\theta_3-\gamma=0$ and $\theta_1+\delta\theta_2+\epsilon\theta_3-\rho=0$. The lattice of isotropy subgroups is given by Fig. 28. One observes (cf. Fig. 20) that the topology of the strata does not allow direct bifurcations U(3)→U(1)xU(1)xU(1).

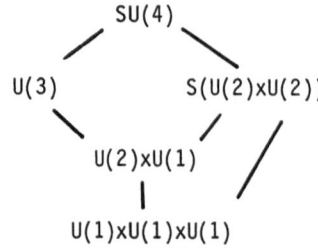

Fig. 28: Lattice of isotropy subgroups of SU(4)

6. Bifurcation Problems in High Energy Physics

We comment here on the role bifurcations may play in high energy physics. Interacting elementary particles are described in terms of a renormalizable Lagrangian L with terms up to fourth order in the scalar fields $\Phi=\{\Phi_i\}$. If L is invariant under a group Γ, the most general Higgs potential is

$$V(\Phi) = (\mu^2 a_1/2)\theta_1(\Phi) + (a_2/3)\theta_2(\Phi) + (a_3/4)\theta_3(\Phi) + (a_4/4)\theta_1^2(\Phi) \qquad (4.13)$$

where the $\theta_i(\Phi)$ are invariants under some irreducible representation of Γ [29]. Equivariant bifurcation equations can be obtained from the gradient of V,

$$G(\Phi) = (a_4\theta_1+a_1\mu^2)\Phi + a_2 G_2(\Phi) + a_3 G_3(\Phi) \qquad (4.14)$$

with equivariant functions $G_n(\Phi)=(n+1)^{-1}d\Phi_n/d\Phi$. It was conjectured [29] that only those nontrivial solutions of G=0 are stable which have maximal isotropy subgroup if the representation is irreducible. Consequently, symmetry breaking occurs only to a maximal isotropy subgroup. In the present case, codimension-0 bifurcation problems and maximal isotropy subgroup solutions of G=0 are related and one may be able to prove [26] the following conjecture. Suppose that Γ acts absolutely irreducibly on R^n and G is a Γ-equivariant bifurcation problem with topological codimension zero (the topological codimension is the codimension minus the number of modal parameters). Then there are only solution branches of G=0 with maximal isotropy subgroup. If, however, the codimension of G is different from zero (as in the examples of Sec. IV.2,3) then solutions with nonmaximal isotropy subgroups exist if the normal forms into which G can be transformed are unfolded.

The physics leading to bifurcations with nonzero codimension relates to energies near the Planck mass because in this case gravitational effects become so essential that they can no longer be regarded as a background field. Then the question arises to which extent such effects also influence the physics below Planck mass. Gravitational effects make themselves felt at lower energies as higher order terms $\sum_{k\geq 4} a_k \Phi^k/\kappa^{k-4}$. Such terms are scaled by inverse powers of the Planck mass κ and are nonrenormalizable but can be disregarded at low energies. Let us assume that $\tilde{V}$ is invariant under, say, the adjoint representation of Γ=SU(n). Let us further suppose that the symmetric solution loses stability at a critical value $\lambda=0$ of a bifurcation parameter λ. Then the symmetry breaking patterns may, near $\Phi=0$, $\lambda=0$ be described by the equivariant bifurcation problem $G(\Phi,\lambda)=0$. In the case of $SU(n)$, G has the form (4.1) and is the gradient of the potential

$$V(\Phi) = \lambda\mu^2\Theta_1(\Phi) + a_2\mu\Theta_2(\Phi) + a_3\Theta_3(\Phi) + \sum_{k=4}^{n} a_k\Theta_k/\kappa^{k-3} \qquad (4.15)$$

where $\Theta_k=\mathrm{Tr}\Phi^{k+1}$, $\dim\Phi=1$, $\dim\mu=1$ and $\dim\kappa=1$, $\kappa=10^{19}$ GeV. Rescaling Φ, $\Phi/\kappa=\phi$, $\tilde{V}$ takes the form

$$\tilde{V}(\phi) = \kappa^4\{\tilde{\lambda}\Theta_1(\phi) + a_2\mu/\kappa\Theta_2(\phi) + a_3\Theta_3(\phi) + \sum_{k=4}^{n} a_k\Theta_k(\phi)\} \qquad (4.16)$$

Here the a_k are dimensionless coupling constants, ϕ is a scalar field, μ is a mass parameter which determines the energy scale at which the $SU(n)$-symmetric solution loses stability and $\tilde{\lambda}=\lambda/\kappa^2$ is a bifurcation parameter to be specified in the context of a specific model (cf., below). The symmetry breaking bifurcations that can be expected are easily visualized if $\Gamma=SU(3)$ and Φ is in the adjoint representation. Without imposing any further restrictions on $\tilde{V}$ one obtains from IV.2 a bifurcation problem with topological codimension one, viz., $F=(\Theta_1-\lambda)x+(\Theta_1+\alpha\Theta_2-\beta)x_vx=0$ where $\beta=a_2\mu/\kappa$ and $\mu\ll\kappa$. If one requires the a_k not to be fine-tuned it follows from Fig. 24 that a secondary bifurcation occurs for $\lambda=\lambda_s$, i.e., for $\Theta_1=\Theta_s\approx\beta$, and at this point the vacuum expectation value is $<\Phi>\approx(\mu\kappa)^{1/2}$.

Physically appealing high energy models are provided by a local extension of supersymmetry (SUSY). While global SUSY avoids some of the deficiencies of grand unified theories (GUTs) [30], local SUSY accounts in a natural way for gravity. In the simplest case, an $N=1$ local SUSY (i.e., supergravity) might be the first step towards a unification of all interactions [31]. $N=1$ supergravity should be interpreted as an effective theory for energies below Planck mass, in analogy to chiral $SU(n)\times SU(n)$ interactions describing hadronic physics below 1 GeV [32]. Therefore $N=1$ supergravity should give rise to nongeneric, i.e., codimension different from zero bifurcation problems considered above. In the following we explain how this comes about. In an $N=1$ supergravity theory all information is condensed into a superpotential [33] $f(\Phi)$ where Φ is a chiral superfield. From this superpotential one derives an effective potential

$$V_{eff} = \exp(\Sigma\Phi_i\Phi_i^*/\kappa^2) \cdot \{\Sigma_i \left|\frac{\partial f(\Phi)}{\partial\Phi_i} + \frac{\Phi_i^*f(\Phi)}{\kappa^2}\right|^2 - \frac{3|f(\Phi)|^2}{\kappa^2}\} + \text{D-terms} \qquad (4.17)$$

Assuming Φ real and acting under a representation of a symmetry group Γ, Equ. (4.17) takes the form of (4.15) as a simplified version of V_{eff}. Since $N=1$ supergravity must be embedded into a more fundamental theory, one can consider V_{eff} to provide an organising center for the bifurcation problem which gives additional information if one analyzes higher-degenerate systems. An example of a topological codimension-2 problem is given by (4.11). This bifurcation problem arises if there is no quartic scalar interaction term present in (4.15). The non-unfolded bifurcation problem (4.10) is structurally unstable, so that certain of its predictions (e.g., spontaneous symmetry breaking patterns) cannot be observed. However, the unfolded problem (4.11) is structurally stable and thus yields the secondary and tertiary bifurcations described in IV.5 leading to symmetry breaking to the generic subgroup. In case of $SU(n)$, the parameters belong to interaction terms of the form $\Theta_k(\Phi)$, $k\leq n-1$. The condition that some of the a_k are zero implies a high topological codimension. One expects that for a sequence of symmetry breaking patterns the topological codimension and therefore the complexity of the problem increases with that of the lattice of isotropy subgroups.

Conditions on the a_k may be due to cosmological constraints which arise naturally in an inflationary universe [34] when one takes into account given bounds on monopole production, density fluctuations and so forth. Inflation is produced, e.g., by a first order phase transition, i.e., by hysteresis (cf., Figs. 23 and 25). Cosmological models also lead to a natural interpretation of the distinguished bifurcation parameter λ. In particular, in an inflationary universe phase transitions are due to a change of temperature T and λ is given by $\lambda=(a\mu^2-bT^2)$ where a and b are dimensionless constants.

We conclude that bifurcation theory provides us with a description of spontaneous symmetry breaking which allows to disregard peculiar interaction details. Namely, while the nongeneric bifurcation problem is structurally unstable, and therefore unobservable, symmetry preserving unfoldings ensure structural stability and induce spontaneous symmetry breaking to ever smaller isotropy subgroups without the need for introducing additional scalar fields.

Acknowledgments

It is a pleasure to acknowledge helpful discussions with D. Armbruster, G. Dangelmayr, D. Lang, E. Knobloch and M. Neveling, and the support of the Stiftung Volkswagenwerk.

References

[1] G. Nicolis and I. Prigogine, "Selforganization in nonequilibrium systems", Wiley (1977).
[2] H. Haken, "Synergetics", Springer (1982); "Advanced Synergetics", Springer (1983).
[3] R. Thom, "Structural Stability and Morphogenesis", Benjamin (1975).
[4] M. Golubitsky and D. Schaeffer, Commun. Pure Appl. Math. 32, 21 (1979).
[5] M. Golubitsky and D. Schaeffer, Commun. Math. Phys. 67, 205 (1979).
[6] D.H. Sattinger, Bull. Am. Math. Soc. 3 (2), 779 (1980) and "Branching in the Presence of Symmetry", CBMS-NSF Conference Notes 40, SIAM, Philadelphia 1983.
[7] M. Golubitsky and W.F. Langford, J. Diff. Eqns. 41, 375 (1981).
[8] D. Armbruster, G. Dangelmayr and W. Güttinger, Physica 16 D, 99 (1985)
[9] D. Armbruster, Z. Phys. B 53, 157 (1983).
[10] M. Golubitsky and B.L. Keyfitz, SIAM J. Math. Anal. 11, 316 (1980).
[11] I. Labouriau, "Applications of singularity theory to Neurobiology", Thesis, University of Warwick (1983).
[12] M. Golubitsky and I. Stewart, "Symmetry and Stability in Taylor-Couette Flow", preprint, University of Warwick (1984)
[13] J. Guckenheimer, SIAM J. Math. Anal. 15, 1 (1984).
[14] J.S. Langer, Rev. Mod. Phys. 52, 1 (1980).
[15] W.W. Mullins and R.F. Sekerka, J. Appl. Phys. 34, 323 (1963); 35, 444 (1964).
[16] D.J. Wollkind and L.A. Segel, Phil. Trans. R. Soc. (Lond.) 268, 351 (1970).
[17] J.S. Langer and L.A. Turski, Acta Metall. 25, 1113 (1977).
[18] M. Kerszberg, Phys. Rev. B 27, 6796 (1983).
[19] L.H. Ungar and R.A. Brown, Phys. Rev. B 29, 1367 (1984).
[20] P. Haug, Thesis (1985).
[21] G. Dangelmayr and P. Haug, in preparation (1985).
[22] J. Bernstein, Rev. Mod. Phys. 46, No. 1, (1974).
[23] P. Becher, M. Böhm, H. Joos, "Eichtheorien der starken und elektro-schwachen Wechselwirkung", Teubner (1981).
[24] H. Georgi, S.L. Glashow, Phys. Rev. Letters 32, 438 (1974).
[25] L. Michel and L.A. Radicati, Ann. Inst. Henri Poincaré, Vol. XVIII, No. 3, 185 (1973).
[26] M. Golubitsky and D. Schaeffer, "The Benard problem, symmetry, and the lattice of isotropy subgroups", in C.P. Borter et al. (eds.), Bifurcation Theory, Mechanics and Physics, Reidel (1983)
[27] C. Geiger, Thesis (1985)
[28] M. Abud and G. Sartori, Phys. Lett. B 104, 447 (1981); see also [25]
[29] L. Michel, CERN-preprint TH 2716 (1979)
[30] S. Ferrara, J. Iliopoulos and B. Zumino, Nucl. Phys. B 159, 420 (1979).
[31] J. Ellis, D.V. Nanopoulos and K.A. Tamvakis, Phys. Lett. 121 B, 123 (1983).

[32] D.V. Nanopoulos, "The Physics of Supersymmetry and Supergravity", in C. Kounnas et al. (eds.), Grand Unification with and without Supersymmetry and Cosmological Implications, World Scientific (1984).

[33] E. Cremmer, B. Julia, J. Scherk, P. Van Nieuwenhuizen, S. Ferrara and L. Girardello, Phys. Lett. 79 B, 23 (1978) and Nucl. Phys. B 147, 105 (1979).

[34] A. Guth and F.J. Weinberg, Phys. Rev. D 23, No. 4, 876 (1981);
A.D. Linde, Phys. Lett. 108 B, 389 (1982).

Pattern Formation and Transients in the Convection Instability

M. Bestehorn and H. Haken

Institut für Theoretische Physik I, Universität Stuttgart, Pfaffenwaldring 57/IV, D-7000 Stuttgart 80, Fed. Rep. of Germany

In this contribution and the following one by Friedrich and Haken we want to show how methods of synergetics, in particular the slaving principle and the order parameter concept, allow us to calculate patterns, including transients, in fluids. The slaving principle enables us to reduce the full equations of fluid dynamics to "generalized Ginzburg Landau equations" for order parameters. These equations can be solved by a mediate size computer in a reasonable time. While in this paper we shall be concerned with regular patterns, the following one will also include chaotic motion.

The formation of convection patterns in fluid layers near threshold has been subject to recent experimental and theoretical studies. (For reviews and a list of references see e.g. [1,2]). A number of experiments [3-6] have shown that the amplitude equation, first derived by Newell and Whitehead [7] from the so-called Boussinesq approximation, is not sufficient to describe all observed stable patterns.
A two-dimensional equation for the horizontal spatial dependence of the temperature field of the fluid was derived by Swift and Hohenberg [8] and solved numerically under rigid horizontal boundary conditions by Greenside et. al.[9]. This equation is invariant against reflection at the vertical symmetry plane of the fluid,and can therefore only explain the formation of rolls.

Our model, which is based on generalized Ginzburg Landau equations [2,10], includes also nonlinear terms resulting from non-boussinesqian effects,or stress effects on a free upper surface (Marangoni-instability). These terms violate the inversion symmetry and give rise to hexagonal structures.

The Model Equation

The generalized Ginzburg Landau equation near threshold derived from the Navier-Stokes-, energy-, and continuity-equations reads in a concise form [10]:

$$\dot{\xi}_{\underline{k}_j}(\underline{x},t) = \varepsilon\, \xi_{\underline{k}_j}(\underline{x},t) - [\Delta - 2i(\underline{k}_j,\underline{\nabla})]^2\, \xi_{\underline{k}_j}(\underline{x},t) + \delta\, \xi_{\underline{k}_n}(\underline{x},t)\, \xi_{\underline{k}_m}(\underline{x},t)$$

$$- \sum_{p=1}^{\infty} c_{j,p}\, |\xi_{\underline{k}_p}(\underline{x},t)|^2\, \xi_{\underline{k}_j}(\underline{x},t)\,, \qquad \underline{k}_j + \underline{k}_n + \underline{k}_m = 0\,, \qquad (1)$$

$$|\underline{k}_j| = |\underline{k}_c|\,.$$

Together with the boundary conditions (B.C.): $\quad \xi_j = \dfrac{\partial \xi_j}{\partial \underline{n}} = 0\,,$ (2)
(n perpendicular to the sidewalls.)

where $\quad \underline{x} = (x_1, x_2), \quad \underline{k} = (k_1, k_2), \quad \varepsilon = (R-R_c)/3R_c\,.$

Δ and $\underline{\nabla}$ are the two-dimensional gradient and the Laplacian and δ is proportional to non-boussinesqian effects or surface effects. The $c_{j,p}$ in Eq (1) can be calculated from the Navier-Stokes equations. (for more details see [2].)

From the slowly varying envelope functions ξ_{k_j} we can construct a time-depending function Ψ which includes the whole horizontal spatial dependence of the fluid pattern :

$$\Psi(\underline{x},t) = \sum_{j=1}^{\infty} \xi_{k_j}(\underline{x},t) \exp(-i\underline{k}_j\underline{x}) , \qquad (3)$$

$$\xi_{\underline{k}} = \xi_{-\underline{k}}^{*} .$$

Ψ serves as order parameter and is related to velocity $\underline{v}$ and temperature Θ of the fluid:

$$v_j = f(x_3) \frac{\partial \Psi}{\partial x_i}, \quad j = 1,2 ,$$

$$v_3 = g(x_3) \Delta\Psi, \qquad f(x_3) = -\frac{dg(x_3)}{dx_3} , \qquad (4)$$

$$\Theta = h(x_3) \Psi .$$

The functions f, g, and h depend only on the vertical B.C. [11].

Numerical Method

To solve Eq (1) on a computer (VAX 11/750) we have to truncate the sum in (3) at j=N. N depends on the accuracy of the numerical procedure and of the geometry of the layer. In our simulation N was between 12 and 24. Under this assumption Eq (1) leads to a system of N partial differential equations for the envelope functions ξ. To solve these equations, we apply a forward integration scheme in time and a finite difference method in space.
The weak spatial dependence of ξ allows us to choose the step size of x_i to the same order as $|\underline{k}_c|$. Consequently, the step size of the forward integration in time may also be fixed to O(1).

Numerical Results

We examined the time development for several parameter values under two different initial conditions (random dot, fig.1, and parallel rolls, fig.2).

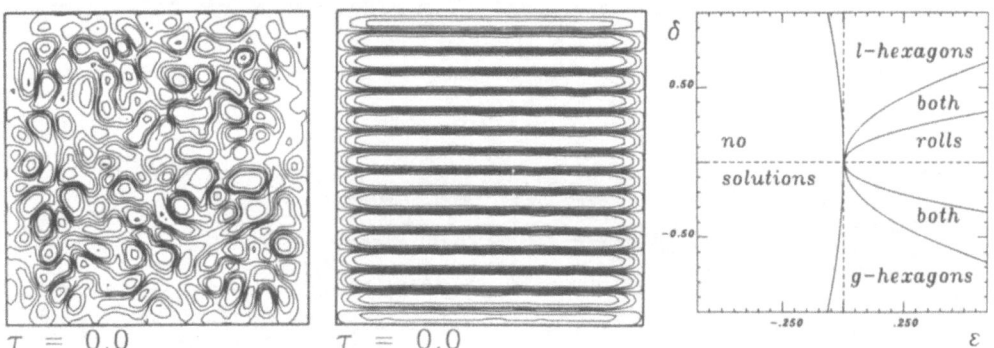

$\tau = 0.0$ $\tau = 0.0$

Fig.1   Fig.2   Fig.3   Phase diagram
Contour plots of the initial conditions used for fig. 4 - 15.   in $\varepsilon - \delta$ - plane

Linear stability analysis leads to a phase diagram as shown in figure 3. Possible stable patterns are rolls, l-hexagons (fluid rises in the center of the hexagon), and g-hexagons (fluid descends in the center).

301

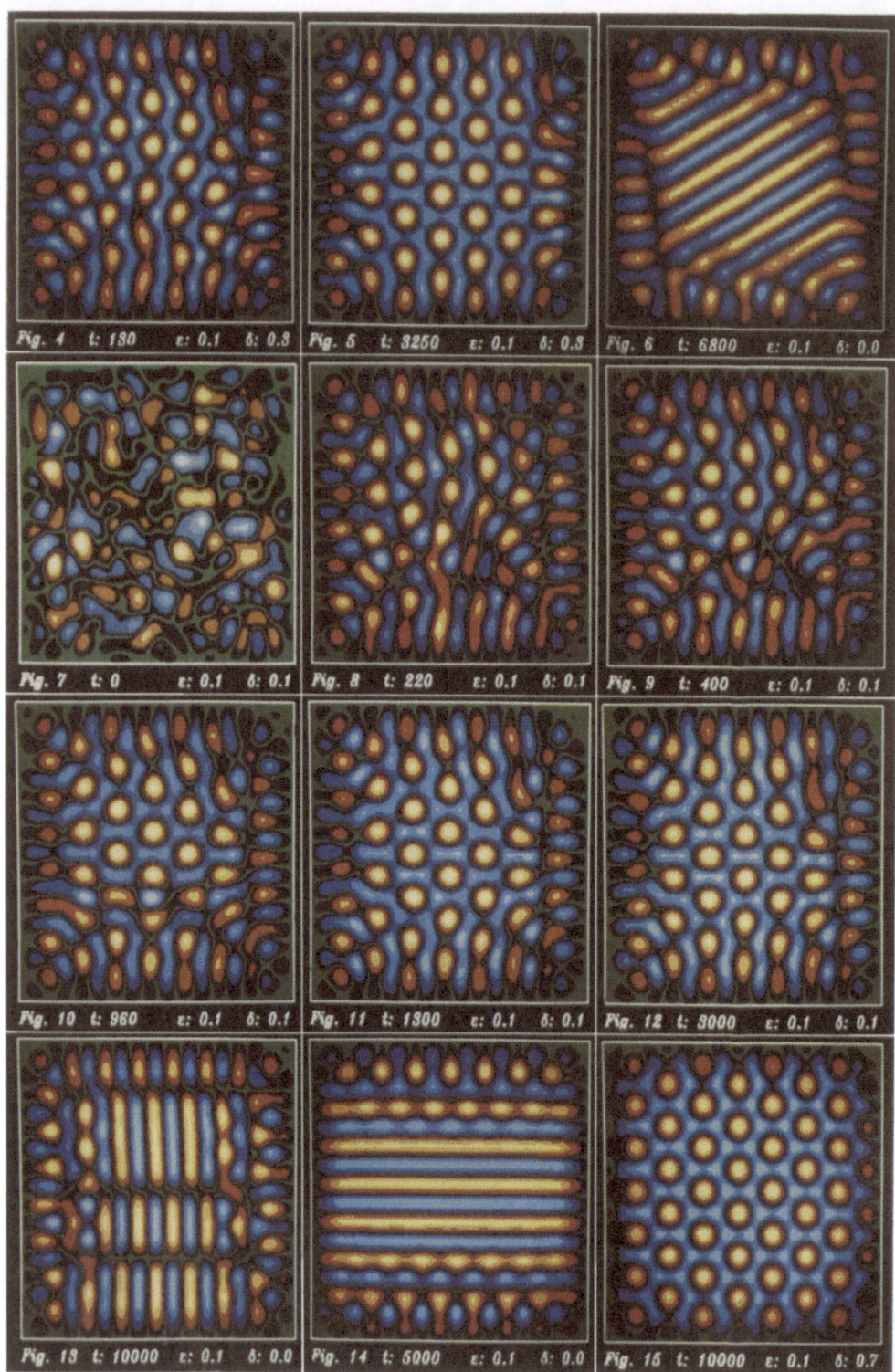

Fig. 4 t: 130   ε: 0.1   δ: 0.3   *Fig. 5 t: 3250   ε: 0.1   δ: 0.3*   *Fig. 6 t: 6800   ε: 0.1   δ: 0.0*

Fig. 7 t: 0   ε: 0.1   δ: 0.1   *Fig. 8 t: 220   ε: 0.1   δ: 0.1*   *Fig. 9 t: 400   ε: 0.1   δ: 0.1*

Fig. 10 t: 960   ε: 0.1   δ: 0.1   *Fig. 11 t: 1300   ε: 0.1   δ: 0.1*   *Fig. 12 t: 3000   ε: 0.1   δ: 0.1*

Fig. 13 t: 10000   ε: 0.1   δ: 0.0   *Fig. 14 t: 5000   ε: 0.1   δ: 0.0*   *Fig. 15 t: 10000   ε: 0.1   δ: 0.7*

Figs.4-15. Caption see opposite page

Figures 4 - 5 show the evolution from a random dot pattern to a hexagonal pattern. After changing δ to zero, hexagons become unstable and rolls influenced by the sidewalls are formed again (fig.6).

The same initial condition but with other parameters leads to an evolution represented in figures 7 - 12. We are now in a parameter region where both patterns can be stable. Hexagons are formed after a relatively long time in the center of the layer. Near the sidewalls, the rolls are the preferred pattern.

Figure 13 shows the steady state obtained again from random dots as initial condition but with δ=0. The resulting pattern shows dislocations which seem to be stable over very long times.

In figure 14, we used the same parameters as in 13 but the initial condition was now a parallel roll pattern (fig.2). The rolls on the sidewalls match the B.C..

In figure 15 the parameters were strongly in the hexagonal region of the phase diagram and a very regular hexagonal pattern was formed. (Initial condition was again random dot.)

References

1. Normand,C., Pomeau,Y., Velarde,M.,: Rev.Mod.Phys. **49** 581 (1977)

2. Haken,H.,: Synergetics, An Introduction. Berlin, Heidelberg, New York, Springer (1978) (3.Edn, 1983)

3. Bergé,P.: in "Chaos and Order in Nature", ed. H.Haken, Berlin, Heidelberg, New York (1981)

4. Koschmieder,E.L.: Adv.Chem.Phys. **26**, 177 (1974)

5. Heutmaker,M.S., Fraenkel,P.N., Gollub,J.P.: Phys.Rev.Lett. **54** 1369 (1985)

6. Ahlers,G., Cannell,D.,S., Steinberg,V.: Phys.Rev.Lett. **54** 1373 (1985)

7. Newell,A.C., Whitehead,J.A.: J. Fluid Mech. **38** 279 (1969)

8. Swift,J., Hohenberg,P.C.: Phys.Rev. **A15** 319 (1977)

9. Greenside,H.S., Coughran,Jr.,W.M.: Phys.Rev. **A30** 398 (1984)

10. Haken,H.,: Advanced Synergetics. Berlin, Heidelberg, New York, Springer(1983)

11. Chandrasekhar,S.: Hydrodynamics and Hydromagnetic Stability. Oxford: Oxford University Press (1961)

Time developments for different parameter values and initial conditions (Compare text). Same colour means same value for the order parameters Ψ. The blue colours correspond to negativ sign of Ψ, the red colours to positive.

Convection in Spherical Geometries

R. Friedrich and H. Haken

Institut für Theoretische Physik I, Universität Stuttgart, Pfaffenwaldring 57/IV,
D-7000 Stuttgart 80, Fed. Rep. of Germany

1. Introduction

Under well-defined conditions systems with many degrees of freedom like lasers , plasmas and fluids show irregular behaviour in their time evolution . These chaotic motions are usually produced by few degrees of freedom only and the basic question arises how to determine these relevant variables and to describe their dynamics . It turns out that the occurrence of low-dimensional chaos in complex systems can qualitatively be explained by using the concepts of synergetics [1] , [2] . The degrees of freedom are drastically reduced due to the slaving principle , i. e. the temporal behaviour of the system is governed by the dynamics of only few order parameters . We applied these ideas to a specific hydrodynamic system . This system especially has the advantage that all steps and conclusions of the above principle can be accomplished explicitly .

2. The model

We consider a fluid contained between two concentric spherical boundaries in a spherical symmetric gravity field . Heat sources on the inner boundary and in the fluid maintain a temperature difference between inner and outer sphere and produce an unstable density distribution, in such a way that above a certain threshold convective motion sets in [4] , [5] .

In the Boussinesq approximation the velocity field $\underline{v}(\underline{r},t)$ obeys the Navier Stokes Equation

$$P^{-1}(\tfrac{d}{dt}\underline{v}(\underline{r},t)+(\underline{v}(\underline{r},t)\cdot \text{grad}) \ \underline{v}(\underline{r},t))=\Delta\underline{v}(\underline{r},t)+R\gamma(r)\underline{r}T(\underline{r},t)-\text{grad } p(\underline{r},t) \qquad (1)$$

and the incompressibility condition

$$\text{div } \underline{v}(\underline{r},t)=0 \ . \qquad (2)$$

The contribution $T(\underline{r},t)$ to the temperature field due to convective fluid motion is determined by the equation of heat conduction:

$$\tfrac{d}{dt}T(\underline{r},t)+(\underline{v}(\underline{r},t)\cdot \text{grad}) \ T(\underline{r},t)=\Delta T(\underline{r},t)+ \tau(r)\underline{v}(\underline{r},t)\cdot\underline{r} \ . \qquad (3)$$

P denotes the Prandtl number , R the Rayleigh number . The function $\gamma(r)$ specifies the gravity field and $\tau(r)$ the distribution of the heat sources inside the fluid [5] .

This model has been discussed in connection with the convection in the earth mantle [4] . It can also be applied to study some aspects of planetary atmospheric motions . Furthermore it can serve , in its own right , to investigate pattern formation as well as selection , instabilities and the occurrence of low-dimensional chaotic motions in a hydrodynamic system .

3. Linear stability analysis

The critical Rayleigh number and the unstable modes for the onset of convection are determined by linear stability analysis [4] . Due to the spherical symmetry of the model there are (2l+1) unstable modes with velocity and temperature fields

$$v(\underline{r},t)=\text{rot rot } \underline{r} \ S_1(r)Y_1^m(\theta,\phi)$$

$$T(\underline{r},t)=T_1(r)Y_1^m(\theta,\phi)$$

(4)

$Y_1^m(\theta,\phi)$ are the spherical harmonics . The functions $S_1(r)$, $T_1(r)$ have to be determined numerically . The actual value of the critical l depends on the aspect ratio ρ , the ratio of the inner sphere to the thickness of the shell .

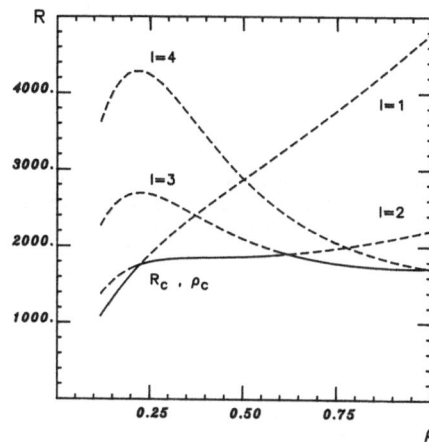

Fig. 1

Critical Rayleighnumbers for disturbances with l=1,2,3,4 as a function of the aspect ratio ρ

Fig. (1) shows the critical Rayleigh numbers for l=1,...,4 . With increasing aspect ratio the value l of the critical modes increases,and the corresponding convection patterns become more and more differentiated.

There are values of ρ where two groups of modes with different values of l simultaneously become unstable . In the neighbourhood of these special values of the aspect ratio one can expect that the nonlinear interactions between the unstable modes lead to interesting and complex dynamical behaviour .

We investigated the model for the case where instability sets in through the two groups of modes with l=1 and l=2 .

4. Generalized Ginzburg-Landau Equations

The application of methods developed in synergetics [1] , [2] , [3] shows that for parameter values in the neighbourhood of R_c , ρ_c (fig.1) the temperature field can be expressed by :

$$T(\underline{r},t)= \sum_{k=-1}^{1} \xi_k(t)T_1(r)Y_1^k(\theta,\phi)$$

$$+ \sum_{k=-2}^{2} \eta_k(t)T_2(r)Y_2^k(\theta,\phi) + \tau(\underline{\xi},\underline{\eta},\underline{r})$$

(5)

where $\tau(\underline{\xi},\underline{\eta},\underline{r})$ is small compared to the first two sums and does not depend explicitly on time . A similar expression holds for the velocity field $\underline{v}(\underline{r},t)$. The time dependence of $T(\underline{r},t)$, $\underline{v}(\underline{r},t)$ is therefore determined by the dynamics of the eight order parameters $\underline{\xi}(t)$, $\underline{\eta}(t)$. The equations of motions for $\underline{\xi}(t),\underline{\eta}(t)$ are Generalized Ginzburg-Landau Equations . They have the following qualitative form :

305

$$d_t \xi_k = \lambda^1 \xi_k + \alpha (\underline{\xi}:\underline{n})_k + a\, \xi_k E_1 + b\, \xi_k E_2 + c\, (\underline{\xi}:\underline{n}:\underline{n})_k$$
$$d_t n_k = \lambda^2 n_k + \beta (\underline{\xi}:\underline{\xi})_k + \gamma (\underline{n}:\underline{n})_k + d\, n_k E_2 + e\, n_k E_1 + f\, (\underline{\xi}:\underline{\xi}:\underline{n})_k$$
$$E_1 = \xi_0^2 - 2\, \xi_1 \xi_{-1}$$
$$E_2 = n_0^2 - 2\, n_1 n_{-1} + 2\, n_2 n_{-2}$$

(6)

E_1 and E_2 are proportional to the total kinetic energies of the convective motions of the modes with $l=1$ and $l=2$. $(\underline{\xi}:\underline{n})$, ... denote special polynomial expressions and λ' are the eigenvalues corresponding to the unstable modes . The coefficients α , β , γ , a , b ... have been determined numerically .

In this way the set of partial differential equations (1) , (2) , (3) is reduced to a set of ordinary differential equations . This enables us to obtain a detailed picture of the convective fluid motion in the neighbourhood of R_c and ρ_c .

5. The phase diagram

Fig. (2) shows an enlargement of the neighbourhood of the parameter values R_c , ρ_c . An increase of the Rayleigh number for $\rho < \rho_c$ first leads to a transition from the purely heat conductive state to an axisymmetric convective state with predominant contribution of the modes with $l=1$. A secondary instability results in a time periodic fluid motion in the form of a rotating wave : the temperature field undergoes a rigid rotation around a constant axis . The next kind of motion is also a rotating wave, but with a different convection pattern . A further increase then leads to a quasiperiodic flow : a time periodic motion is superposed on the rotating wave, and the kinetic energy of the fluid oscillates between the two groups of modes . This flow finally undergoes a transition which leads to an aperiodic time behaviour . For $\rho \geqslant \rho_c$ there is a transition to axisymmetric convection with $l=2$. This convection pattern may , for aspect ratios close to ρ_c , lose its stability also, and one again observes chaotic fluid motion .

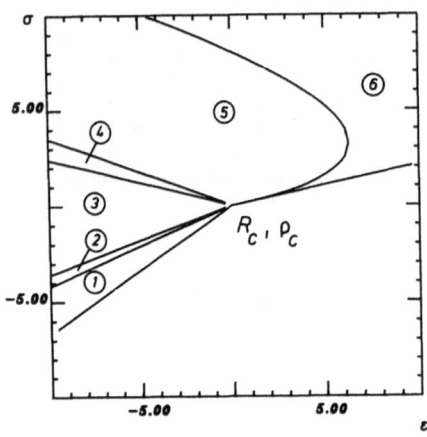

Fig. 2

Phase diagram :

(1) stationary convection ($l=1$) ,
(2) rotating wave ,
(3) rotating wave ,
(4) quasiperiodic convection ,
(5) chaotic convection,
(6) stationary convection ($l=2$) ;

$\sigma = 100(R-R_c)/R_c$, $\varepsilon = 100(\rho - \rho_c)/\rho_c$

Typical patterns of convection are represented in figs. (3) and (4) . They show the temperature field on a spherical surface in between the fluid shell . The red colour corresponds to a hot region or to upwelling fluid motion . The blue colour marks a cold area . The plates of fig. (3) show the patterns of a rotating wave . Patterns of chaotic convection are shown in fig. (4) : Plates 1 – 3 show the fluid in a state mainly determined by modes with $l=2$. A change of the pattern

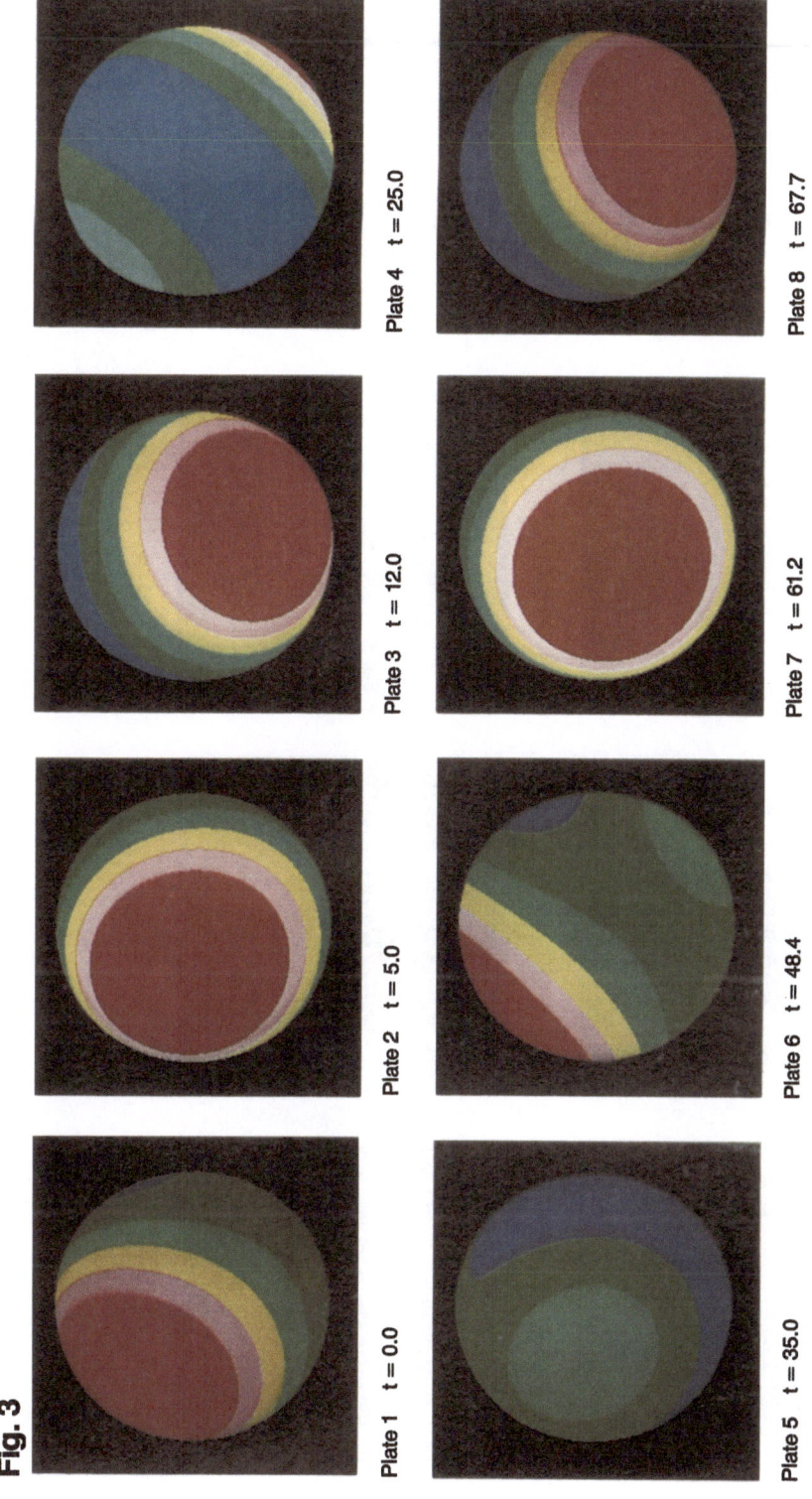

Fig. 3

Plate 1 t = 0.0

Plate 2 t = 5.0

Plate 3 t = 12.0

Plate 4 t = 25.0

Plate 5 t = 35.0

Plate 6 t = 48.4

Plate 7 t = 61.2

Plate 8 t = 67.7

Fig. 3 , plates 1 – 8   Patterns of convection in the case of a rotating wave

Fig. 4

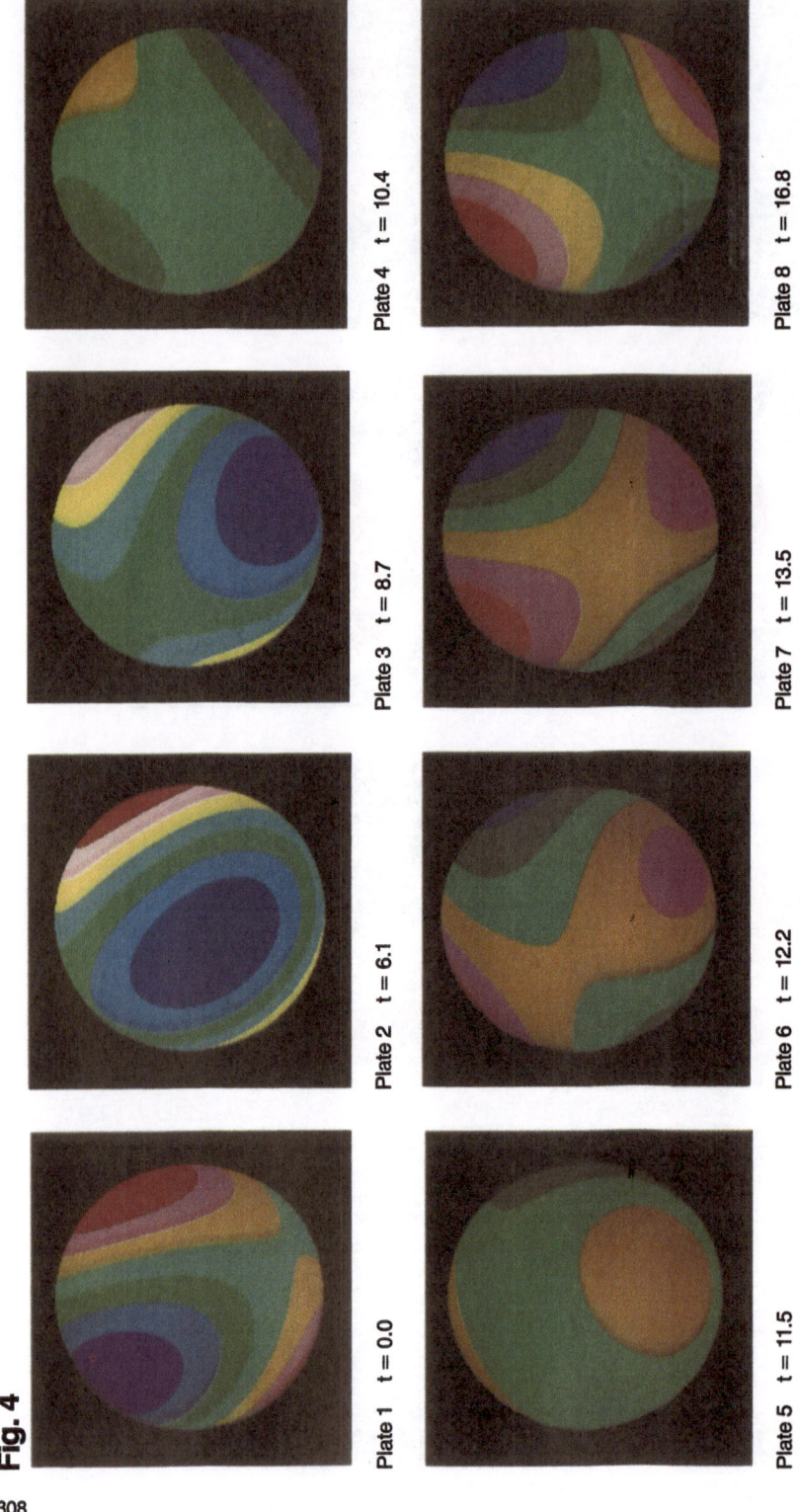

Plate 1   t = 0.0

Plate 2   t = 6.1

Plate 3   t = 8.7

Plate 4   t = 10.4

Plate 5   t = 11.5

Plate 6   t = 12.2

Plate 7   t = 13.5

Plate 8   t = 16.8

Plate 9   t = 23.6

Plate 10   t = 28.7

Plate 11   t = 41.5

Plate 12   t = 91.7

Plate 13   t = 95.2

Plate 14   t = 97.5

Plate 15   t = 98.5

Plate 16   t = 99.9

Fig. 4 , plates 1 - 16   Patterns of convection in the case of chaotic fluid motion

occurs due to the influence of modes with l=1 (plates 4 - 8) . This change leads nearly to the same pattern as shown in plate 1 but with a different orientation . A longer period with a pulsation of this pattern follows (plates 9 - 11) until the next change sets in (plates 12 - 16) .

References

[1] H. Haken : Synergetics . An Introduction ,
         Springer-Verlag , Berlin , Heidelberg , New York (1978)
[2] H. Haken : Advanced Synergetics ,
         Springer-Verlag , Berlin , Heidelberg , New York (1983)
[3] A. Wunderlin , H. Haken : Z . Physik . B - Condensed Matter
         44 , 135-141 ,(1981)
[4] S. Chandrasekhar : Hydrodynamics and Hydrodynamic Stability ,
         Clarendon , Oxford (1961)
[5] F.H. Busse : J . Fluid Mech. 72 , 65-85 , (1975)
    F.H. Busse , N. Riahi : J . Fluid Mech. 123 , 283-291 , (1981)

How Does Low-Dimensional Chaos Arise
in Complex Systems with Infinite Degrees of Freedom

E. Meron and I. Procaccia

Department of Chemical Physics, The Weizmann Institute of Science,
Rehovot 76100, Israel, and
The James Franck Institute and the Department of Chemistry,
University of Chicago, Chicago, IL 60637, USA

I. Introduction

The prediction that chaos should arise as a low dimensional phenomenon, even in systems that are described by partial differential equations, goes back to the seminal paper of Ruelle and Takens from 1971 [1]. However, most physicists became comfortable with low dimensional models only after the success of Feigenbaum in providing the theoretical framework for period doubling [2], a framework that was based on low dimensional iterative maps, but that applied to experiments on hydrodynamic systems [3]. In the last 2-3 years we have seen a growing number of experiments that proved directly that chaos sets in with low dimensional attractors [4]. These experiments measured dimensions of attractors, Kolmogorov entropies and Lyapunov exponents, and established that the description of the onset of chaos and its development can be modelled with low dimensional ordinary differential equations (or one-time maps). What has been lacking however is an example where both theory and experiment exist in parallel. In no case to date there has been (to our knowledge) a theory which described how the reduction from the infinite to the finite occurs in practice for an example on which detailed experiments have been conducted. The aim of this paper is to report on such a theory.

The experiment that seemed ideal for analysis has been conducted by Ciliberto and Gollub and dealt with chaos in surface waves [5,6]. The advantage of this experiment is that there is a point in parameter space where the transition to chaos occurs essentially straight from the quiescent state (see section II for details). This is in contrast to all previous experiments where the chaotic motion set in after a series of bifurcations that resulted in a space and time dependent state before the onset of chaos. The theoretical analysis called therefore for working around a complicated state.

In section II we review briefly the experiment. Section III sets the pde's description of the problem. In section IV we describe the center manifold theory which allows the reduction of the mathematical description to low dimensional ode's. In section V we discuss briefly the normal form theory that allows for reaching the minimal (simplest) nonlinear description of the dynamics. Section VI summarizes some of the numerical results and offers a brief comparison of theory and experiment. The paper is sketchy in terms of mathematical details; the aim is to review the ideas that allow the reduction from the infinite to the finite and the reason for the appearance of normal forms (which are the basis of universality) in the context of the transition to chaos.

II. The experiment

The experimental set up consists basically of a plexiglass cylinder containing water of depth 1 cm, mounted on a cone of a loudspeaker oscillating accurately in the vertical direction [5,6]. When the amplitude of the oscillations exceed some frequency dependent threshold value, surface wave patterns appear on the free surface of the water. The patterns were studied by refraction. The basic modes that span the surface deformation are $J_\ell(k_{\ell,m}r) \begin{smallmatrix} \cos\ell\theta \\ \sin\ell\theta \end{smallmatrix}$ where r is the radial coordinate, the azymuthal coordinate, J_ℓ are the Bessel functions of order ℓ and the allowed wavenumbers $k_{\ell,m}$ are determined by the boundary condi-

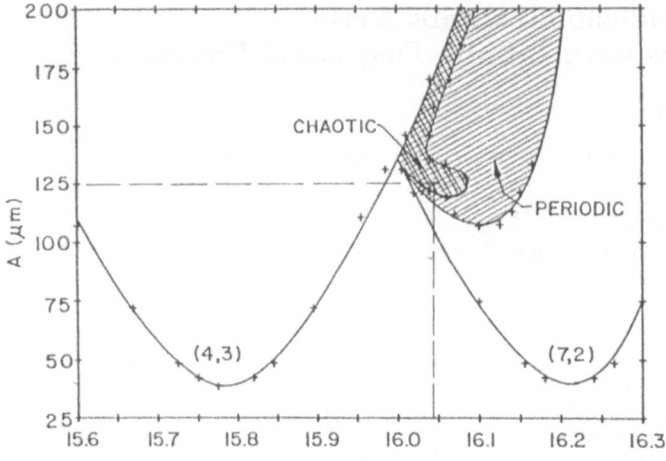

Figure 1.
Experimental phase diagram. The regions denoted (4,3) and (7,2)
display stable patterns. In the shaded regions one sees slow periodic
and chaotic motions.

tions. The modes can be denoted by the double index ℓ, m. A portion of the
experimental phase diagram is shown in Fig. 1. The crosses are the experimentally determined points on the stability boundaries of the (4,3) and (7,2) modes. In the regions labelled (4,3) and (7,2) the system displays pure stable oscillations of these modes at half the driving frequency. In a stroboscopic
measurement synchronized at this frequency these signals appear stationary. In
the shaded regions periodic (in the stroboscopic measurement) and chaotic responses were observed.

Clearly, the most interesting point in this phase diagram is the point where
the neutral stability curves of the two modes meet. It appears there that the
system becomes chaotic straight from the quiescent state. It thus seems worth-
while to develop the theory around this point.

The experimental study of this system has been extensive [6]. Ample evidence
for the low dimensionality of the signals has been presented. The reader is referred to ref. 6 for full details. One additional result will be reproduced
here, however, since it seemed very perplexing at first. Figure 2 is the result
of a time resolved Fourier analysis in the region of competition between the =4
and $\ell = 7$ modes. Shown is the angular power spectrum $P(\ell)$, at two different

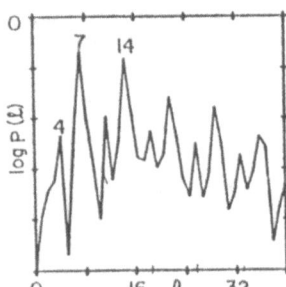
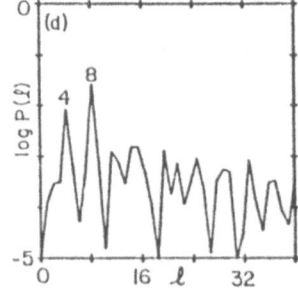

Figure 2. Time resolved Fourier spectra showing the ℓ modes found experimentally to be involved in the deformation pattern.

312

times. It is clear that in addition to the (4,3) and the (7,2) modes, many other modes with ℓ =3,8,11,14,18,21,25,28 etc. are present. Concerning the fact that the dimension of the strange attractor found here was smaller than 3, it is difficult at first sight to rationalize the active participation of so many modes. It will be seen that the theory resolves this riddle.

We are thus looking for a theory that besides giving a basis for the low dimensionality of the phenomenon and the appearance of the phase diagram, would provide an approximate solution of the space-time dependent deformation field such that we would be able to rationalize most of the details seen in the experiment. Naturally we begin with the hydrodynamic description.

III. Hydrodynamics

Consider a fluid layer of height h in a cylinder of radius R. Fix a Cartesian coordinate system such that xy is the horizontal plane and the free surface of the water is at z=0. Allow the coordinate system to move with the vessel. For simplicity we shall neglect (at this point) the viscosity of the fluid. The equations of motion for the velocity field $\vec{v}$ are then [7]

$$\frac{\partial \vec{v}}{\partial t} + (\vec{v} \cdot \vec{\nabla}) \vec{v} = - \frac{1}{\rho} \vec{\nabla} p + (g - \tilde{A}\cos\omega t)\hat{z} \tag{3.1}$$

where ρ, p, g, $\tilde{A}$ and ω are the density, pressure, gravitational acceleration, and the amplitude and frequency of the oscillation respectively. Since the velocities involved are all subsonic we can assume that the fluid is incompressible:

$$\vec{\nabla} \cdot \vec{v} = 0 \quad . \tag{3.2}$$

The advantage of neglecting the viscosity is that one can define [7] a velocity potential ϕ, $\vec{v}=\vec{\nabla}\phi$, or

$$\nabla^2 \phi = 0. \tag{3.3}$$

Integrating Eq. (3.1) we find [7]

$$\frac{\partial \phi}{\partial t} + \frac{1}{2} \vec{v} \cdot \vec{v} = - \frac{p}{\rho} + (g - \tilde{A}\cos\omega t)z \quad . \tag{3.4}$$

We are interested in the motion of the free surface, $z=\xi(x,y,t)$. Since the pressure at the free surface is

$$p = \gamma(\frac{1}{R_1} + \frac{1}{R_2}) \tag{3.5}$$

where R_1 and R_2 are the principal radii of curvature, we can estimate p (for small ξ) as

$$p = \gamma(\nabla^2 \xi - \frac{1}{2}\vec{\nabla} \cdot (\vec{\nabla} \xi |\vec{\nabla} \xi|^2) \tag{3.6}$$

where $\vec{\nabla}=\partial_x \hat{X}+\partial_y \hat{Y}$. Using Eq. (3.6) in Eq. (3.4) and expanding around z=0 to second order we arrive finally to the equation of motion

$$\frac{\partial \phi}{\partial t} = (g - \tilde{A}\cos\omega t)\xi - \frac{\gamma}{\rho} \nabla^2 \xi - N_2(\xi,\phi) \quad ; \quad z=0 \tag{3.7}$$

where

$$N_2(\xi,\phi) = \frac{1}{2} |\nabla\phi|^2 + \frac{1}{2} \xi \frac{\partial |\nabla\phi|}{\partial z} + \ldots \tag{3.8}$$

To close this equation we need an equation for which is provided by the kinematic surface condition D/Dt($\xi(x,y,t)-z)=0$. After expansion around z=0 this equation reads

$$\frac{\partial \xi}{\partial t} = \frac{\partial \phi}{\partial z} = N_1(\xi,\phi) \quad ; \quad z=0 \tag{3.9}$$

313

where

$$N_1(\xi,\phi) = \vec{v}\,\xi\cdot\vec{v}\,\phi + \xi v^2\phi + \frac{1}{2}\,\xi^2 v^2\,\frac{\partial\phi}{\partial z} + \xi\vec{v}\,\xi\cdot\vec{v}\,\frac{\partial\phi}{\partial z} \qquad (3.10)$$

The boundary conditions that we choose are $\partial\phi/\partial n=0$ on the walls, $\partial\phi/\partial z=0$ at $z=h$ and $\partial\xi/\partial n=0$ on the wall at the free surface. The last condition means that the contact angle is 90°. (For further discussion of these boundary conditions see section IV).

In dimensionless units the equations (3.7) and (3.9) read

$$\frac{\partial\xi}{\partial t} = \frac{\partial\phi}{\partial z} - N_1(\xi,\phi) \qquad (3.11a)$$

$$\frac{\partial\phi}{\partial t} = -\frac{1}{h\omega^2}\,[\frac{\gamma}{\rho h^2}\,v^2 - g + \tilde{A}\cos t]\xi - N_2(\xi,\phi) \ . \qquad (3.11b)$$

3.1 The Linear Problem

Some of the gross features of the phase diagram in Fig. 1 can be understood on the basis of the linearized problem. This is obtained by setting $N_1(\xi,\phi)$ and $N_2(\xi,\phi)$ to zero, and expanding ξ and ϕ in the eigenfunctions of v^2 (i.e. the functions $f_\ell \equiv J_\ell(k_{\ell,m}r)\,{\sin\theta \choose \cos\theta}$. We denote the amplitudes of these modes by ξ_ℓ and ϕ_ℓ with a single index. For example

$$\xi(x,y,t) = \sum_\ell \xi_\ell(t)f_\ell(x,y) \ . \qquad (3.12)$$

The linear problem then takes the form

$$\begin{pmatrix}\dot{\xi}_\ell \\ \dot{\phi}_\ell\end{pmatrix} = \begin{pmatrix}0 & -\chi_\ell \\ \Gamma_\ell + \Lambda\cos t & 0\end{pmatrix}\begin{pmatrix}\xi_\ell \\ \phi_\ell\end{pmatrix} \qquad (3.13)$$

where

$$\chi_\ell = hk_\ell\tanh(hk_\ell)\sim 1 \ , \qquad \Gamma_\ell = \frac{1}{h\omega^2}\,[\frac{\gamma k_\ell^2}{\rho} + g]\sim10^{-1}, \qquad \Lambda = -\frac{\tilde{A}}{h\omega^2}\sim 10^{-3} \ .$$

The numerical estimates pertain to the experimental conditions of Fig. 1.

The system (3.13) is a Floquet problem which is equivalent to the Mathieu equation

$$\ddot{\xi}_\ell + (\omega_\ell^2/\omega^2 + \chi_\ell\Lambda\cos t)\xi_\ell = 0 \qquad (3.14)$$

where $\omega_\ell = \omega\sqrt{\chi_\ell\Gamma_\ell}$ is the natural frequency of the ℓ'th mode. The solutions of such Floquet problems can be written in the general form

$$\psi_\ell = z_\ell(t)e^{L_\ell t} \ ; \ z_\ell(t) = z_\ell(t+2\pi) \ . \qquad (3.15)$$

By examining the real part of the eigenvalues of L_ℓ (the "Floquet exponents") we can delineate the stability of the spatial mode ℓ. In our case this can be done analytically since Λ is so small ($\sim 10^{-3}$). We can thus find the neutral stability curves by perturbation theory. After some algebra we find the neutral stability curves for 4π periodic solutions (i.e. half the driving frequency) to be (further details can be found in ref. 8)

$$\Gamma_\ell(\Lambda) = \frac{1}{4\chi_\ell} \pm \frac{1}{2}\Lambda + O(\Lambda^2) \ . \qquad (3.16)$$

To compare with Fig. 1 we have to return back from the dimensionless description. Rather than in equations we present in Fig. 3 a sketch of the neutral stability curves predicted by Eq. (3.16) in the range of parameters of Fig. 1. The modes (4,3) and (7,2) have stability curves precisely in the region of Fig. 1. However the mode (11,1) which is not seen experimentally, is found between them. The reason for this is the boundary conditions chosen in section III. In

314

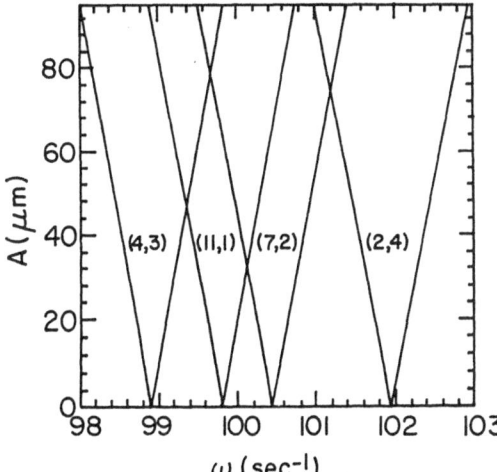

Figure 3. Theoretical stability boundaries as predicted by the linearized theory without damping. The mode (10,1) has a tongue to the left of the mode (4,3). The ordinate is $A=2h|\Lambda|$.

fact, if we pick a boundary condition $\xi=0$ at the walls, the (11,1) mode is pushed outside of the region of interest, and the modes (2,4) and (10,1) creep in. Our conclusion is that the experimental boundary conditions are probably neither, but a mixture of the two. We thus disregard the (11.1) mode from now on.

Due to the neglect of viscosity the stability curves start at zero amplitude, but the point of intersection is very well predicted (in terms of frequency) by the linear theory. Other modes of comparable wavelength have their instability at different values of ω. There are additional modes that fall between $2\omega_{4,3}$ and $2\omega_{7,2}$, but these are of considerably higher wavevectors and therefore are still strongly damped in the range of amplitudes which is scanned experimentally. It is therefore safe to conclude that there are only two spatial modes that become unstable in the region of interest of parameter space.

IV. Center Manifold Theory

Since we neglected viscous dissipation, all the neutral stability curves start at zero value of the amplitude of oscillation. However with viscous damping the spectrum of Floquet exponent at the point of intersection of the stability curves of the (4,3) and (7,2) modes takes the form sketched in Fig. 4. All the modes besides (4,3) and (7,2) have Floquet exponents whose real parts are well bounded away from the imaginary axis. When one has such a situation in a case of autonomous equations of motion one can use the center manifold theorem [9] to reduce the nonlinear dynamics to a set of ordinary differential equations [10]. We shall review first this theorem and then rectify the fact that in our case we have explicit time dependence such that the theorem would become useful for us as well.

The Center Manifold Theorem

Consider the autonomous system

$$\dot{x} = Ax + f(x,y) \qquad x \in R^n \qquad R \cdot \sigma(A) = 0 \qquad (4.1a)$$

$$\dot{y} = By + g(x,y) \qquad y \in R^m \qquad R \cdot \sigma(B) < 0 \qquad (4.1b)$$

315

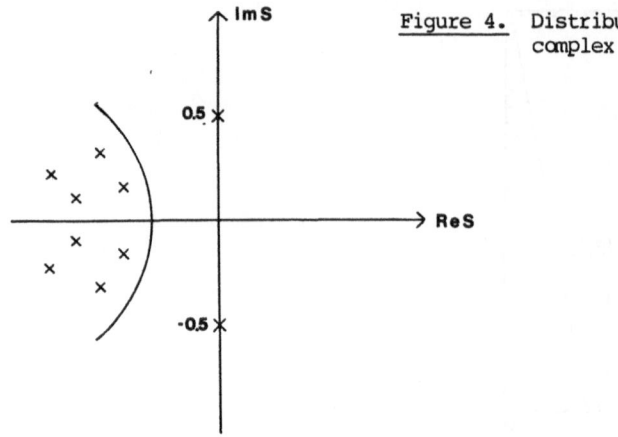

Figure 4. Distribution of Floquet exponents in complex plane.

with f,g and their first partial derivative ϵC^2, and $f(0)=g(0)=0$. Then there exists an invariant manifold $y=h(x)$ where $h=\nabla h=0$ at $x=0$, and where h is C^2. The significance of this theorem is that the long time properties of the equation

$$\dot{x} = Ax + f(x,h(x)) \tag{4.2}$$

provide a good approximation to the long time properties of the set (4.1). This is the basis of the reduction from infinite degrees of freedom to a small number of ode's. For extension to the case $\sigma(A)\neq 0$ but positive and small see for example ref. 10.

As mentioned earlier, our problem is not autonomous. If we denote the amplitudes of the critical modes by α_i, whereas those of the stable modes by β_j, then Eqs. (3.11) lead to a set

$$\dot{\alpha}_i = K_i(t)\alpha_i + g_i(\alpha,\beta) \tag{4.3a}$$

$$\dot{\beta}_j = K_j(t)\beta_j + g_j(\alpha,\beta) . \tag{4.3b}$$

where g_i, g_j are uncomputed functions at this point. We can remove the explicit time dependence on the RHS of (4.3) by introducing another "critical" degree of freedom α_c,

$$\alpha_c = \frac{\Lambda}{2} e^{it} \tag{4.4}$$

such that $\dot{\alpha}_c=i\alpha_c$, $\dot{\bar{\alpha}}_c=-i\bar{\alpha}_c$. In ref. 8 we argue that after this substitution there is a center manifold $\beta=h(\alpha)$. Inserting $\Lambda\cos t=\alpha_c+\bar{\alpha}_c$ and $\beta=h(\alpha)$ we get a system

$$\dot{\alpha}_i = K_i\alpha_i + G_i(\alpha) \tag{4.5}$$

where G_i is to be calculated and K_i is

$$K_i = \begin{pmatrix} 0 & -\chi_1 \\ \Gamma_i & 0 \end{pmatrix} \tag{4.6}$$

α contains six amplitudes now: $\xi_{4,3}, \xi_{7,2}, \phi_{4,3}, \phi_{7,2}$, and $\alpha_c, \bar{\alpha}_c$. It is convenient to diagonalize K. Denoting the amplitudes in the (linear) diagonal representation by $\tilde{\alpha}$, we find the equations of motion

$$\dot{\tilde{\alpha}} = J\tilde{\alpha} + G(\tilde{\alpha}) \tag{4.7}$$

where $G(\tilde{\alpha})$ is to be computed and J has the form

316

$$J = \begin{pmatrix} i\Omega_a & 0 & & & & \\ 0 & -i\Omega_a & i\Omega_b & 0 & & \\ & & 0 & -i\Omega_b & i & 0 \\ & & & & 0 & -i \end{pmatrix} \qquad (4.8)$$

We thus see that at this stage of analysis the point of intersection of the stability curves of Fig. 1 acts effectively like a codimension-3 point, where the linear problem is of the form of a triple Hopf bifurcation, leading to a 3-torus. Evidently this is not a true codimension-3 problem since one of the modes is the artificial mode α_c which is always critical. Notwithstanding, one is less surprised that chaos can set immediately close to this point in parameter space.

V. Normal Form Theory

For evaluating the term $G(\tilde{\alpha})$ in Eq. (4.7) one can pick a number of standard methods. However since we have six amplitudes and we have to go to at least cubic order there is a large number of nonlinear terms that might result. It becomes essential therefore to construct the simplest or minimal nonlinear equation. Such a construction is based on the theory of Normal Forms [11] which we review here briefly. The two fundamental theorems are due to Poincare and Poincaré-Dulac, and the key concept is that of resonance:

Resonances: Let

$$\dot{\underline{X}} = A\underline{X} + \ldots \qquad (5.1)$$

where the dots represent nonlinear terms. The linear operator A has eigenvalues λ. The n-tuple $\underline{\lambda} \equiv (\lambda_1, \ldots, \lambda_n)$ is called resonant if there is vector $\underline{m} \equiv (m_1, \ldots, m_n)$ such that one of the eigenvalues, say λ_s, obeys

$$\lambda_s = (\underline{m}, \underline{\lambda}) \qquad (5.2)$$

Here $m_k > 0$, $\Sigma m_k > 2$.

Poincare Theorem: Let $\dot{X} = AX + \ldots$ If λ are non-resonant, there exists a formal change of variables $\underline{X} = \underline{Y} + \ldots$ such that

$$\dot{\underline{Y}} = A\underline{Y} . \qquad (5.3)$$

Poincaré-Dulac Theorem: In the case of resonances, Eq. (5.3) becomes

$$\dot{\underline{Y}} = A\underline{Y} + \omega(\underline{Y}) \qquad (5.4)$$

where the only nonlinear terms in $\omega(y)$ are resonant monomials (i.e. $Y_1^{m_1} Y_2^{m_2} \ldots Y_n^{m_n}$ such that $\lambda_s = (\underline{m}, \underline{\lambda})$.

The meaning of these theorems is that by a change of variables one can reduce any nonlinear equation of motion to the simplest or minimal one, containing only resonant monomials.

In practice we follow the method described by Coullet and Spiegel [10] which derives the nonlinear equations automatically in normal form. The derivation involves some cumbersome algebra that will be reported somewhere else [8].The final results are the nonlinear equations of motion in addition to a solution of the deformation field itself (i.e. an approximate solution of the pde's (3.11)).

The equations of motion take the form

$$\dot{\alpha}_a = i\Omega_a \alpha_a + i\gamma_1 e^{it}\bar{\alpha}_a + i\gamma_2 |\alpha_a|^2 \alpha_a + i\gamma_3 |\alpha_b|^2 \alpha_a + i\gamma_4 \bar{\alpha}_a \alpha_b^2 \qquad (5.5a)$$

$$\dot{\alpha}_b = i\Omega_b \alpha_b + i\delta_1 e^{it}\bar{\alpha}_b + i\delta_2 |\alpha_b|^2 \alpha_b + i\delta_3 |\alpha_a|^2 \alpha_b + i\delta_4 \bar{\alpha}_b \alpha_a^2 \qquad (5.5b)$$

where $a \equiv (4,3)$, $b \equiv (7,2)$, $\Omega_{a,b} = \omega_{a,b}/\omega$, and γ_i, δ_i are numerical coefficients that were calculated theoretically for the boundary conditions of section III.

The deformation field is found in the form

$$\xi(x,y,t) = \xi^{(1)}(x,y,t) + \xi^{(2)}(x,y,t) + \xi^{(3)}(x,y,t) \tag{5.6}$$

where

$$\xi^{(1)}(x,y,t) = \frac{i}{\sqrt{2}} \alpha_{4,3} f_{4,3} + \frac{i}{\sqrt{2}} \alpha_{7,2} f_{7,2} + C.C. \tag{5.7}$$

where as $\xi^{(2)}(x,y,t)$ is quadratic in the amplitudes and contains modes with $\ell = 0,3,4,7,8,11,14$. $\xi^{(3)}(x,y,t)$ is cubic in the amplitudes and contains modes with $\ell = 1,4,7,10,12,15,18$ and 21. All the time dependent coefficients are known once Eqs. (5.5) are solved. One sees that to lowest order only the (4,3) and (7,2) modes are expected to exist. However to higher order the stable modes appear as dictated by the Center Manifold theorem. We see that these are precisely the modes seen experimentally via the time-resolved spatial Fourier transform (cf. Fig. 2). There is of course no contradiction between the low dimensionality of the dynamics (eq. 5.5) and the fact that many modes are seen in the field, since the latter are enslaved by the fundamental linearly unstable modes.

VI. Numerical Results and Comparison With the Experiment

As noted before, the boundary conditions used in the theory do not correspond to the experimental ones. Accordingly, the numerical coefficients in Eq. (5.5) cannot be used to compare theory and experiment. The structure of the equations, however, being of a normal form, is correct. We thus base further analysis on these equations.

As a first step we wish to eliminate the trivial, fast time dependence that results from the periodic forcing and the response at 1/2 the frequency. To do that we use the two-time scales method [12]. One writes $\alpha_i(t)$ as

$$\alpha_i(t) = \alpha_i^{(0)}(t,\tau) + \varepsilon\alpha_i^{(1)}(t,\tau) + 0(\varepsilon^2) \qquad \text{where} \tag{6.1}$$

$$\tau = \varepsilon t . \tag{6.2}$$

Here we pick

$$\varepsilon = 2(\Omega_b - \Omega_a) \sim 0(10^{-2}) . \tag{6.3}$$

From Eq. (6.1) we find

$$\frac{d\alpha_i}{dt} = \frac{\partial\alpha_i}{\partial t} + \varepsilon[\frac{\partial\alpha_i^{(0)}}{\partial\tau} + \frac{\partial\alpha_i^{(1)}}{\partial t}] + 0(\varepsilon^2) . \tag{6.4}$$

Inserting Eqs. (6.1), (6.3) and (6.4) in Eqs. (5.5), equating terms of the same order in ε, and using the condition to remove secular terms we finally arrive to the results

$$\alpha_a^{(0)} = a(\tau)e^{\frac{1}{2}t} \tag{6.5a}$$

$$\alpha_b^{(0)} = b(\tau)e^{\frac{1}{2}t} \qquad \text{where} \tag{6.5b}$$

$$\frac{da}{d\tau} = (-L_a+i\phi_a)a+i\Gamma_1\bar{a}+i\Gamma_2|a|^2a+i\Gamma_3|b|^2a+i\Gamma_4\bar{a}b^2 \tag{6.6a}$$

$$\frac{db}{dt} = (-L_b+i\phi_b)b+i\Delta_1\bar{b}+i\Delta_2|b|^2b+i\Delta_3|a|^2b+i\Delta_4\bar{b}a^2 \tag{6.6b}$$

where Γ_i and Δ_i are defined as γ_i/ε and δ_i/ε respectively. The constants $L_{a,b}$ are phenomenological damping constants added to Eqs. (6.6). Since the normal form (5.5) is invariant to the addition of damping, we can still use the same equations. $\phi_{a,b}$ is defined as $(\Omega_{a,b}-1/2)/\varepsilon$.

Equations (6.6) can be integrated numerically to explore the phase diagram. The values of the parameters used are displayed in Table 1. We discuss here the part of the phase diagram that pertains to the immediate vicinity of the point of intersection of the neutral stability curves in Fig. 1. The results of the numerical investigations are displayed in Fig. 5. The main characteristics of Fig. 1 are reproduced: (i) The asymmetry between the modes, i.e. the fact that the (4,3) mode damps the (7,2) mode whereas the (7,2) mode pumps the (4,3) mode; (ii) The existence of a region with periodic competition between the modes and (iii) the existence of a chaotic competition. Notice that the boundaries between these regions converge close to the point of intersection, as seen experimentally. We note however that a careful search reveals periodic windows in the chaotic regions, and also very small chaotic regions in the periodic regime, close to the "boundary" with the chaotic regime. Further results will be published elsewhere.

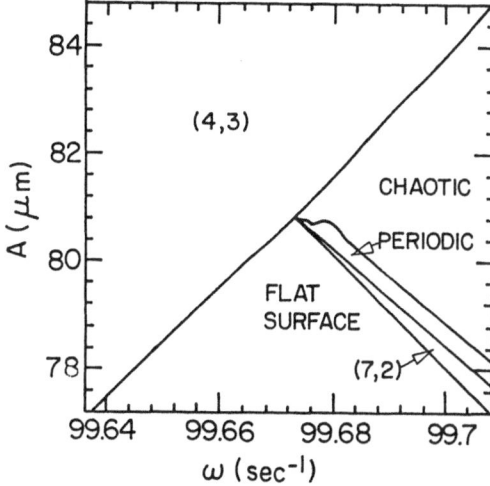

Figure 5. The theoretical phase diagram. A is $2h|\Lambda|$.

Table 1. Parameters used to reproduce the phase diagram

ω_a = 49.4490 sec^{-1}	γ_2 = -5.0x10^{-3}
ω_b = 50.2265 sec^{-1}	γ_3 = 8.5x10^{-2}
χ_a = 1.9249	γ_4 = 0
χ_b = 1.9684	δ_2 = 6.5x10^{-3}
L_a = 5x10^{-4}/ε	δ_3 = 8.5x10^{-3}
L_b = 5x10^{-4}/ε	δ_4 = 0

Acknowledgements
We are indebted to Ed Spiegel for some very useful conversations. We express our gratitude to Jerry Gollub for discussing his beautiful results with us freely long before publication, and for sharing with us his insights. Parts of this work were done when both authors were at the University of Chicago and enjoyed the wonderful hospitality of R.S. Berry and L.P. Kadanoff. This work was supported in part by the MRL of the University of Chicago, an NSF grant and the Minerva Foundation, Munich, Germany.

References

1. D. Ruelle and F. Takens: Comm.Math.Phys. **20**, 167(1971).

2. M.J. Feigenbaum:J.Stat.Phys. **19**, 25 (1978); **21**, 669 (1979).

3. A. Libchaber and J. Maurer: J.Phys. **41**, 13 (1980).

4. I. Procaccia: Physics Scripta, **59**, 40 (1985), and references therein.

5. S. Ciliberto and J.P. Gollub: Phys.Rev.Lett. **52**, 922 (1984).

6. S. Ciliberto and J.P. Gollub: J.Fluid.Mech. in press.

7. T.B. Benjamin and F. Ursell, Proc.Roy.Soc.London Ser.A **225**, 505 (1954).

8. E. Meron and I. Procaccia: Phys.Rev.A, to be published.

9. J. Guckenheimer and P. Holmes: Nonlinear Oscillations, Dynamical Systems and Bifurcations of Vector Fields (Springer, New York 1983).

10. P. Coullet and E.A. Spiegel: SIAM J.Appl.Math. **43**, 776 (1983).

11. V.I. Arnold: Geometrical Methods in Ordinary Differential Equations (Springer, New York, 1983).

12. See for example, C.M. Bender and S.A. Orszag: Advanced Mathematical Methods for Scientists and Engineers (McGraw Hill, New York 1978).

Chaos and Turbulence in an Electron-Hole Plasma in Germanium

G.A. Held and C.D. Jeffries

Department of Physics and Lawrence Berkeley Laboratory,
University of California, Berkeley, CA 94720, USA

1. Introduction

In this experiment we study the spatio-temporal behavior of an electron-hole (e-h) plasma. The plasma is produced by injecting both electrons and holes into a rod-shaped crystal of germanium at liquid nitrogen temperatures as shown in Fig. 1. The crystal is placed in a magnetic field parallel to its axis, and an adjustable electric field is also applied along the length of the sample. The plasma can absorb energy from the applied fields and, beyond some threshold (typically a few volts/cm at a few kilogauss), an unstable travelling helical density wave develops within the plasma, as shown in the Appendix, Fig. A-1. Several nonlinearly coupled modes can be excited within the boundaries of the crystal.

Experimentally, we measure the total current $I(t)$ through the crystal and the potential across it, $V(t)$, as the driving parameter V_{dc} is increased. By also recording the voltages $V_i(t)$ across pairs of probe contacts formed along the length of the sample, we can observe spatial variations in the plasma density. At the onset of the helical instability, spontaneous current oscillations are observed. As V_{dc} is increased further, we find that this simple physical system exhibits complex nonlinear dynamics, including a period doubling route to chaos when only one mode is excited. More generally, when more modes are excited, we observe quasi-periodicity (i.e., simultaneous oscillations at two incommensurate frequencies); self-entrainment (i.e., frequency ratio becomes a simple fraction); temporal chaos; and a partial loss of spatial coherence -- indicating the spatial breakdown of the helical density wave and the onset of "turbulence" in this solid state system.

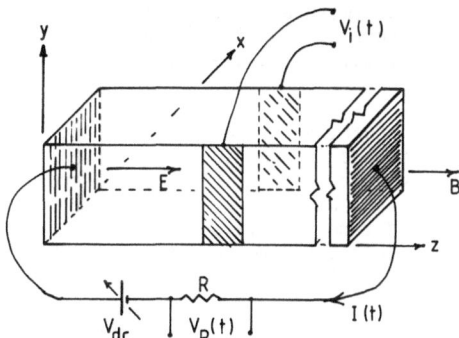

Fig. 1. Ge crystal in electric field E_0 and magnetic field B_0 at temperature T = 77°K. Electrons and holes injected from n^+ and p^+ contacts, respectively, generate plasma, density $\approx 10^{13}$ cm^{-3}. On increasing V_{dc} helical plasma density waves form and become chaotic. Plasma monitored by current $I(t)$, voltage $V(t)$ between end contacts, and voltage $V_i(t)$ from eight pairs of n^+ side probe contacts, each 0.5 mm wide, spaced by 1 mm along z-axis. Only one such pair shown in figure.

Our experiments are, of course, related to some hydrodynamic experiments on fluids, e.g., Rayleigh-Benard convection and Coette-Taylor flows, as well as other experiments on nonlinear dynamical systems [1]. Such experiments are partly motivated by the conjecture that in dissipative nonlinear media the dynamics may be modeled by a strange attractor of relatively low dimension [2], in contrast to a very large number of degrees of freedom associated with ergodic systems. Coherent oscillations of the type we study were originally observed by Ivanov and Ryvkin [3] in Ge and were subsequently studied both theoretically [4,5] and experimentally [5,6] in a number of other semiconductors. It is possible that chaotic behavior was observed earlier but not recognized as such, owing to the lack of the mathematical framework now available [1,2]. Our physical system is well characterized, and the equations of motion well known (see Appendix). This appears to be a good system for detailed study of spatio-temporal plasma turbulence and, in fact, it is the first plasma system found to exhibit the universal period doubling and quasi-periodic transitions to chaos [7].

2. Experimental Procedures

Our experimental apparatus is depicted in Fig. 1. We cut a $1 \times 1 \times 10$ mm³ sample from a large single crystal of Ge having a net donor concentration of 3.7×10^{12} cm⁻³. A lithium-diffused n^+ contact (electron injecting) and a boron-implanted p^+ contact (hole injecting) were formed on opposite 1×1 mm² ends. Phosphorous-implanted n^+ contacts were formed on two opposite 1×10 mm² faces. Onto these two faces we etched a pattern of eight pairs of contacts along the z-axis; one such pair is shown in Fig. 1. The voltage $V_i(t)$ between a pair of contacts is a measure of the local variation in the plasma density [5]. The sample was lapped, etched, and then stored for 72 hours in dry air to allow the surfaces to passivate.

When taking data, the control parameters include the applied dc magnetic field B_0, the applied dc voltage V_{dc}, and the angle θ between the sample axis and the magnetic field; typically, $\theta = 0 \pm 3°$. In practice, we fix B_0 and θ and sweep V_{dc}, while recording the dynamical variables $I(t)$, $V(t)$, and $V_i(t)$ which characterize the plasma behavior.

Perhaps the single feature most useful in characterizing the plasma is the power spectrum of the current, $|I(\omega)|^2$, from which we can detect the onsets of spontaneous oscillations, period doubling, quasiperiodicity, and chaos. However, observation of only power spectra does not enable us to distinguish between deterministic chaos and stochastic noise; both result in broadened spectral peaks. To uniquely identify the observed spectral broadening as deterministic chaos, we observe in real time the phase portrait, a plot of $V(t)$ vs. $I(t)$; and the first return map, a plot of I_n vs. I_{n+1}, where $\{I_n\}$ is the set of local current maxima. The return map is topologically equivalent to a Poincaré section of the attractor [8]. When the return map does not fill an entire area within 2-dimensional space, the motion of the system is confined to a low-dimensional strange attractor [8]. However, a system in which the return map does fill an entire area within 2-dimensional space may still be characterized by low-dimensional chaos (with attractor dimension typically $\gtrsim 2.5$). In these cases even a return map cannot distinguish between chaos and stochastic noise, and one must consider more quantitative measures of the dimensionality of the system.

The fractal dimension [9] provides just such a quantitative measure, and thus an approximate measure of the number of degrees of freedom needed to characterize the plasma at any instant of time. We use the following procedures [10] to measure the fractal dimension d of our plasma instabilities: we begin by recording a data set of N values of the current at uniformly spaced time intervals [i.e., $I(t+mT) \rightarrow I(mT)$, $m = 1,2,...,N$] using a fast 12-bit analog-to-digital converter and an LSI-11/23 computer. From the data set $\{I(T), I(2T), ..., I(NT)\}$ we construct $N - D + 1$ vectors $\vec{G}_m = [I(mT), I((m+1)T), ..., I((m+D-1)T)]$ in a D-dimensional phase space; D is referred to as the embedding dimension of the reconstructed phase space $\vec{G}$. Next, we compute the number of points on the attractor, $N(\varepsilon)$, which are contained within a D-dimensional hypersphere of radius ε centered on a randomly selected vector $\vec{G}_m$. One expects scaling of the form

$$N(\varepsilon) \propto \varepsilon^d \tag{1}$$

where d is the fractal dimension of the attractor. Thus, a plot of $\log \overline{N(\varepsilon)}$ vs. $\log \varepsilon$ is expected to have a slope d, where $\overline{N(\varepsilon)}$ is the average for hyperspheres centered on many different $\vec{G}_m$. This procedure is carried out for consecutive values of D = 2, 3, 4, ..., until the slope has converged. This is done to ensure that the embedding dimension chosen is sufficiently large [11] (important if the dimension of the phase space is not known) and to discriminate against high-dimensional stochastic noise, not of deterministic origin.

To determine whether or not a plasma density wave is spatially coherent, we compare the fluctuations in plasma density at different points along the sample. We obtain a crude measure of the degree of coherence by using a fast two-channel digital storage oscilloscope and comparing the voltages $V_i(t)$ across pairs of contacts located at different positions along the z-axis of the sample. If the temporal behavior of $I(t)$ is periodic, we observe only a phase-shift in $V_i(t)$ along z, which indicates a coherent travelling wave.

To obtain a more quantitative measure of the degree of spatial coherence, applicable for nonperiodic behavior, we calculate a spatial correlation function $C(r)$, defined as

$$C(r) = \sqrt{\frac{2}{N} \left| \sum_{n=1}^{N} V_i(nT) \ V_j(nT) \right|^{\frac{1}{2}}}, \qquad (2)$$

where $V_i(t)$ and $V_j(t)$ are the voltages across two pairs of contacts separated by a distance r, T is the sampling time interval, and N is a number large enough that $C(r)$ has converged, typically 20,000. We find that $C(r)$ is independent of T. From these spatial measurements we are able to determine whether temporally chaotic states are spatially coherent or disordered.

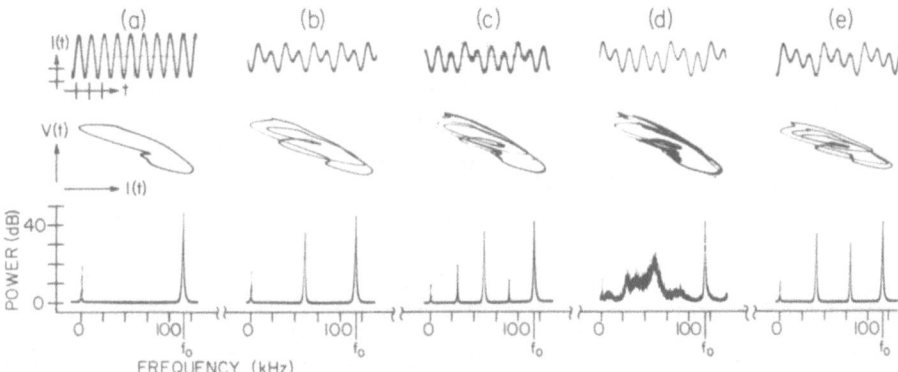

Fig. 2. Observed period doubling sequence for increasing drive voltage V_{dc}, (a) to (e), for plasma wave in Ge at B_o= 4 kG. (a) Current $I(t)$, phase portrait $V(t)$ vs. $I(t)$, and power spectrum of $I(t)$ show periodic behavior at $f_o \approx$ 118 kHz. (b) Period doubling to $f_o/2$. (c) Period doubling to $f_o/4$. (d) Onset of chaos. (e) Period 3 window. Behavior consistent with nonlinear wave-wave coupling model (Appendix). $I(t)$ scales: 8 μsec/div and 0.05 mA/div. For time series, power spectra and return maps in Figs. 2, 3, 4, the current $I(t)$ was ac coupled and also filtered to remove harmonics of f_o.

3. Results: Temporal Routes to Chaos

In different regions of parameter space (B_o, θ, V_{dc}) we observe different types of transitions to chaos. Two paths through parameter space will be discussed on this section, first a period doubling route to chaos and later a quasiperiodic route.

The first sequence was taken with B_o = 4 kG, as V_{dc} was increased from 0 to 25 V. The overall behavior of $I(t)$ was found to be as follows: For V_{dc} < 6 V, $I(t)$ has only a dc component. At V_{dc} = 6 V, $I(t)$ spontaneously becomes periodic. Regions of chaotic dynamics occur in the intervals 7.0 ≤ V_{dc} ≤ 7.4 V; 10.0 ≤ V_{dc} ≤ 10.7 V; and 14.9 ≤ V_{dc} ≤ 18 V; otherwise, $I(t)$ is periodic.

The clearest of these three chaotic sequences is shown starting in Fig. 2(a) at $V_{dc} = 10.0$ V: $I(t)$ is oscillating at a fundamental frequency $f_0 \approx 118$ kHz, i.e., at period 1. The phase portrait, $I(t)$ vs. $V(t)$ in Fig. 2(a), shows that the oscillation has a small spectral component at a *harmonic* of f_0. However, there is no *subharmonic* component, as seen in the power spectrum of $I(t)$ of Fig. 2(a). As V_{dc} is increased, $I(t)$ shows a period doubling bifurcation; the data shown in Fig. 2(b) display the emergence of a spectral component at $f_0/2$. At larger V_{dc}, another period-doubling bifurcation occurs [Fig. 2(c)] with new spectral components at $f_0/4$, $3f_0/4$, $5f_0/4$, ... At slightly larger V_{dc} $I(t)$ becomes nonperiodic [Fig. 2(d)] and its power spectrum enters a region of broadband "noise". For further increases of V_{dc} there appear noise-free windows of periods 3, 4, 5, ..., within this region of broadband noise; see Fig. 2(e) for period 3. This sequence ends at $V_{dc} = 10.7$ V with a return to period 1 oscillations.

The broadband "noise" of Fig. 2(d) is characteristic of chaotic behavior. The fact that this state is approached through a series of period doubling suggests that the spectral broadening is a result of deterministic nonlinear dynamics. However, to firmly establish that it is not due to some stochastic process, a return map of the current is recorded for the same parameters as Fig. 2(d) and is shown in Fig. 3. If the broadening of the spectral peaks were due to stochastic noise, then the height of one peak in $I(t)$ would be unrelated to the next, and the return map would be a uniformly filled square area. That this is not the case indicates that the evolution of the system is indeed governed by low-dimensional nonlinear dynamics; i.e., it is chaotic.

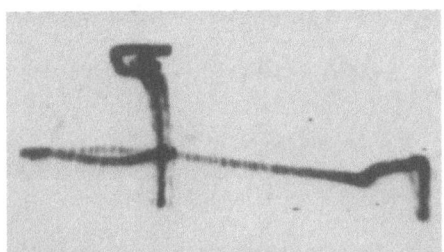

Fig. 3. Observed return map I_n (horizontal) vs. I_{n+1} (vertical) for the total plasma current, corresponding to the conditions of Fig. 2(d). Here, $\{I_n\}$ is the set of local current maxima. That the points do not fill the space shows that the "noise" in the power spectrum, Fig. 2(d), is not due to very high dimensional stochastic processes.

The second type of transition which we have observed is the quasiperiodic route to chaos: as V_{dc} is increased, the onset of a quasiperiodic state is followed by a transition to chaos. One such sequence of power spectra taken at $B_0 = 11.15$ kG, is shown starting in Fig. 4(a) with $V_{dc} = 2.865$ V: $I(t)$ is spontaneously oscillating at a fundamental frequency $f_1 = 63.4$ kHz. At $V_{dc} = 2.907$ V, the system becomes quasiperiodic: a second spectral component appears at $f_2 = 14$ kHz, incommensurate with f_1 [Fig. 4(b)]. At $V_{dc} = 2.942$ V, the system is still quasiperiodic; however, the two modes are interacting,and the nonlinear mixing gives spectral peaks at the combination frequencies $f = mf_1 + nf_2$, with m,n positive and negative integers.

As V_{dc} is increased further, we observe a series of frequency lockings [12], i.e., (f_1/f_2) = rational number, until the onset of chaos is reached, indicated by a slight broadening of the spectral peaks [Fig. 4(d)]. As V_{dc} is increased further, the e-h plasma exhibits increasingly turbulent behavior [Fig. 4(e)]. This is followed by a return to quasiperiodicity at $V_{dc} = 3.125$ V and, subsequently, simple periodicity at $V_{dc} = 3.442$ V.

Figure 4 also shows a sequence of return maps, topologically equivalent to Poincaré sections. Periodic motion corresponds to a closed 1-dimensional orbit in phase space; the Poincaré section in this case is simply a point [Fig. 4(a)]. Similarly, when the system is quasiperiodic, corresponding to motion on a 2-dimensional torus, the Poincaré section is approximately a circle [Fig. 4(b) and (c)]. However, as the system becomes chaotic, we find that the Poincaré section begins to wrinkle and to occupy an extended region. This does not *necessarily* imply that the behavior

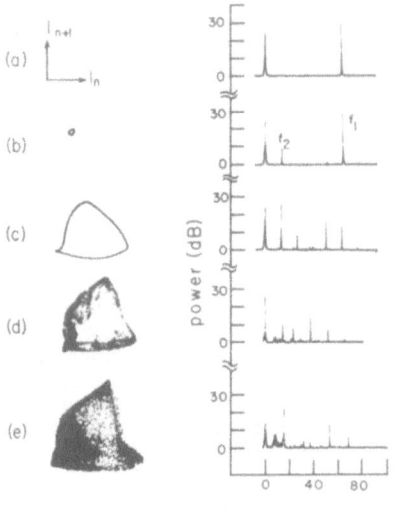

Fig. 4. Poincaré sections, I_n vs. I_{n+1}, and power spectra of the total plasma current $I(t)$ at $B_0 = 11.15$ kG with increasing V_{dc}: (a) 2.865 V, periodic at $f_1 = 63.4$ kHz. (b) 2.907 V, quasi-periodic with second frequency $f_2 = 14$ kHz. (c) 2.942 V, quasiperiodic with combination frequency components. (d) 3.033 V, onset of chaos. (e) 3.058 V, more chaotic; the fractal dimension of this attractor, $d = 2.7$, is measured in Fig. 5(a).

is stochastic, but rather that the dimension of the strange attractor (which is one greater than the dimension of the Poincaré section) is too large to be determined by visual inspection of the Poincaré section. For these attractors we calculate the fractal dimension d of Eq. (1) using the algorithm outlined above. Figure 5(a) shows the results for $V_{dc} = 3.058$ V with the embedding dimension, D = 2, 4, 6, and 8; for D ≥ 6 the slope has converged to 2.7. The fractal dimension for all the states shown in Fig. 4, as well as several states not shown, are plotted in Fig. 5(b). Within the chaotic regime, the fractal dimension of the attractor varies between 2 and 3. This demonstrates that the plasma turbulence shown in Fig. 4(d) and (e) may be described with only a few degrees of freedom; the behavior of the system remains largely deterministic. Were the observed turbulence due to thermal or stochastic processes, the slope would not have converged for small embedding dimension D. The number of degrees of freedom could then have been on the order of the number of electrons and holes in the crystal.

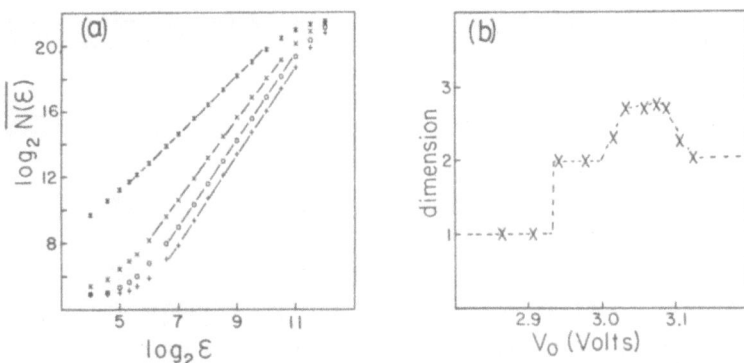

Fig. 5. (a) Plots to determine the fractal dimension d of the chaotic attractor at $V_{dc} = V_0 = 3.058$ V, $B_0 = 11.15$ kG, using method discussed in text and Eq. (1). Embedding dimension D = 2, 4, 6, and 8 correspond, respectively, to symbols *, x, o, and +; for D ≥ 6, the slope converges to d = 2.7. (b) Dependence of measured dimension d on applied voltage $V_{dc} = V_0$. Values d = 1 and d = 2 correspond to periodic and quasiperiodic orbits, respectively.

4. Results: Spatial Behavior

Having established that the e-h plasma instabilities do indeed exhibit chaos, we turn attention to the question of spatial coherence within the instabilities. In particular, we ask whether the chaotic states we observe correspond to a temporally chaotic yet still spatially coherent helical plasma density wave (e.g. of a form depicted in the Appendix, Fig. A-1) or whether the onset of chaos corresponds to a breakup of spatial order within the density wave. We consider two different observed transitions to chaos, the first being a quasiperiodic transition to "weak" turbulence, such as the sequence shown in Fig. 4, and the second being a transition to "strong" turbulence to be discussed below.

For our system we define a transition to "weak" turbulence to be one in which the transition from periodicity to chaos is followed by a transition back to periodicity as V_{dc} is increased further. All such transitions that we have observed occur over a small range of V_{dc} (i.e., a few volts) and in all such chaotic states there exists at least one fundamental peak that stands out clearly above the broadband "noise" level of the power spectrum. The two scenarios discussed in the previous section both correspond to transitions to "weak" turbulence.

To determine whether or not a weakly turbulent state is spatially coherent, we calculate the spatial correlation function $C(r)$, Eq. (2), for periodic, quasi-periodic, and chaotic (weakly turbulent) states of the system. The transition which we consider (taken with B_0 = 11.15 kG) is periodic at V_{dc} = 5.50 V. The correlation function $C(r)$ for different spacings r between the pairs of probe contacts is plotted in Fig. 6(a). For each pair of contacts, the voltage difference, $V_i(t)$, is periodic. By noting the phase-shift between pairs of probes as a function of distance, we estimate that the spatial wavelength $\lambda \approx 4.9$ mm. The theoretical correlation function for a travelling wave,

$$S(r) = \left[\frac{2}{T} \int_0^T \sin\omega t \, \sin(\omega t - 2\pi r/\lambda)dt \right]^{\frac{1}{2}} \tag{3}$$

is also shown in Fig. 6(a); the periodic data points lie close to the theoretical curve. Thus we conclude that the periodic oscillations are spatially coherent, not surprising, and consistent with previous experimental work [5,13]. In Fig. 6(b)

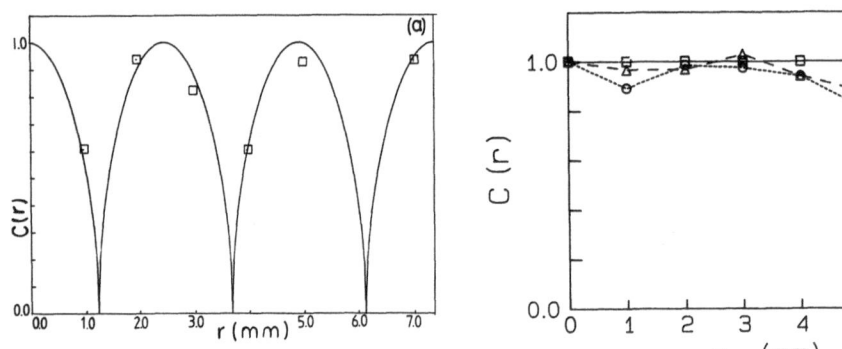

Fig. 6. (a) Comparison of measured values of spatial correlation function $C(r)$, Eq. (2), with theoretical correlation function for a travelling wave. The data ($\square$) were taken for a periodic state at B_0 = 11.15 kG, V_{dc} = 5.50 V using voltages $V_i(t)$ from pairs of side probes separated by distance r. The solid line is $S(r)$, Eq. (3), computed for a wavelength λ = 4.9 mm. (b) Normalized comparison of measured values of spatial correlation function $C(r)$ for three data sets at B_0 = 11.15 kG: ($\square$) periodic, V_{dc} = 5.50 V; ($\triangle$) quasiperiodic, V_{dc} = 5.59 V; (o) chaotic, V_{dc} = 5.71 V. In this case of "weak turbulence" the wave is essentially spatially coherent while exhibiting temporal chaos.

we plot the correlation function for quasiperiodic (V_{dc} = 5.59 V) and chaotic (V_{dc} = 5.71 V) states. In these plots C(r) is normalized with respect to the periodic case at each distance r. We find that the quasiperiodic and chaotic states both have correlation functions that follow the periodic case, which has been normalized to one. Therefore, we conclude that this weakly turbulent instability is chaotic in the temporal domain only. Even while exhibiting chaotic behavior, it remains essentially a spatially coherent plasma density wave.

With sufficiently large applied electric and magnetic fields, we find we can drive the plasma into a turbulent state which will not become periodic again as V_{dc} is increased further. Instead, all of the frequency peaks in the power spectrum merge into a single, broad, noiselike band. We classify this as a transition to "strong" turbulence. Such a transition is shown in Fig. 7. When V_{dc} = 10.4 V, I(t) is simply periodic at f_o = 321 kHz, with higher harmonics present as well -- Fig. 7(a). At V_{dc} = 11.6 V, I(t) is quasiperiodic, and at V_{dc} = 12.1 V the onset of broadband "noise" can be observed. At V_{dc} = 13.8 V, [Fig. 7(b)] only a few of the peaks can be seen above the noise, and when V_{dc} = 21.8 V [Fig. 7(c)] only a very broad peak remains.

We find that this transition to "strong" turbulence is characterized by a partial loss of spatial coherence. In the right column of Fig. 7 we plot the voltage traces across two pair of probe contacts which are separated by r = 4 mm, for V_{dc} = 10.4, 13.8, and 21.8 V. In the periodic case, the wave is spatially coherent with a wavelength of approximately 8 mm (i.e., a 4 mm separation corresponds to a 180° phase-shift). At V_{dc} = 13.8 V we find onset of the breakup of spatial order -- the basic oscillatory pattern and the 180° phase-shift are approximately maintained between the two traces, but changes in the shapes and spacings of the peaks can also be observed. For V_{dc} = 21.8 V, the wavelike structure of the traces, as well as the readily observable spatial correlation, are no longer present. We have not found a

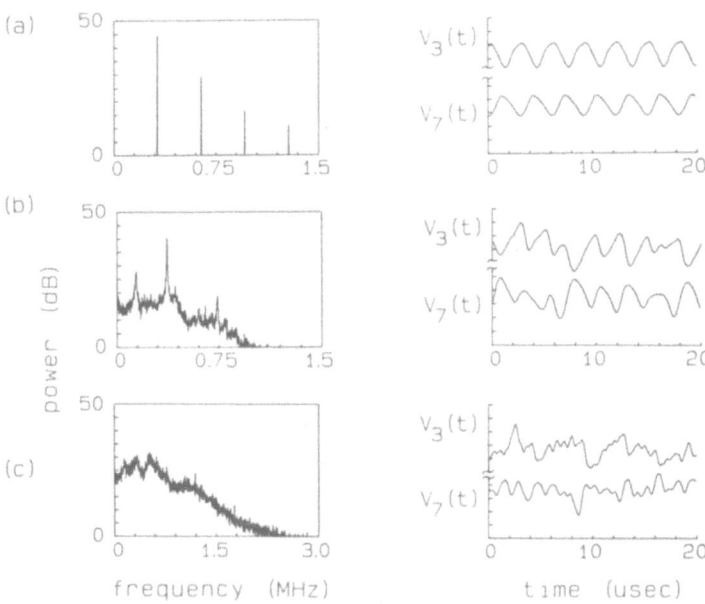

Fig. 7. Left, measured power spectra of I(t); right, measured voltages for two pairs of probe contacts separated by r = 4 mm: V(z = 3 mm) and V(z = 7 mm). B_o = 11.15 kG. (a) V_{dc} = 10.4 V, periodic at f_o = 321 kHz. At V_{dc} = 12.1 (not shown) temporal chaos has set in, with measured fractal dimension d ≈ 2.6, Fig. 8(a). (b) V_{dc} = 13.8 V, power spectrum has broad base and peaks; comparison of V_3(t) with V_7(t) shows beginning of spatial incoherence; measured fractal dimension d > 8. (c) V_{dc} = 21.8 V, power spectra very broad, more marked loss of spatial coherence, measured fractal dimension d > 8.

327

correlation function of analytic form which fits the data, in contrast to that of Fig. 6.

We would like to determine whether this breakup of spatial order can be characterized by chaotic dynamics: do the spatially uncorrelated states still correspond to motion in phase space along a low-dimensional strange attractor? We have yet been unable to definitively answer this question. Just prior to the breakup of spatial coherence, V_{dc} = 12.1 V, the total current $I(t)$ of the system is characterized by a low-dimensional attractor; measurements of the fractal dimension of the system yield d = 2.6 [Fig. 8(a)]. However, just after the onset of spatial disordering, V_{dc} = 12.9 V, the fractal dimension has increased to the point where we cannot calculate its value -- we can only set a lower limit: d $\gtrsim$ 8. this is shown in Fig. 8(b) where the slope has not converged either with respect to embedding dimension D or number of data points N. Figure 8(b) was taken with N = 884,000 and required 50 hours of CPU time on a Sun Microcomputer. For V_{dc} = 21.8 V, N = 884,000 points and embedding dimension D = 18, the slope is 14 and has definitively not converged. This difficulty in calculating large fractal dimensions is a problem encountered whenever one works with a very chaotic system [14]; the number of data points required for convergence increases exponentially with the fractal dimensions of the system. At present, although we know that our system experiences a large jump in dimensionality at the onset of spatial incoherence, we have not yet determined whether this onset is characterized by chaotic dynamics of an attractor of fractal dimension many orders smaller than the number of degrees of freedom of the particles in the system.

5. Concluding Remarks

We find that unstable helical density waves in an e-h plasma, when excited, will undergo transitions to low-dimensional chaotic turbulence. The observed behavior includes both period doubling and quasiperiodic transitions to chaos. Following a quasiperiodic route, we find that the resulting chaotic state of the system is characterized by a strange attractor which has a fractal dimension of between two

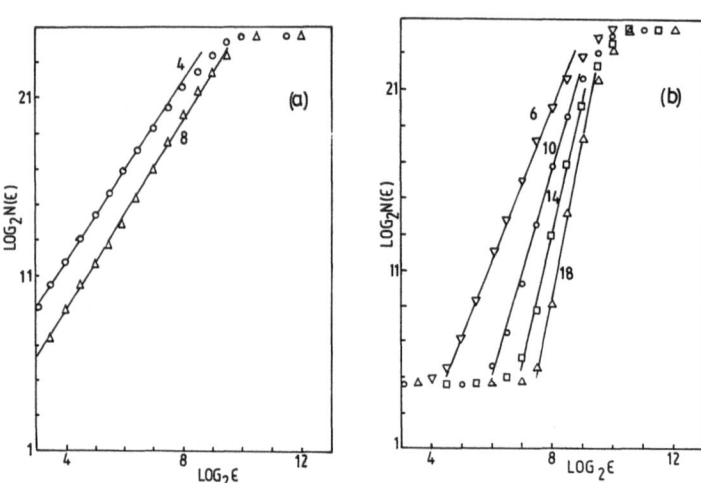

Fig. 8. Plots of $\log_2 N(\epsilon)$ vs. $\log_2\epsilon$ used to compute fractal dimension d at B_0 = 11.15 kG. $N(\epsilon)$ is the sum over 25 randomly selected vectors $\vec{G}_m$ in the reconstructed space. (a) V_{dc} = 12.1 V, N = 490,000 data points, embedding dimension D = 4, 8. The slope converges to d = 2.6. (b) V_{dc} = 12.9, N = 884,000, embedding dimension D = 6, 10, 14, and 18. The maximum slope is 8.8 but has not converged, suggesting that d > 8. The horizontal line of data points at small ϵ corresponds to hyperspheres for which there is only one vector $\vec{G}_m$; at large ϵ the horizontal line corresponds to all $\vec{G}_m$ vectors within the hypersphere.

and three. By measuring the local variations in plasma density, we find that these "weakly" turbulent states correspond to a temporally chaotic, spatially coherent plasma density wave.

Further, we find that, when more strongly excited, these plasma density waves undergo a loss of spatial coherence. The onset of the loss of spatial order is characterized by a sudden jump in the dimension of the attractor -- from less than three for the spatially coherent states to at least eight after the onset of spatial disorder.

We wish to thank E. E. Haller and the members of his laboratory for the Ge samples and assistance in the sample preparation. This work was supported by the Materials Sciences Division of the U.S. Department of Energy.

Appendix: Equations of Motion for Electron-Hole Plasma

For times and distances of interest, the following partial differential equations describe the motion of conduction electrons and holes in the crystal [5]:

$$\vec{J}_{e,h} = n_{e,h}q\mu_{e,h}\vec{E} \pm qD_{e,h}\vec{\nabla}n_{e,h} \mp \mu_{e,h}\vec{J}_{e,h} \times \vec{B} \tag{A1}$$

$$\frac{\partial n_{e,h}}{\partial t} = \pm\frac{1}{q}(\vec{\nabla} \cdot \vec{J}_{e,h}) + \gamma \tag{A2}$$

$$\vec{\nabla} \cdot \vec{E} = -q(n_e - n_h)/\varepsilon \tag{A3}$$

where the subscript e(h) and the upper (lower) signs refer to electrons (holes); n is the carrier density, $\vec{J}$ is the current density, q is the magnitude of the electronic charge, μ is the mobility, D is the diffusion constant, ε is the dielectric constant of the sample, γ is the net carrier generation rate. At surfaces perpendicular to the applied electric field, these equations are subject to the boundary condition $J_{e\perp} = J_{h\perp} = qsn_e$, where s is the surface recombination rate. By expanding the carrier densities and the electric field about their equilibrium values [$n_e = n_{eo} + n_{e1}(t)$, $n_h = n_{ho} + n_{h1}(t)$, $\vec{E} = \vec{E}_0 - \nabla\psi_1(t)$], and substituting these expressions into Eq. (A1), it has been shown [5] that the first-order terms lead to a helical density wave, $n_{e1} \approx n_{h1} = N_1(r)\exp[i\omega t - ikz - im\phi]$ and $\psi_1 = \psi_1(r)\exp[i\omega t - ikz - im\phi]$; m is an integer. Beyond certain thresholds of the applied electric and magnetic fields, this helical density wave is absolutely unstable [Im(ω) < 0], and spontaneous oscillations occur, coincident with the onset of nonlinear behavior. An illustration of this unstable helical density wave is shown in Fig. A-1, adapted from [17].

To incorporate nonlinear behavior into a model which explains the observed chaotic dynamics, we consider a *superposition* of waves in which the time-dependence is not assumed to be periodic,

$$n_{h1} \approx n_{e1} = \sum_{k,m} C_{km}(t)N_{km}(r)e^{-ikz-im\phi} + c.c. \tag{A4}$$

$$\psi_1 = \sum_{k,m} C_{km}(t)\psi_{km}(r)e^{-ikz-im\phi} + c.c. \tag{A5}$$

resulting in a differential equation of the form

$$\partial C_{km}/\partial t = M_{km}C_{km} + \sum_{\substack{k_1,k_2:k_1+k_2=k \\ m_1,m_2:m_1+m_2=m}} M'_{k_1k_2m_1m_2}C_{k_1m_1}C_{k_2m_2}$$

$$+ \sum_{\substack{k_1,k_2:k_1-k_2=k \\ m_1,m_2:m_1-m_2=m}} M'_{k_1k_2m_1m_2}C_{k_1m_1}C^*_{k_2m_2} \tag{A6}$$

where M and M' are independent of time. This equation describes a wave-wave interaction [15] in which a wave with wave vector k can couple nonlinearly to many different pairs of waves (k_1,m_1), (k_2,m_2). It turns out that a special case of Eq.
(A6) had been considered by Wersinger et al. [16] who studied numerically the evolution of an undamped wave coupled bilinearly to two damped waves:

$$\frac{\partial C_1}{\partial t} = \gamma_1 C_1 + MC_2 C_3 \exp(-i\delta t) \tag{A7}$$

$$\frac{\partial C_{2,3}}{\partial t} = \gamma_{2,3} C_{2,3} - MC_1 C^*_{3,2} \exp(i\delta t) \tag{A8}$$

For $\gamma_2 = \gamma_3$ and $\Gamma \equiv |\gamma_2|/\gamma_1$ they found that C_1 undergoes a period doubling cascade
to chaos as Γ is increased. The computed Poincaré section can be reduced to a uni-
modal return map. This is known to give rise to a period doubling cascade to chaos
[1,2]. It seems that Eqs. (A7) and (A8) may also exhibit a Hopf bifurcation to
quasiperiodicity but no numerical solutions are yet available.

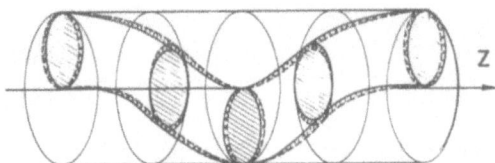

Fig. A-1. Model of helical plasma density wave in semiconductor cylinder, after
HOH and LEHNERT [17]. Electric field E_0 and magnetic field B_0 are applied along
the z-axis. The shaded sections represent regions of enhanced electron and hole
density.

References

1. For a review of chaotic behavior in physical systems, see H. L. Swinney:
    Physica (Utrecht) 7D, 3 (1983). See also general references, e.g., H. Haken:
    Advanced Synergetics (Springer-Verlag, Berlin, 1983); John Guckenheimer and
    Philip Holmes: Nonlinear Oscillations, Dynamical Systems, and Bifurcations
    of Vector Fields (Springer-Verlag, New York, 1983).
2. D. Ruelle and F. Takens: Comm. Math. Phys. 20, 167 (1971); J.-P. Eckmann: Rev.
    Mod. Phys. 53, 643 (1981); E. Ott: Rev. Mod. Phys. 53, 655 (1981).
3. I. L. Ivanov and S. M. Ryvkin: Zh. Tekh. Fiz. 28, 774 (1958) [Sov. Phys. Tech.
    Phys. 3, 722 (1958)].
4. R. D. Larrabee and M. C. Steele: J. Appl. Phys. 31, 1519 (1960); R. D. Larrabee:
    J. Appl. Phys. 34, 880 (1963); J. Bok and R. Veilex: Compt. Rend. 248, 2300
    (1958).
5. C. E. Hurwitz and A. L. McWhorter: Phys. Rev. 134, A1033 (1964).
6. M. Glicksman: Phys. Rev. 124, 1655 (1961); M. Shulz: Phys. Status Solidi 25, 521
    (1968).
7. G. A. Held, Carson Jeffries, and E. E. Haller: Phys. Rev. Lett. 52, 1037 (1984).
8. N. H. Packard, J. P. Crutchfield, J. D. Farmer, and R. S. Shaw: Phys. Rev. Lett.
    45, 712 (1980).
9. See, for example, J. D. Farmer, E. Ott, and J. A. York: Physica (Utrecht) 7D,
    153 (1983).
10. This is the method used by A. Brandstäter et al.: Phys. Rev. Lett. 51, 1442
    (1983); see Peter Grassberger and Itamar Procaccia: Phys. Rev. Lett. 50,
    346 (1983).
11. For the embedding theorem to be applicable, it is required that D ≥ 2d+1, where
    D is the embedding dimension and d is the fractal dimension of the attractor;
    see Ref. 1.

12. See, e.g., J. P. Gollub and S. V. Benson in: Pattern Formation and Pattern Recognition, edited by H. Haken (Springer-Verlag, Berlin, 1979), p. 74.
13. T. Misawa and T. Yamada: Jpn. J. Appl. Phys. 2, 19 (1963).
14. H. S. Greenside, A. Wolf, J. Swift, and T. Pignataro: Phys. Rev. A 25, 3453 (1982).
15. R. Z. Sagdeev and A. A. Galeev: Nonlinear Plasma Theory (Benjamin, New York, 1969).
16. J.-M. Wersinger, J. M. Finn, and E. Ott: Phys. Fluids 23, 1142 (1980).
17. F. C. Hoh and B. Lehnert: Phys. Rev. Lett. 7, 75 (1961).

Turbulent Motion. The Structure of Chaos

Yu.L. Klimontovich

Department of Physics, Moscow State University, Moscow 117234, USSR

> If we close the door before Fallacy,
> how is then Verity to come in?
>
> Rabindranath Tagore

1. Synergetics and the statistical theory of nonequilibrium processes

Two entirely dissimilar classes can be distinguished among various
processes in physical, chemical and biological systems. First, there
are time-dependent processes in closed systems. They lead to the
establishment of an equilibrium state, which under any given external
conditions corresponds to the highest possible degree of disorder.
This state can be appropriately called physical chaos. The current
view on the equilibrium state is based on remarkable works of Ludwig
Boltzmann and Josiah Willard Gibbs. They have shown entropy, the
quantity introduced in thermodynamics by Rudolf Clausius, to be one
of the major characteristics in the statistical theory. Entropy is
the measure of the disorder — tne measure of chaoticity of a state
of the system. Boltzmann's H-theorem and Gibbs theorem are the primary
tools in the construction and development of the contemporary statistical theory of nonequilibrium processes.

Secondly, there are processes in open systems, by which structures
evolve from physical chaos, i.e. the system arrives at a state of
higher order. These are the processes of self-organization.

The problem of self-organization in various systems is by no means
a new one. Many outstanding works were dedicated to its different
aspects; a special place among them belongs to Charles Darwin's
theory of natural selection in the course of evolution.

In the past Darwin's theory seemed to challenge the second law
of thermodynamics. Indeed, according to Darwin, biological evolution
leads to higher order and increasing complexity of structures. At
the same time the second law of thermodynamics states that evolution
in a closed system proceeds towards higner disorder and is associated with increasing entropy. This apparent controversy was solved
as soon as it was recognized that there are entirely different kinds
of processes. Some of them (in closed systems) lead to thermal equilibrium, while otners (in open systems) represent self-organization.

332

The basics of the theory of self-organization have evolved rather recently; the crucial years were 1977-78 when isolated (however important)   investigations were brought together to form the basis of what Ilya Prigogine named the theory of self-organization in nonequilibrium systems, while Hermann Haken coined the term synergetics to emphasize the fundamental role of cooperative, concerted action in the processes of self-organization. It is also significant that the Solvay conference of 1978 convened under the motto "Order and fluctuations in equilibrium and nonequilibrium statistical mechanics".

Obviously, there is a need for the theory capable of describing both these classes of processes. Such theory must be efficient on all levels of description: kinetic, hydrodynamic, diffusive, thermodynamic. A theory of this kind, the so-called statistical theory of nonequilibrium processes, is successfully being developed by the efforts of many researchers and allows to deal with a very wide scope of problems in various branches of science.

All this inspired the author to write a concise book entitled "Turbulent motion. The structure of chaos" in an attempt to give a unique interpretation of the basic ideas of this theory, to reveal its general structure and to discuss some particular results on different levels of description.

We were determined to make the treatment comprehensible for a wide range of university-educated readers. The particulars of proofs are often skipped, as long as they can be found in easily accessible sources, such as the Springer series in Synergetics edited by H.Haken, as well as other recent books on the kinetic theory, dissipative structures and self-organization. Many of these are listed in the References.

The interpretation is based on the method of construction of the statistical theory pursued by the author for over twenty years and already used in a number of books: "Statistical theory of non-equilibrium processes in plasma" (KLIMONTOVICH, 1964, 1967), "Kinetic theory of nonideal gases and nonideal plasmas" (KLIMONTOVICH, 1975, 1982), "The kinetic theory of electromagnetic processes" (KLIMONTOVICH, 1980, 1983), "Statistical physics" (KLIMONTOVICH, 1982, 1985).

The new book is certainly not a synopsis of previous publications. It contains a good deal of fresh material, which has been only partly included in the English version of "Statistical physics" (KLIMONTOVICH, 1985) and in the book "Selforganization and turbulence in liquids" (EBELING, KLIMONTOVICH, 1984).

From a very extensive material accumulated so far in the statistical theory of nonequilibrium processes we have selected only those facts which form the basis of our views on the structure of chaos and turbulent motion. The general approach and the unique standpoint in dealing with a very wide scope of problems combine here with the most plain examples used for illustrating the results, which brings the book within grasp of the general audience.

Let us also give some more explanations to the choice of the title.

When going from microscopic reversible equations to irreversible ones (on the kinetic, hydrodynamic, diffusive and thermodynamic levels of description), we introduce macroscopic collective characteristics. Three types of motions can be distinguished depending on the number of macroscopic degrees of freedom.

The first of these is, essentially, chaotic thermal motion, when the macroscopic parameters (thermodynamic functions) are constant. In other words, thermal chaotic motion is not associated with any processes in time and space characterized by macroscopic degrees of freedom. Fluctuations of thermodynamic functions are low, and themselves are a manifestation of thermal chaotic motion.

The second group includes coherent motions and laminar flows. They appear against the background thermal motion and are characterized by a small number of macroscopic degrees of freedom.

Finally, the third group includes turbulent motions, characterized by the involvement of a large number of macroscopic degrees of freedom. Of course, turbulent motions are quite diverse and may also include coherent states (space-time structures), which appear this time, however, against the background of turbulent motion, i.e., against the background of many other degrees of freedom.

The nonequilibrium spatial-temporal structures will also be called, after I. Prigogine, dissipative structures.

It follows that the transition from equilibrium (thermal) motion to laminar, and further on to turbulent motion is the transition from the most chaotic movement to the most structurally complicated turbulent motion. In other words, turbulent motion is viewed as highly intricate space-time structures arising from chaos in open systems, and hence the title of the book.

It is a sequence of ever increasingly complex dissipative structures that shapes the process of self-organization, if under given external conditions the creation of structure depends on the inherent properties of the system. Hermann Haken was one of the first to treat self-organization as a sequence of nonequilibrium phase transitions. This approach proved very fruitful.

As we have already mentioned, the new unifying approach, aimed at establishing the general laws and principles of self-organization in most diverse systems was christened Synergetics by H.Haken. It can be said that the new book exposes the ideology, structure and the main techniques of the contemporary statistical theory, which is the foundation and the principal tool of Synergetics.

The problems we are to deal with are well known to be quite complex. As yet, few of them can be handled in a comprehensive and consistent fashion. In many cases one has to rely more on intuition than on figure-work. In this plight we draw encouragement from the words of Rabindranath Tagore, taken here for an epigraph. The following section we shall also begin with a quotation, which illustrates how complicated the theory of turbulence seemed even to Theodor von Karman.

2. Chaos and Order

> I recall that von Karman, in his opening address, said that, when he finally came face to face with his Creator, the first revelation he would supplicate would be an unfolding of the mysteries of turbulence.
>
> H. Moffatt

The concept of chaos was essential in the doctrines of ancient philosophers, especially the representatives of Platonic school. Without going into the details, whose value now is mainly historical, let us recall only two definitions formulated by the ancient thinkers, which are still useful for understanding the physical sense of "chaos".

According to Plato and his followers, chaos is that state of a system which persists as the possibilities for the system to manifest its properties are removed.

On the other hand, all that constitutes the essence of Universe emerges from initially a chaotic state. The role of creative force (the Creator) Plato ascribed to Demiurge, who transformed the primeval Chaos into Cosmos. All existing structures are thus born from chaos.

The concept of structure is also quite general. Structure is a certain form of organization and linkage of elements in the system. It is not the particular form of the elements that is essential but the totality of their interrelations.

In spite of their fundamental importance, the concepts of chaos and chaotic motion in physics are not defined strictly enough.

Indeed, chaotic is the movement of atoms in any system that is in a state of thermal equilibrium. Chaotic is the movement of Brownian particles, which are small but macroscopic bodies. Applied to motion, "thermal" and "chaotic" appear to be synonyms. Thus, we speak of chaotic/thermal oscillations of charge and current in thermostated electric circuit, of chaotic/thermal electromagnetic radiation etc.

In all these cases we are dealing with motion in the state of thermal equilibrium. The terms chaos and chaotic motion, however, are often employed for characterizing the states which are far from thermal equilibrium - for instance, in the description of turbulent motion.

It is very hard to answer, what is turbulence. Opinions differ, for instance, whether the turbulent motion is more chaotic (less ordered) than the laminar. In the "Handbook of turbulence" FROST and MOULDEN (1977) write, for example, that turbulent motion is chaotic. "Chaotic" in this context is almost synonymous to "turbulent". Chaoticity constitutes the main property of such motion.

This definition derives from the utmost complexity of turbulent motion, from the supreme intricacy of its structure. At the same time, this view is not constructive. It fails to acknowledge the difference between the motion in the state of thermal equilibrium and the turbulent motion, which is very far from equilibrium.

A similar view is adopted by H.HAKEN in his "Synergetics" (1983, 1985). In Sect.1.2.1, dealing with dynamic structures in liquids (Benard cells), we read:
"When... the Rayleigh number is increased further, the rolls can start an oscillation, and at a still higher Rayleigh number an oscillation of several frequencies can set in. Finally, an entirely irregular motion, called turbulence or chaos, occurs."

More or less the same we find in connection with coherent oscillations in lasers (Sect.1.2.2):
"Under different conditions the light emission may become "chaotic" or "turbulent", i.e. quite irregular."

It seems almost self-evident that the transition from laminar to turbulent motion is the transition from order to chaos. A different approach is used in our "Statistical physics" (KLIMONTOVICH, 1982, 1985). It can be summarized with the help of the following example.

Imagine motionless liquid. It has no singled-out macroscopic degrees of freedom, and hence no <u>macroscopic structure</u>. There only is thermal/chaotic movement of atoms.

A laminar flow of incompressible liquid in a tube gives rise to a macroscopic structure imposed on the thermal motion of atoms. This structure is determined by the spatial-temporal distribution of the mean flow velocity. On the hydrodynamic level of description the thermal motion (atomic structure) is reflected only in the existence of small hydrodynamic fluctuations.

As the Reynolds number approaches its critical value, the intensity of hydrodynamic fluctuations increases. This is a harbinger of the change in motion, a forerunner of the change in the macroscopic structure.

When the laminar flow finally gives way to turbulence, the macroscopic structure becomes much more complicated. The motion is now characterized by a large number of macroscopic degrees of freedom.

Turbulent motion is associated with a highly increased withdrawal of energy from the averaged flow towards other degrees of freedom: turbulent viscosity sets in. This implies that in the turbulent flow the liquid exhibits a much ampler capacity for transferring the momentum from one layer to another. Such a large amount of momentum cannot be exchanged via the ordinary mechanism of molecular transfer, which is characterized by common viscosity.

We see that with the onset of turbulence the microscopic (molecular) mechanism of momentum transfer is replaced by the macroscopic mechanism. Metaphorically speaking, the system switches from "individual" resistance to "organized" (collective) resistance. This is reflected in the change of drag when laminar flow becomes turbulent.

Calculations which corroborate this view have been published by the author; some of them can be found in the English version of "Statistical physics" (KLIMONTOVICH, 1985) and in "Selforganization and turbulence in liquids" (EBELING, KLIMONTOVICH, 1984).

Thus, turbulent motion is characterized by a large number of macroscopic degrees of freedom.

Today, however, we also know examples of highly complicated and seemingly chaotic motions which occur with a small number of macroscopic degrees of freedom. This kind of motion is described, for instance, by a well-known Lorentz system of three ordinary nonlinear differential equations.

The Lorentz system is obtained by replacing the hydrodynamic equations for convective motion by a set of equations of three modes. These equations helped to discover the strange attractor, which is a restricted region in three-dimensional phase space which attracts all trajectories from the neighborhood. The trajectories in the neighborhood of a strange attractor are dynamically unstable; this means that the initially close trajectories spread over the entire phase space of the strange attractor.

The motion in a strange attractor is so complex and tangled that it resembles turbulence. This provides some justification for considering the Lorentz system as a most simple model of turbulent motion.

There are also other examples of dynamic systems of relatively few equations, whose solutions with appropriately chosen parameters are seemingly chaotic. It is not yet clear, to what extent such systems may be viewed as models of real turbulence.

Now let us draw some conclusions.

We started with the philosophic definition of chaos. The philosopher's (or, for that matter, layman's) chaos has no definite measure, which prevents comparing the "chaoticity" of different types of motion.

The concept of chaos in physics is no less ambiguous, since both the equilibrium thermal motion and the highly nonequilibrium turbulent motion can be referred to as "chaotic".

As we have already indicated, turbulence may be viewed as, in a certain sense, a state of higher order compared with both thermal and laminar motions. In this sense, as we shall see, the transition to turbulence can be interpreted as a nonequilibrium phase transition, or, to be more precise, as a series of phase transitions which constitute the process of self-organization. The same view is expressed by PRIGOGINE and STENGERS in "Order out of chaos" (1984):

"For a long time turbulence was identified with disorder or noise. Today we know that this is not the case. Indeed, while turbulent motion appears as irregular or chaotic on the macroscopic scale, it is, on the contrary, highly organized on the microscopic scale. The multiple space and time scales involved in turbulence correspond to the coherent behavior of millions and millions of molecules.

Viewed in this way, the transition from laminar flow to turbulence is a process of selforganization. Part of the energy of the system, which in laminar flow was in the thermal motion of the molecules, is being transferred in macroscopic organized motion."

Identifying turbulence with chaos is not constructive. Insofar as turbulence is produced under the action of external agents (i.e., the kinetic head in a tube), the problem of the theory of turbulence in our "Statistical physics" (KLIMONTOVICH, 1982,1985) was formulated as revealing the "structure of chaos".

Before approaching this problem, however, we must introduce a measure of chaoticity of the system's state. This measure is furnished by the entropy of Boltzmann and Gibbs.

It must be observed that the concept of entropy was considerably broadened in connection with the study of complicated chaotic (or, as they are sometimes called, stochastic) regimes in dynamic systems. For such systems A.N.Kolmogorov introduced the concept of dynamic entropy, also called K-entropy. It is closely related to Lyapunov indices. The basis of the theory of dynamic chaos was laid also in the works of N.S.KRYLOV (1950, 1979). Essential is also the relationship of K-entropy with Boltzmann and Gibbs entropy, and with the concept of entropy production.

3. Changes in entropy in the processes of self-organization. S-theorem

The possible use of Boltzmann-Gibbs entropy for estimating the degree of order in the processes of self-organization is not self-evident. There are two ways of stating the problem.

(1) We consider the time evolution of entropy as the system approaches a stationary state, characterized by a certain value of control parameter a. The control parameter can be the amount of feedback or pumping in classical and quantum generators, the Reynolds and Rayleigh numbers in hydrodynamics etc. The problem is then reduced to finding the Lyapunov function, which stands in place of Boltzmann's H-function when studying the time-development in open systems. The Boltzmann-Gibbs entropy can also be used, on condition that the values of entropy for all intermediate states are renormalized to the same values of "mean energy" $\langle E \rangle$. The "energy" here is the effective Hamilton function, which characterizes the stationary states. Thus defined, the entropy in the course of evolution towards stationary state may either increase or remain the same. The latter is only possible when the initial state coincides with the stationary one.

(2) We consider a set of stationary states which correspond to different values of the control parameter. Here we also renormalize the values of entropy to the given value of mean energy. If, as the value of the control parameter is increased and the system recedes from the equilibrium state, the entropy decreases, then the process of self-organization is under way. We called this statement S-theorem.

According to this criterion, the averaged turbulent flow in a channel or tube is more ordered than a laminary one. As we have already indicated, the higher order in the averaged turbulent flow manifests itself, for one thing, by collective organized resistance, which replaces the chaotic molecular transfer in the laminar flow.

This attempt of a consistent and unified treatment of the contemporary statistical theory of nonequilibrium processes, which allows to assess the degree of order of different structures arising from chaos seems to us very well-timed and useful. The spectrum of systems which require assessing the relative order of different states in the course of evolution is quite wide: from the most simple physical systems to the Universe, from a gas of structureless particles to physical vacuum. Vacuum apparently exhibits the highest degree of chaos; at the same time it gives birth to all that constitutes Cosmos. The problem of finding the right control parameters is one of the most important in the study of complex systems. If there are more than one control parameters, the evolution may take different routes, which implies the existence of the most favorable way to self-organization.

References

Balescu R. Statistical mechanics of charged particles. Wiley, New
        York 1963; Mir, Moscow 1967
Bogolubov N. Problems of the dynamic theory in statistical mechanics.
        Gostekhizdat, Moscow 1946; North Holland, Amsterdam 1962
Ebeling W., Klimontovich Yu. Selforganization and turbulence in
        liquids. Teubner, Berlin 1984
Frost W., Moulden T. Handbook of turbulence, v.1. Fundamentals and
        applications. Plenum Press, New York, London 1977
Haake F. Statistical treatment of open systems by generalized master
        equation. Springer, Berlin, Heidelberg, New York 1973
Haken H. Synergetics. Springer, Berlin, Heidelberg, New York 1978
Haken H. Advanced synergetics. Springer, Berlin, Heidelberg, New York,
        Tokyo 1983; Mir, Moscow 1985
Haken H., Light V. Waves, photons, atoms. North Holland, Amsterdam,
        New York, London 1981
Klimontovich Yu.L. Statistical theory of non-equilibrium processes
        in plasma. MGU, Moscow 1964; Plenum Press, New York, London
        1967
Klimontovich Yu.L. Kinetic theory of nonideal gases and nonideal
        plasmas. Nauka, Moscow 1980; Springer, Berlin, Heidelberg,
        New York 1982

Klimontovich Yu.L. The kinetic theory of electromagnetic processes. Nauka, Moscow 1980; Springer, Berlin, Heidelberg, New York 1983

Klimontovich Yu.L. Statistical physics. Nauka, Moscow 1982; Gordon and Breach, Harwood Academic publishers, New York 1985

Klimontovich Yu.L. Brownian motion and turbulence; entropy, entropy production with laminar and turbulent motions. In: Nonlinear and turbulent processes in physics. Gordon and Breach, Academic publishers, New York 1984

Krinsky V.I. Self-organization. Autowaves and structures far from equilibrium. In: Proceedings of an International Symposium, Pushchino, USSR, July 18-23, 1983. Springer, Berlin, Heidelberg, New York, Tokyo 1984

Krylov N.S. Works on the foundations of statistical physics, Nauka, Moscow 1950; Princeton University Press, Princeton 1979

Lifshits E.M., Pitaevsky L.P. Physical kinetics. Nauka, Moscow 1979; Pergamon Press, Oxford, New York 1982

Moffatt H. Some developments in the theory of turbulence. J.Fluid Mech. <u>106</u>, 27-47 (1981)

Monin A.S., Yaglom A.M. Statistical fluid mechanics. Nauka, Moscow 1965, 1967; MIT Press, Cambridge 1973

Nicolis G., Prigogine I. Selforganization in nonequilibrium systems. Wiley, New York 1977; Mir, Moscow 1979

Nicolis G., Dewel G., Turner J. Order and fluctuations in equilibrium and nonequilibrium statistical mechanics. (XVIIth International Solvay conference on physics). Wiley, New York 1981

Prigogine I. Non-equilibrium statistical mechanics. Wiley, New York, London 1962

Prigogine I. From being to becoming. Freeman, San Francisco 1980; Nauka, Moscow 1985

Prigogine I., Stengers I. Order out of chaos. Bantham Books, Toronto, New York, London 1984

Fully Developed Turbulence as a Complex Structure in Nonlinear Dynamics

S. *Grossmann*

Fachbereich Physik der Philipps-Universität, Renthof 6,
D-3550 Marburg/Lahn, Fed. Rep. of Germany

1. Introduction

This talk aims at presenting (i) the idea why a fully developed turbulent fluid flow is a deterministic chaotic state that has to be described statistically, (ii) how turbulent convective transport happens,
which gives the most direct picture of the selfsimilarly nested hierarchical structure of the turbulent flow field, and (iii) to describe a
random walk on a fractal with a *space-scale* as well as a *time-scale*
hierarchy imaging the turbulent eddy hierarchy, but which is of rather
general interest also in itself. In these notes I concentrate on the
basic physical ideas and the discussion of results. The reader who is
interested in details of the formulations, derivations, and more applications is referred to the original references. Some additional aspects
are discussed here for the first time.

2. Strong turbulence as a chaotic state

Many systems when driven far away from equilibrium by appropriate external conditions enter a chaotic state. Its characteristic is a temporal permanently varying signal, non-periodic, irregular, unpredict-
able, with broad band continuous spectrum and correlation decay. This
stems from the coexistence of several ($\gtrsim 3$) competing collective modes
or order parameters coupled by a nonlinear set of equations of motion,
that has *no stable* time independent solution *nor any stable* periodic
solution. The nonlinearities, at the other hand, guarantee global stability, i.e. there is an increase of one collective amplitude at the
cost of others, which then implies feeding of other modes at the cost
of the own strength, providing thus an upper bound, followed by a decrease, etc. again and again.

Such behaviour is typically preceded by certain regular states,
such as steady, periodic, quasi-periodic states, entered with or without a period-doubling sequence, with or without intermittency. It is
called *weak* turbulence and characterized (i) by a strange attractor
of low dimension in the phase space of the order parameters (= collective modes or variables), (ii) by a positive Kolmogorov-Sinai entropy
indicating the permanent creation of information, (iii) by a few positive Lyapunov exponents, indicating the sensitive dependence on the
initial conditions, i.e. the strong divergence of nearby starting trajectories, (iv) by ergodic behaviour, i.e. the existence of a smooth
invariant measure, describing the relative frequency with which the
available phase space is continuously covered, (v) by temporal correlation decay, either exponential or algebraically.

Are these notions applicable to understand also *fully developed strong turbulence?* Turbulence is displayed by the Eulerian velocity field $\vec{u}(\vec{x},t)$ of a fluid in a high Reynolds number state. It satisfies the Navier-Stokes equation

$$\partial_t u_i = - u_j \partial_j u_i - \partial_i p + \nu \Delta u_i \quad , \quad x \in \Omega. \tag{1}$$

Here p is the pressure divided by the (constant) mass density ρ_0, ν is the kinematic viscosity, and Ω the volume occupied by the turbulent flow. The velocity field is solenoidal, div $\vec{u}$ = 0, reflecting incompressibility. The solutions of (1) are selected by conditions to be satisfied at the boundary $\partial\Omega$ with typical linear extension ℓ_b and typical imposed velocity u_b. The Reynolds number is Re = $\ell_b u_b / \nu$ or, equivalently, $Re_T = u'\lambda_T/\nu$, the Taylor-Reynolds number. The relation holds $Re_T \cong 0.9 \sqrt{Re}$; for atmospheric turbulence $Re_T \cong 4300$. u' = $\sqrt{\langle \Delta u_i \Delta u_i \rangle /3}$ denotes the degree of turbulent velocity fluctuations $\Delta u = u - \langle u \rangle$ and λ_T the Taylor microscale $\lambda_T^{-1} = \sqrt{\langle u_{1|1} u_{1|1} \rangle}/u'$.

According to the simultaneously developed ideas of Kolmogorov, Obukhov, v.Weizsäcker, Heisenberg, and Onsager a high Reynolds number flow is sustained by permanent input of energy into the large scale motion, which cascades down to smaller and smaller eddies, feeding finally the viscous degrees of freedom, Richardson [1]. The mean rate of energy per unit mass, ε, fed into the system is $\varepsilon \cong u_b^3/\ell_b$ and the one which is taken out by viscosity is $\varepsilon = \langle \varepsilon(x) \rangle$,

$$\varepsilon(x,t) = \nu u_{i|j}(x,t) u_{i|j}(x,t). \tag{2}$$

$\varepsilon(x)$ is the operator of local energy dissipation. The reader is referred to textbooks (e.g. [2]).

It is only now that we slowly begin to understand how the intuitively appealing cascade idea finds its support in the underlying equation of motion (1), in particular how energy dissipation enters from theory and not simply by intuition and scaling arguments. To exhibit this is one main emphasis of this and the following section.

Is there sensitive dependence on initial conditions? To answer this the linear variational equation is needed.

$$\partial_t \delta u = L(u(t))\delta u \quad , \quad \delta u = 0 \text{ on } \partial\Omega. \tag{3}$$

The linear operator L(u) depends on the reference trajectory u(t) due to the nonlinearity of (1). δu is the distance to a neighbouring trajectory. "Trajectory" here means a hole flow field in space and time.

The essential question is whether L has a *positive* real part. This would imply, evidently, exponential growth of a finite initial disturbance $\delta u(x,t_0)$. We therefore have to consider the Hermitian part of L,

$$H(u(t)) = - (u_{i|j} + u_{j|i})/2 + \delta_{ij}\nu\Delta. \tag{4}$$

What does its spectrum look like? First, due to the very existence of

343

boundary conditions at all, it is discrete, giving rise to a discrete set of Lyapunov exponents. Second, it contains the "kinetic energy operator" but with "negative mass" $m \hat{=} - \nu$; hence from this term the spectrum is bounded from *above* (instead of below as for quantum particles). Third, there is an "external potential" in addition, whose largest local eigenvalue is (at given t)

$$w(x) = [(1/4) \sum_{i,j} (u_{i|j} + u_{j|i})^2]^{1/2}. \tag{5a}$$

This introduces the local energy dissipation based on the equation of motion (1) via (3), since

$$\sqrt{2}w(x) = [\varepsilon(x)/\nu]^{1/2}. \tag{5b}$$

We note that the largest eigenvalue is connected with $\nu\Delta \hat{=} 0$, i.e. the positive eigenvalues range up to $\sqrt{\varepsilon/\nu}$ (which is the Kolmogorov viscous time scale $\tau_\eta = \sqrt{\nu/\varepsilon}$). The more $\nu\Delta$ contributes, the smaller the eigenvalues a_α of $H(u)$ are. They get arbitrarily large negative (indicating stability) under the sole government of viscosity ν via "kinetic energy".

This mathematical frame has been formulated more rigorously as an estimate for upper bounds by Ruelle (1982/4) and Lieb (1984) [3,4]. The Kolmogorov-Sinai entropy has the upper bound

$$h = \sum_{\lambda_\alpha > 0} \lambda_\alpha \le K_1 \nu^{-11/4} \int d\mu(u) \int d^3x\, \varepsilon(x)^{5/4} , \tag{6}$$

and the Hausdorff-Lyapunov dimension of the strange attractor in the infinite dimensional phase space has the *finite* upper bound

$$N \le K_2 \nu^{-9/4} [\int d\mu(u) \int \frac{d^3x}{|\Omega|} \varepsilon(x)^{5/4}]^{3/5} \cdot |\Omega|. \tag{7}$$

$\mu(u)$ is the (unknown) invariant measure of the stationary flow field, $K_1 = 0.030\,624$, $K_2 = 0.174\,906$. These are the results for 3-dimensional flow, extension to arbitrary space dimension d is given in [3,4].

The reader acquainted with turbulence physics will notice that neglecting the spatial fluctuations of $\varepsilon(x)$, i.e. without intermittency, one recovers an old conjecture [2] for the number of degrees of freedom of a turbulent flow from (7):

$$N \cong |\Omega|/\eta^3 \cong Re^{9/4} , \quad \eta \equiv (\nu^3/\varepsilon)^{1/4} , \quad \text{dissipative length.} \tag{8}$$

The old dimensionality argument thus finds its clarifying reinterpretation as the dimension of the strange attractor in phase space.

In atmospheric turbulence $Re \cong 2.3 \cdot 10^7$; then one evaluates from (7) $N \cong 10^{-3} Re^{9/4} \cong 10^{13}$. This seems pretty large as compared with weak turbulence in the onset range, where N is of order unity. But if the volume covered by atmospheric turbulence is estimated as 10km height times $(100\text{km})^2$ area, expressed by the molvolume of 22.4ℓ, one finds $|\Omega| \cong 10^{16}$ molvolumes. So the attractor owns hardly one degree of freedom per mol compared with 10^{23} mol^{-1} on the microscopic level (or ∞ on the continuum limit level). This exemplifies how huge a reduction in the effective number of variables is achieved by viscosity but still sufficiently chaotic, unpredictable motion left.

Instead of only a few, one has a whole spectrum of Lyapunov exponents, since the eigenvalues of H(u) are fairly close if $|\Omega|$ is large. This suggests to introduce a distribution function $dN(a)/da$ of eigenvalues (and correspondingly of the Lyapunov exponents) per a-range [3]. It can be shown to be monotonously decreasing with increasing a, ranges up to the viscous decay rate $\tau_\eta^{-1} = (\varepsilon/\nu)^{1/2}$ (cf.(5)), and has a finite value at $a = 0$ if there is no intermittency. Figure 1 shows the estimates for the β-model in classical approximation for the density of states calculated on the basis of Ruelle's general formulae.

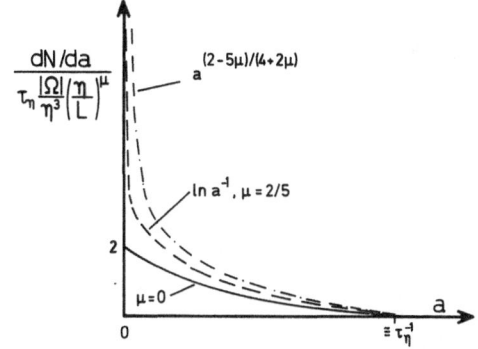

Fig.1 Upper bound for the density of characteristic exponents vs. a. It is finite at $a = 0$ for intermittency exponent $0 \le \mu < 2/5$, gets logarithmic singular at selfsimilarity dimension $D = 3 - \mu = 2.6$ and is infinite if $\mu > 2/5$

3. Turbulent convective transport

The most direct way to consider the chaotic motions of fluid elements in strong turbulence is to observe convection. This leads to a diffusive broadening of an initially small cloud of particles (or, more generally, passive scalars). Since Richardson's work [1] it is clear that it is particle *pair* separation which has to be considered in order to understand fluid dynamics. While single particle motion depends on the flow's gross structure, i.e. on the geometry of the boundaries, a pair's relative diffusion has a chance to be universal, since large eddies only advect both particles simultaneously without separating them and very small eddies are too weak in energy. So two particles separated by a distance r probe the r-eddy dynamics.

One of the great puzzles of turbulent diffusion was its anomalously large extension velocity. If particles are released at a certain time t_0 and watched over a time *interval* t the variance increases algebraically

$$\sigma_t \propto t^\Theta. \tag{9}$$

The astounding point is the magnitude of the exponent Θ. In common Brownian motion [5] one has $\Theta = 1$ due to a fast correlation decay. If successive steps of the particles are correlated, one has Θ = 2, as observed e.g. if a cloud of birds swarms off when startled by an explosion, say. But turbulent spreading is even stronger, namely, Θ ≅ 3. This is experimentally supported beyond doubt; examples of measurements are in Figures 2 and 3. The large value Θ ≅ 3 is equivalent to Richardson's famous finding that the turbulent diffusivity is $d\sigma_t/dt \propto r^{4/3}$ [1]. It is found in measurements of air and water pollution, in governmental control rules which use anomalously large Θ, and in the spreading of rain clouds [6,7].

How can we understand the physical mechanism of the anomalously strong diffusion and, even more important, how can it be based on the Navier-Stokes equation. Let me explain the physical idea. Details are given in a unified theory [11] and its extension to incorporate the difference of transverse and longitudinal spreading in [12]. The key is the observation how energy dissipation ε enters via correlation decay of Lagrangian motion directly from the Navier-Stokes equation [13].

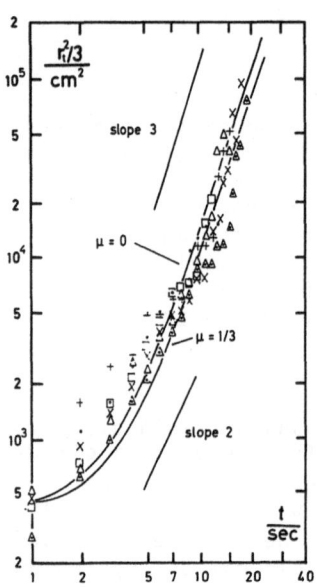
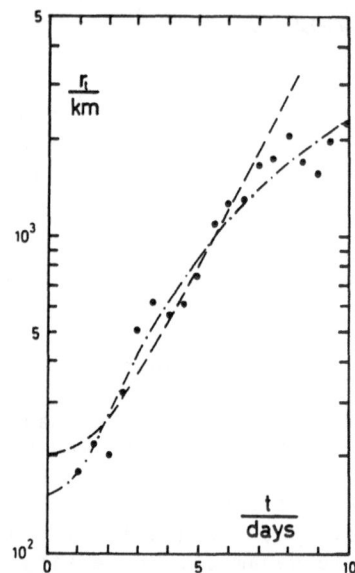

Fig.2 Mean square particle separation of gun powder explosion clouds vs time (Frenkiel, Katz [8], Gifford [9]). Curves from theory are explained later (eq.(12))

Fig.3 Mean separation of balloon pairs vs time of flight in about 14km height (TWERLE-data, evaluated by Lundgren [10]). Dashed curve Lundgren's theory, dash-dotted present theory (see later (12))

The variance can be described as usual by a Kubo-type time integral over the correlation function of the velocities. But note, it is the *Lagrangian* and relative velocity that is responsible for pair separation while both particles are advected, cf.Fig.4. The variance (tensor) for $\delta\vec{R} = \vec{R} - \langle\vec{R}\rangle = \vec{R} - \vec{r}$ is

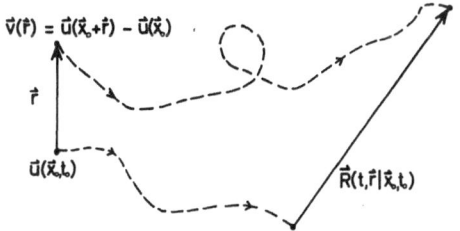

$$\vec{v}(\vec{r}) = \vec{u}(\vec{x}_o + \vec{r}) - \vec{u}(\vec{x}_o)$$

$\vec{r}$

$\vec{u}(\vec{x}_o,t)$

$\vec{R}(t,\vec{r}|\vec{x}_o,t_o)$

Fig.4 Lagrangian motion of a particle pair, released at distance r, observed after time lapse t at distance $\vec{R}(t,\vec{r}) = \vec{r} + \int_o^t d\tau \vec{v}(\tau,\vec{r})$. Statistical averaging over x_o, t_o, position and/or time of release

$$\sigma_t = \langle \delta R \delta R \rangle = \int_0^t d\tau_1 \int_0^t d\tau_2 \ \langle v(\tau_1 r) v(\tau_2 r) \rangle \ . \tag{10}$$

If the correlation decays fast enough, one gets $\sigma_t \propto t$ with a coefficient reflecting the mean energy $\langle v^2 \rangle$ ($\cong \kappa T/m$ for thermal diffusion) *and* the decay rate γ (Einstein: $D = \kappa T/m\gamma$). If the correlation persists, one obtains $\sigma_t \propto t^2$, the coefficient being $\langle v^2 \rangle$. In turbulence this happens since the decay rate turns out to decrease with eddy size [13] and hence to decrease with cloud size. The increase of t and the decrease of correlation decay rate just compensate; a quantitative analysis [11,12] (using relation (11) given later) shows that the v-correlation (10) practically never decays, i.e. the particles are always correlated by that eddy's motion which is appropriate to the respective distance $r_t = \sqrt{\langle R^2(t,r) \rangle}$.

Still, the diffusion exponent derived so far is $\Theta < 2$. The unique feature of turbulent transport which implies the anomalous enhancement is that the *magnitude* of the correlation function, $\langle v^2 \rangle$, itself depends on the extension which the cloud has reached after time t. It is $\langle v^2(t,r) \rangle \propto r_t^{2/3} \propto (t^3)^{1/3} \propto t$, yielding the 3rd power in (9). All this can be thrown in a quantitative closure scheme to calculate σ_t including prefactors and physical flow parameters, cf. [11,12]. The results are the theoretical curves in figures 2,3.

The key observation is how Lagrangrian velocity correlations decay according to the Navier-Stokes equation. In the single exponent approximation (lowest order continued fraction) the decay rate turns out to be [13]

$$\gamma(r) = - \langle v|Lv \rangle / \langle vv \rangle = (2\varepsilon - \nu \Delta_r D(r))/D(r) \ , \tag{11}$$

with the structure function $D(r) = \langle v^2(r) \rangle$. This holds for arbitrary r, in the viscous as well as in the inertial subrange. In the former, $D(r) \propto r^2$, so ε precisely cancels, and $\gamma \to \gamma_o \cong 10^{-1} \tau_\eta^{-1}$. In the inertial subrange $\nu \Delta_r D(r) \ll \varepsilon$, so ε takes over in (11), so determining γ and consequently σ_t. Note that (11) is a consequence of the Navier-Stokes equation (1), no scaling argument needs to be introduced. The following expression summarizes the results, giving a closed expression for the variance.

$$\sigma(r,t) = (r^{4/3 - \mu/9} + Bt^2)^{18/(12-\mu)} - r^2 \ , \tag{12a}$$

$$\text{with } B \equiv (22/9) b_u \varepsilon^{2/3} (Re_\lambda \eta)^{-\mu/9} \ . \tag{12b}$$

347

(log-normal ε fluctuations according to Kolmogorov and Oboukhov were used.) Expressions (12) yield the theoretical curves in the figures 2,3. Molecular viscosity can be included as well [14], i.e. this theory is not restricted to the scaling (inertial) subrange. Work is in progress to finally derive also the structure function using (11) the very root how *the Navier-Stokes equation introduces quite naturally the energy dissipation.*

To close this section a by-product of interest shall be given. The closure leading to (12) implies for the equal time velocity variance $D(r,t) \equiv \langle v^2(r,t) \rangle \equiv D(\sqrt{r^2+\sigma_t})$, cf. Fig.5. Note that it is the *velocity* which asymptotically diffuses "regular", $D(r,t) \propto \varepsilon t$, instead of the position variable's variance as in common Brownian motion.

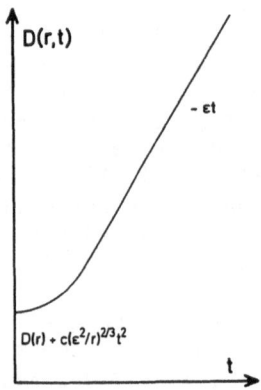

Fig.5 Variance increase with time t of the relative velocity of a particle pair released with distance r

4. Random walk on a fractal

Turbulent relative diffusion was shown to be anomalously enhanced since the medium helps twofold: It sustains the correlation by reducing the time scale for increasing distance scale and it provides the larger eddies with more energy, so the wider-spread particles are favorably spread further in contrast to thermal Brownian motion, where the relevant energy is equal for all pairs, namely $\kappa T/m$. This mechanism reflects the self-similar hierarchical structure of turbulent eddies. In order to study the dynamics of a random walk on such a phase space for the relative distances (which, of course, is no longer translational invariant, al-though the Eulerian flow field $u(x_o,t_o)$ is), we investigated a model, whose geometry and transition rates were stimulated by the imagined properties of turbulence eddy fields, but which has great interest in its own and numerous applications. I briefly describe it now and summa-rize some main results. Details can be found in a series of papers de-voted to the analytical solution of this model [15], [16], [17]. The first mention how turbulent diffusion might be modeled in a scale inva-riant hierarchical manner by a selfsimilar broken linear map implying deterministic (chaos-induced) diffusion was given in [18]. A similar hierarchical model but with thermally activated transitions that have scaling self-similarity is considered in another context in [19]; its be-haviour is investigated by renormalization group transformation.

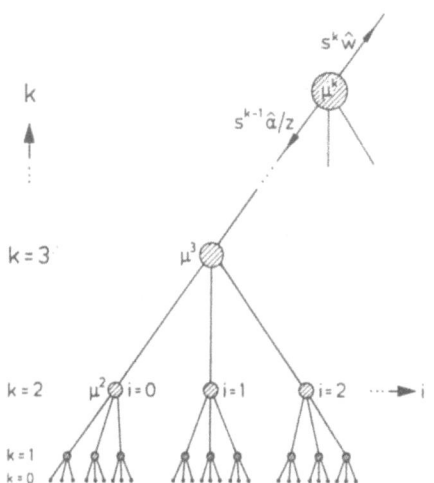

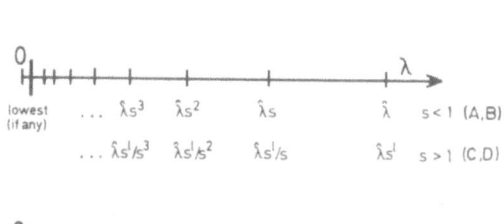

Fig.6 Hierarchical model with lower cut-off and branching number z = 3

Fig.7 Eigenvalues in the various parameter regions, indicated in the inset of table 1

The model is presented in Figure 6. Imagine a cloud of phase points (particles) in the initial interval i = 0 at level k = 0 whose spatial extension is put equal to unity. The particles may climb upwards, becoming spatially more spread on each higher level by an additional factor μ. (In this section μ denotes a length scale factor and not the intermittency exponent as in section 3.) Downward transitions with a branching ratio z serve to homogenize the cloud on the successively reached levels. The fractal dimension of the geometrical structure is $d_F = \ell n z / \ell n \mu$. The dynamics is characterized (i) by the growth ratio $r = \hat{w}/\hat{a}$ of the upward transition rate $\hat{w}$ to the downward rate $\hat{a}$ (each single branch: $\hat{a}/z$) and (ii) by s which scales the transition rates between adjacent levels. In turbulence the constancy of energy dissipation on all levels of eddy size implies length-scale^2/time-scale^3 = const, i.e. $\mu^2/s^{-3} = 1$. We do not restrict to this particular relation but treat μ, s, r as parameters to be chosen differently for different applications.

The striking consequence of the time scaling s ≠ 1 is that the eigenvalues accumulate geometrically towards zero (Fig.7). If s = 1 the bulk of eigenvalues is bounded and approximately equally distributed. Whenever r ≤ 1 and rs ≤ 1 there is an additional separate smallest eigenvalue dominating the moment growth as well as the correlation decay. Each finite set of $\ell + 1$ eigenvalues describes the eigenstates in the space of all trajectories that never exceed the finite level ℓ; of course t → ∞ needs ℓ → ∞.

We found various kinds of moment growth or diffusion behaviour which is summarized in table 1. It may be exponential (rate a = (μ^m-1) $(1-r^{-1}\mu^{-m})$), algebraical with exponents Θ that depend differently on the dynamical parameters r, s, and on the geometry μ, and sometimes logarithmic corrections show up. The exponents of the anomalous m-th moment growth are

Table 1 Eigenvalues and diffusion laws in the various parameter regions, which are defined in the inset (from [16])

Region	Eigenvalues $\lambda_i^{(\ell)}$ lowest eigenvalue ($i = 0$)	other eigenvalues ($i > 0$)	Diffusion behaviour of $\langle x^m \rangle (t)$ for $\mu^m rs < 1$	$\mu^m rs = 1$	$\mu^m rs > 1$
A				t^{Θ_v}	
AB	$\sim s^{\ell}/\ell$	$\sim s^{\ell-i}$	$(t/\ell nt)^{\Theta_v}$	$t/\ell nt$	$t^{\Theta_v}/\ell nt$
B				$t\,\ell nt$	$t^{\Theta_>}$
BC	$c(rs)^{\ell}$	bounded	$t^{\Theta_<}$	t^2	e^{at}
C					
CD	$\sim 1/\ell$	$\sim s^{i}$			∞
D					
DA	bounded				e^{at}

Inset diagram (regions A, B, C, D; axes $\ell n\,r$ vertical, $\ell n\,s$ horizontal, line $rs=1$).

$$\Theta_v(m) = m\,\ell n\mu / \ell ns^{-1} , \tag{13}$$

$$\Theta_<(m) = \ell n\mu^m / \ell n(rs)^{-1} , \quad \text{always} < 1 , \tag{14}$$

$$\Theta_>(m) = \ell n(r\mu^m)/\ell ns^{-1} , \quad \text{always} > 1 . \tag{15}$$

Application to a turbulence-like system ($\mu^2 s^3 = 1$) yields quite different behaviour for the variance growth (i.e. for $m = 2$) depending on the growth ratio r. The relevant exponents Θ are the following. If $1 < r$, $\Theta_v = 3$; if $s^2 < r < 1$, $\Theta_> = 3 - \ell nr^{-1}/\ell ns^{-1} \in (1,3)$; if $r < s^2 < 1, \Theta_< = 3/(1 + \ell nr^{-1}/\ell ns^{-1}) < 1$. Note that the dynamical fluctuations which are fully included in the model's solution, tend to reduce the exponent if the transitions are dominantly downwards, $r < 1$. This may be an indication that in real turbulence these fluctuations might counteract the effects of intermittency in the static structure function $D(r) \propto r^{2/3+\mu/9}$, which in turn enhances the moment's exponent. It may be this kind of counter-balance that makes the influence of fluctuations so small that the experimental values of Θ are very near 3, the value of the mean field theory.

We have also studied [16] the decay of correlations, in particular the probability to find the particle in the initial set ($k = 0$, $i = 0$) again after time t provided it started here at time 0.

$$p_{0,0}(t) \propto t^{-\nu}, \quad \nu = \ell n z / \ell n(rs)^{-1}, \quad r < 1 \text{ and } rs < 1 . \tag{16}$$

Applied to a system with activation type transitions as for instance for glassy systems we may choose the following interpretation. Let $k = 0$ denote states of local minima, $k > 0$ saddle point states between adjacent valleys, $rs = Ke^{-\varepsilon/T}$ the rates, with ε activation energy, T temperature, K effective number of saddle points. This leads to a temperature dependent algebraic correlation decay, also derived in [19], and quite unique for these hierarchical systems

$$\nu = T\ell n z / (\varepsilon - T\ell n K) . \tag{17}$$

The Fourier spectrum in such systems is of $1/f$-type; more precisely $\tilde{p}_{0,0}(f) \propto 1/|f|^\kappa$ with $\kappa = 1 - \nu$. In particular for vanishing temperature $T \to 0$ there is long lasting correlation $\nu \propto T \to 0$ and $1/|f|$ noise. If T increases, correlations decay increasingly faster, $\nu \to \infty$, and at $T_c = \varepsilon/\ell n K$ the long-lasting correlation disappears. That may be interpreted as the glass transition.

There is a universal, temperature independent relation between moment growth and correlation decay if $\mu^m rs < 1$, i.e. if $\Theta_<(m) < 1$.

$$\nu = d_F \Theta_<(m)/m , \quad d_F \text{ fractal dimension.} \tag{18}$$

This generalizes the well-known relation between random walk dimensionality $\Theta(\equiv\Theta_<(2))$, spectral dimension d_s, and fractal dimension d_F as given in [20,21,22]

$$2\nu \equiv d_s = d_F \cdot \Theta. \tag{19}$$

This was derived originally for fractals with reduced random walk, $\Theta < 1$. As the more general results (17) and (14), (15) show, the universal relation (18) must be substituted by a connection between correlation and diffusion that contains more details of the dynamical process and depends on temperature in activation-type systems if T exceeds $T_1 \equiv \varepsilon/\ell n(zK)<T_c$.

This general relation has the following form: (i) If $\mu^m rs = 1$, we have $\nu = d_F/m$ and $\langle x^m \rangle$ is $t\ell nt$ in B, t^2 in BC, ∞ in C. (ii) If $\mu^m rs > 1$ we have $\nu/\Theta_>(m) = \zeta d_F/m$ with $\zeta = (1+\ell nr/\ell n(rs)^{-1})/(1+\ell nr/\ell n\mu)$ explicitly depending on the dynamics, and $\langle x^m \rangle \propto t^{\Theta_>}$ in B, $\propto e^{at}$ in BC, infinite in C.

Experimental tests of the relations between algebraic correlation decay and anomalous variance growth of diffusion, $\Theta > 1$, thus would provide much insight into the geometrical structure as well as the dynamical process that both characterize the hierarchical structure.

1. L.F.Richardson, Proc.Roy.Soc. London, Ser. A110,709(1926)
2. L.D.Landau, E.M.Lifshitz, Fluid Mechanics, Oxford etc: Pergamon Press, 1959;1975
3. D.Ruelle, Comm.Math.Phys. 87,287(1982);93,285(1984)

4. E.Lieb, Comm.Math.Phys. 92,473(1984)
5. A.Einstein, Ann.Phys. (Leipzig) 19,289(1906)
6. S.Lovejoy, Science 216,185(1982)
7. H.G.E.Hentschel, I.Procaccia, Phys.Rev. A29,1461(1984)
8. F.N.Frenkiel, I.Katz, J.Meteorol. 13,388(1956)
9. F.Gifford, J.Meteor. 14,410(1957)
10. T.S.Lundgren, J.Fluid Mech. 111,27(1981)
11. S.Grossmann, I.Procaccia, Phys.Rev. A29,1358(1984)
12. H.Effinger, S.Grossmann, Phys.Rev.Lett. 53,442(1984)
13. S.Grossmann, S.Thomae, Z.Phys. B49,253(1982)
14. S.Grossmann, I.Procaccia, P.S.Stern, Phys.Letters 104A,140(1984)
15. F.Wegner, S.Grossmann, Z.Phys. B59,197(1985)
16. S.Grossmann, F.Wegner, K.-H.Hoffmann, J.Phys.Lettres 46,L-575(1985)
17. K.-H.Hoffmann, S.Grossmann, F.Wegner, Preprint 1985, Random walk
    on a Fractal: Eigenvalue Analysis, Z.Phys. B 60, 3/4 (1985)
18. S.Grossmann, S.Thomae, in: Multicritical Phenomena, Geilo Adv.
    Stud.Inst. April 10-21,1983, eds R.Pynn, A.T.Skjeltorp, pp.423-
    450, New York, London: Plenum Press, 1984
19. B.A.Huberman, M.Kerszberg, J.Phys. A18, L331(1985)
20. R.Rammal, G.Toulouse, J.Physique Lettres 44,L-13(1983)
21. S.Alexander, R.J.Orbach, J.Physique Lettres 43,L-625(1982)
22. B.O'Shaughnessy, I.Procaccia, Phys.Rev.Lett. 54,455(1985)

Cooperative Effects and Superradiance in Compton Scattering and Their Relevance to Free Electron Lasers

R. Bonifacio and F. Casagrande

Dipartimento di Fisica dell'Università, Via Celoria 16,
I-20133 Milano, Italy

1. Relativistic Mirrors, Collective Compton Effect and Free Electron Lasers

The problem of obtaining a source capable of producing coherent radiation in the whole electromagnetic spectrum, which would be an ideal tool for research in physics and biology, could be easily solved if we had *relativistic mirrors* at our disposal; that is, mirrors moving at a velocity v_0 close to the velocity of light c. Let a monochromatic laser field, of frequency ω_i, impinge on such a mirror. The radiation would be reflected back with the same frequency, $\omega'_r = \omega'_i$, in the reference frame S' moving with the mirror; however, a Lorentz transform to the laboratory frame shows that the reflected frequency ω_r would appear much higher than the incident frequency ω_i :

$$\omega_r = [(1+\beta)/(1-\beta)]\omega_i \simeq 4\gamma^2\omega_i \ , \quad \text{where} \tag{1}$$

$$\beta = v_0/c \simeq 1, \quad \gamma = (1-\beta^2)^{-1/2} \gg 1 \ . \tag{2}$$

Hence (1) describes a very simple, though ideal, up-conversion process which would allow the generation of coherent radiation tunable on a very broad range just by varying the incident frequency and the velocity of the mirror.

Actually, if we consider the well-known Compton effect, that is the electron-photon scattering, an electron can play the role of a *microscopic* relativistic mirror, since the backscattered radiation has the same frequency of the incident one in the electron rest frame, provided that the Compton shift can be neglected. However, the radiation backscattered from a *beam* of N relativistic electrons produced by an accelerator is *incoherent*, with intensity proportional to N, due to the random distribution of the particle positions in a radiation wavelength. On the other hand, if the beam is spatially modulated so that the electrons are bunched, a *coherent* backscattering can take place, with intensity as N^2. Most interesting is the case in which the electrons *self-bunch*, without any preparation but only by virtue of their interaction with the common radiation field. This synergetic behaviour is just what occurs in a novel source of coherent radiation, the *free-electron laser* (FEL) [1].

In the basic configuration of this device a beam of relativistic electrons interacts with a radiation field and a transverse, spatially periodic magnetostatic field in such a way that the radiation gets amplified at the expenses of the kinetic energy of the particles. Let us consider a beam injected into the magnetic field array (undulator or wiggler), parallel to its axis. The magnetic field bends the electron trajectories so that the particles radiate even in the absence of pre-existing radiation: this is the FEL *spontaneous* emission. This radiation is strongly peaked in the forward direction and sharply centered on a frequency

$$\omega_{sp} = \omega_0/(1-\beta_{\parallel}) \simeq 2\gamma_{\parallel}^2\omega_0 \ , \tag{3}$$

353

that is the frequency associated with the spatial periodicity of the wiggler, $\lambda_0 = 2\pi c/\omega_0$, as "seen" by the relativistic electron. In (3)

$$\gamma_\parallel^2 = \gamma^2/(1 + K^2) \ , \quad K = eB_0\lambda_0/2\pi m_0 c^2 \ , \tag{3'}$$

B_0 being the magnetic field amplitude. Hence ω_{sp} can be changed by varying the electron energy or the wiggler parameters (FEL tunability).

The transverse velocity acquired by the electrons due to the magnetostatic field can couple to a radiation (or laser) field of frequency ω_L provided that this field is nearly resonant, that is

$$\omega_L \simeq \omega_{sp} \simeq 2\gamma_\parallel^2\omega_0 \ . \tag{4}$$

Thus an energy-transfer occurs between the particles and the field, leading under suitable conditions to FEL *stimulated* emission.

This is the picture in the laboratory frame. On the other hand, in the electron rest frame the magnetostatic field is felt by the particles as a radiation field, the so-called pseudoradiation field (PRF) (Weizsäcker-Williams approximation [2]). Hence the FEL process can be described as a Compton scattering of PR photons into laser photons. A Lorentz transform shows that the PRF frequency in the laboratory frame is

$$\omega_{pr} = \omega_0/2 \ . \tag{5}$$

It follows that the FEL resonance condition (4) can be written

$$\omega_L \simeq 4\gamma_\parallel^2\omega_{pr} \ , \tag{4'}$$

namely, it takes exactly the form (1), with bunched electrons playing the role of the mirror. In the next sections we show that indeed in the FEL process the electrons can self-organize so that a collective, stimulated Compton effect takes place.

2. The Hamiltonian Model

Let us consider the relativistic Hamiltonian for one electron interacting with an electromagnetic field

$$H = [c^2(\boldsymbol{p} - e\boldsymbol{A}/c)^2 + m_0^2 c^4]^{1/2} \ , \tag{6}$$
$$\boldsymbol{A}(z,t) = \boldsymbol{A}_L(z,t) + \boldsymbol{A}_w(z,t) \ ,$$

where the vector potential $\boldsymbol{A}$ is the sum of a "laser" contribution $\boldsymbol{A}_L$ and a "wiggler" contribution $\boldsymbol{A}_w$. The p_x, p_y components of the electron canonical momentum are constants of motion; we can set them equal to zero if we assume that initially $\boldsymbol{p} \equiv p_z$. It follows that $\boldsymbol{A} \cdot \boldsymbol{p} \equiv \boldsymbol{A}_\perp \cdot \boldsymbol{p}_\parallel = 0$ in (6). Also, we neglect $|\boldsymbol{A}_L|^2$ with respect to $|\boldsymbol{A}_w|^2$ since the wiggler field plays the role of an intense pump field. Furthermore, the transverse motion can be eliminated by introducing a renormalized electron mass [3]

$$m = m_0(1 + K^2)^{1/2} \ , \tag{7}$$

thus obtaining a one-dimensional Hamiltonian $H(z, p \equiv p_z)$

$$H = \{p^2 c^2 + m^2 c^4 + 2e^2 \boldsymbol{A}_L \cdot \boldsymbol{A}_w\}^{1/2} \ . \tag{8}$$

By a Lorentz transform in the z-direction we can describe the dynamics in a reference frame, moving at a velocity close to the initial electron velocity $v_0 \simeq c$, in which the nonrelativistic approximation holds and quantization is easily performed. If we consider two counterpropagating, circularly polarized field modes

$$
\begin{aligned}
A_L(z,t) &= \hat{e} Q_L \exp[ik_L(z - ct)] + \hat{e}^* Q_L^+ \exp[-ik_L(z - ct)] , \\
A_w(z,t) &= i\{\hat{e} Q_w \exp[-ik_w(z - ct)] - \hat{e}^* Q_w^+ \exp[ik_w(z + ct)]\} ,
\end{aligned}
\tag{9}
$$

and if we pass from one to N electrons setting the velocity of the moving frame equal to the mean initial electron velocity $\langle p \rangle_0 / m$, we obtain the many-particle, one-mode quantized Hamiltonian

$$
H = \sum_{i=1}^{N} p_i^2 / 2m + i\hbar g [Q_w Q_L^+ \sum_{J=1}^{N} \exp(-ikz_J) - \text{h.c.}] - \hbar \Delta Q_L^+ Q_L ,
\tag{10}
$$

$$
[Q_L, Q_L^+] = [Q_w, Q_w^+] = 1 , \quad [z_i, p_J] = i\hbar \delta_{ij} ,
$$

where $k = k_L + k_w$, g is a coupling constant and Δ a detuning parameter [4]. Hamiltonian (10) admits the constants of motion

$$
Q_L^+ Q_L + Q_w^+ Q_w = \text{const}_1 ,
\tag{11a}
$$

$$
\sum_{i=1}^{N} p_i + \hbar k_L Q_L^+ Q_L - \hbar k_w Q_w^+ Q_w = \text{const}_2 .
\tag{11b}
$$

Equation (11a) describes the exchange of laser and wiggler photons, (11b) the momentum conservation in the Compton scattering process; (11a) and (11b) can be combined into a single equation,

$$
\sum_{i=1}^{N} p_i / \hbar k + Q_L^+ Q_L = \text{const} ,
\tag{12}
$$

which illustrates the basic interaction process in the model, namely the transfer of momentum between the particles and the laser field. In particular, gain (loss) for the field occurs simultaneously with deceleration (acceleration) of the particles. A realistic assumption is to treat the wiggler field as a classical, constant field. By a suitable scaling, Hamiltonian (10) can be written in dimensionless form [4]

$$
H = \sum_{i=1}^{N} \bar{p}_i^2 / 2 + iw [Q_L^+ \sum_{J=1}^{N} \exp(-iO_J) - \text{h.c.}] - \bar{\Delta} Q_L^+ Q_L ,
\tag{13}
$$

$$
[O_i, \bar{p}_J] = i\delta_{iJ}, \quad [Q_L, Q_L^+] = 1 ,
$$

where O_i are the electron phases (positions in a radiation wavelength), $\bar{p}_i$ the normalized electron momenta, $\bar{\Delta}$ the detuning parameter and w the coupling constant which includes the wiggler field amplitude. From Hamiltonian (13) one derives the Heisenberg equations of motion ($\bar{t} = 2w_0 t$) :

$$dO_i/d\bar{t} = \bar{p}_i \ ,$$

$$d\bar{p}_i/d\bar{t} = -w[Q\exp(iO_i) + Q^+\exp(-iO_i)] \ ,$$

$$dQ/d\bar{t} = w\sum_{J=1}^{N}\exp(-iO_J) + i\Delta Q \ . \tag{14}$$

In the classical limit, that is operators replaced by c-number variables, (14) turn into a set of Hamilton equations, which have been investigated by us a few years ago [5]. The analysis can be simplified by a further scaling, which reduces the relevant parameters from two to one (the detuning parameter). In this way (14) become formally identical to the classical equations derived in the laboratory frame in [6], provided that we assume $(\gamma_i - \gamma_0)/\gamma_0$, $(\gamma_0 - \gamma_R)/\gamma_R \ll 1$, γ_R being the resonance electron energy in units m_0c^2, and space-charge effects are neglected [7]. These equations read

$$dO_i/d\tau = \eta_i \ , \tag{15a}$$

$$d\eta_i/d\tau = -[A\exp(iO_i) + A^*\exp(-iO_i)] \ , \tag{15b}$$

$$dA/d\tau = N^{-1}\sum_{J=1}^{N}\exp(-iO_J) + i\delta A \ , \tag{15c}$$

where

$$O_i = (k + \kappa_0)z_i - (\omega/2\omega_0 + \delta)\tau \ ,$$
$$\eta_i = \varrho^{-1}(\gamma_i - \gamma_0)/\gamma_0 \ , \quad |A|^2 = \varrho^{-1}(|E_0|^2 V/4\pi)/N\gamma_0 m_0c^2 \ ,$$
$$\tau = 2\omega_0\varrho t \ , \quad \delta = \varrho^{-1}(\gamma_0 - \gamma_R)/\gamma_0 \tag{16}$$
$$\varrho = [(K/4)(\Omega_p/\omega_0)]^{2/3} \ , \quad \Omega_p = (4\pi e^2 N/m_0 V\gamma_0^3)^{1/2}.$$

In (16) ϱ is a generalized Pierce parameter and Ω_p the relativistic plasma frequency. A further approximation used to derive (15c) will be discussed in Sect. 4. The constant of motion reads

$$\langle\eta\rangle + |A|^2 = \text{const} \ , \tag{17}$$

that is the electron+field energy conservation

$$m_0c^2\sum_{i=1}^{N}\gamma_i + |E_0|^2 V/4\pi = \text{const} \ . \tag{17'}$$

Let us remark that if the field amplitude remains nearly constant during the interaction, $A(\tau) \simeq A_0$, (15c) can be neglected while (15a,b) become the equations for N uncoupled pendula,

$$d^2O_i/d\tau^2 = -2A_0\cos O_i \ . \tag{18}$$

The first FEL experiments can be described fairly well in the approximation of weakly coupled pendula [8]. In this regime the electrons radiate almost independently so that the emitted intensity is proportional to the number N of radiating particles. Clearly, cooperative effects are to be expected in the opposite limit of sensible field variations. In this case, since the complex field amplitude A depends on the phases of all electrons (15c), any electron can "talk" to all others via the common radiation field (15a,b).

356

3. Collective Instability and Amplified Spontaneous Emission Regime

The initial condition for the evolution equations (15a−c) with no field excitation and a monokinetic, unbunched electron beam

$$A(0) = 0, \quad \bar{\eta}_i(0) = 0, \quad \sum_{J=1}^{N} \exp(-i O_J^0) = 0 \tag{19}$$

is an equilibrium condition. The linear stability analysis is remarkably simplified by the introduction of two suitably defined electron collective operators, which allow the reduction of the number of linear equations from $2N + 2$ to $2 + 2$, independently of N [5,6]. It turns out that the system is stable for values of the detuning parameter

$$\delta > \delta_T = 3/2^{2/3} \simeq 1.89 \tag{20}$$

and unstable for $-\infty < \delta < \delta_T$. In the former case, the system is "below threshold" in the sense that the radiated intensity is merely oscillating in time; in this regime one recovers the picture of weakly coupled pendula which leads to the FEL small-signal gain. In the latter case, the system is "above threshold" and exhibits an exponential gain before saturation effects set in. One can describe these different regimes in terms of the *electron bunching parameter*

$$b = N^{-1} \sum_{J=1}^{N} \exp(-i O_J), \quad 0 \leq |b| \leq 1 . \tag{21}$$

The quantity (21) is the order parameter of this problem and plays a role analogous to that of polarization in atomic lasers [9], acting as a source for the field (15c). In the small-signal regime the field does not correlate the particles, so that an initially unbunched beam, $b_0 \simeq 0$, remains almost completely unbunched during the whole process, i.e., $|b(t)| \ll 1$. By contrast, in the collective unstable regime the particles are correlated through the field, and self-bunching occurs such that b reaches peak values close to one; this is the regime of *amplified spontaneous emission* (ASE). Recently, a high-gain FEL has been operated in this regime by a Livermore-Berkely collaboration [10]. In this experiment, exponential gain has been demonstrated with a peak value of 80 MW for operation in the millimeter range ($\lambda = 9$ mm). FEL operation in the ASE regime should provide a source of coherent radiation even in the VUV or soft X-ray regions; note that the single-passage amplifier configuration does not require any optical element, whereas the oscillator mode of operation is affected by the poor efficiency of mirrors at such short wavelengths. Furthermore, very-high gain in this range should be achieved due to the recently predicted optical guiding effect, that is the trapping of radiation in the coherent interaction with the electron beam like in an optical fiber [11].

The equations of motion (15a−c) have been numerically integrated starting close to the equilibrium condition (19), e.g., by introducing an initial value of the bunching parameter $b_0 \neq 0$ which simulates noise. Indeed, the main source of fluctuations is the electron shot-noise, that is the randomness of the electron positions in the beam. In the instability region, $\delta \lesssim \delta_T$, the bunching of particles makes them radiate in such a way that the normalized peak intensity $|A|_p^2$ reaches values on the order of one. Going back to the original variables, this means that the peak intensity I_p is proportional to $N^{4/3}$, whereas $I_p \propto N$ in the small-signal regime; this proves the cooperative nature of the process.

The initial noise level b_0 plays a relevant role in the ASE regime. In fact, the radiation build-up time (or delay time) t_D, that is the time of the first radiated peak, turns out to depend critically on b_0; clearly, it is crucial that t_D does not exceed the transit time in the wiggler, $t_i = L_0/c$. In [12] it is shown that this condition leads to the threshold condition

$$4\pi\varrho N_0 \gtrsim \ln\sqrt{N_\lambda} \, , \tag{22}$$

where $N_0 = L_0/\lambda_0$ is the number of wiggler periods and N_λ the number of electrons per radiation wavelength.

While the short-time behaviour is relevant in connection with the experiments on high-gain FEL amplifiers, the long-time behaviour is an interesting problem in nonlinear Hamiltonian dynamics. We had already shown the occurrence of Hamiltonian chaos in the present FEL model [5]. In particular, the system undergoes a stochastic transition from ordered to chaotic motion for decreasing values of the detuning parameter δ, that is the control parameter in (15a–c), as proven both by the criterion of the local instability of motion and by the Poincaré surfaces of section. Now we are reconsidering this problem due to the more mature status-of-art of the experiments. Some numerical results are reported in Figs. 1a–d. While the underlying chaotic electron dynamics is fully confirmed, as shown e.g. by the behaviour of the electron bunching parameter (Fig. 1b) or the momentum spread (Fig. 1d), the time evolution of macroscopic quantities, such as the radiated intensity (Fig. 1a) or the mean electron momentum (Fig. 1c) looks less irregular, for a sufficiently high number of particles, than it appeared from previous numerical results with few electrons. In particular, for $N \gtrsim 50$, one finds pulsations of these quantities (Figs. 1a–c) with a nearly well-defined period, though the amplitudes are not regular at all. While the meaning of this quasi-periodicity deserves further investigations, this behaviour confirms the occurrence of a process of self-organization in the system when it operates in the ASE regime.

4. Dissipative Model and Superradiant Regime

In the previous section we have seen that in the ASE regime the electrons exhibit a strong self-bunching, and the radiated power from N bunched charges is proportional to $N^{4/3}$. On the other hand, according to the interpretation of the FEL process as a many electron, stimulated Compton scattering (see Sect. 1), one is led to expect that under suitable conditions the peak power should scale as N^2, that is, superradiant emission. In this section we present the basic ideas and some analytical and numerical results on the novel concept of *superradiance* (SR) in the interaction between radiation and "free" electrons [13].

In the derivation of the evolution equations (15a–c) it is assumed that the difference between the velocity of light and the velocity of the copropagating electrons, or slippage effect, is negligible during the interaction time t_i [6]. It follows that a photon emitted by an electron can be reabsorbed by another one before it may leave the undulator. Indeed, Figs. 1a,c show the continuous energy exchange between particles and field in the saturation regime. However, the reabsorption process should be drastically reduced if the photons were allowed to escape from the electron beam at a rate much faster than the energy-transfer rate. In order that this fast removal of photons may occur, first of all a short-pulse regime is required for the electron beam so that the slippage cannot be neglected. Furthermore, the model equations (15a–c) must be generalized to describe the photon escape from an electron bunch length L_e; this implies to include dissipation in the dynamics. The simplest way to do that is by adding a linear damping term in the field equation (15c), which

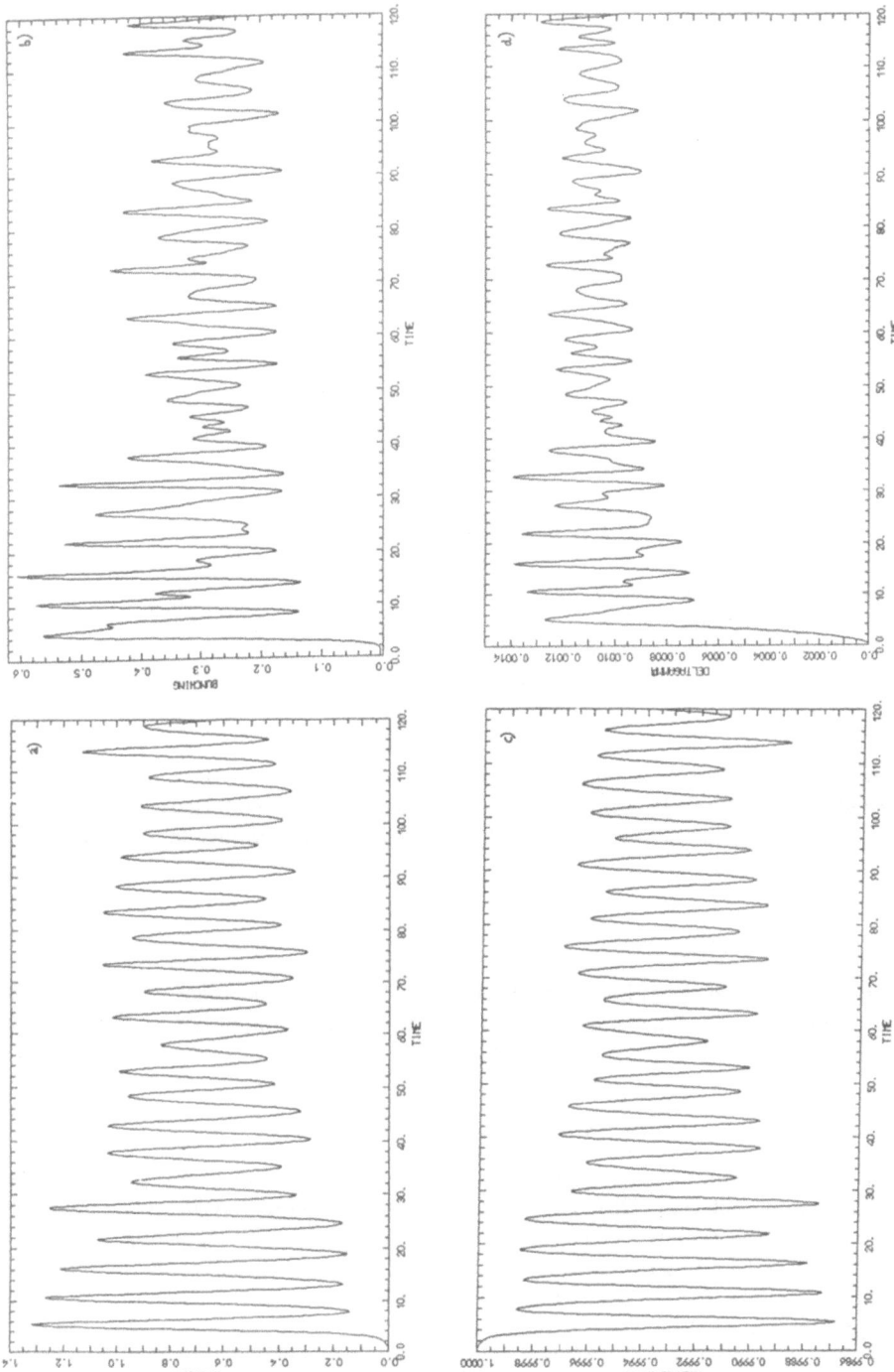

Fig. 1a–d. Time evolution of **(a)** field intensity $|A|^2$, **(b)** electron bunching parameter $|b|^2$, **(c)** mean electron energy $\langle\gamma\rangle/\langle\gamma\rangle_0$, **(d)** electron energy spread $\Delta\gamma/\langle\gamma\rangle$, from (15a–c), for $N = 100$, $\delta = 0$, $A(0) = \bar{\eta}_i(0) = 0$, $|b|_0 \simeq 0.07$

becomes

$$dA/d\tau = b + (i\delta - \bar{k})A \ , \tag{23}$$

where (21) has been used and the normalized field amplitude decay-rate is

$$\bar{k} = k/2\omega_0\varrho = c(1 - \beta_{||}^0)/2\omega_0\varrho L_e \equiv S/G \simeq \lambda_L/4\pi\varrho L_e \ . \tag{24}$$

In (24) the field damping constant k is taken simply equal to the inverse of the photon time-of-flight through the electron bunch; furthermore, we have introduced the slippage parameter S and the total gain G defined as

$$S = c(1 - \beta_{||}^0)L_0/L_e, \quad G = 4\pi\varrho N_0 \ . \tag{24'}$$

Note that S is the ratio of the slippage distance $c(1 - \beta_{||}^0)L_0 \simeq \lambda_L N_0$ to the electron bunch length L_e. From (15a,b) and (23) one derives the dissipation law

$$(d/d\tau)(\langle\eta\rangle + |A|^2) = -2\bar{k}|A|^2 \ , \tag{25}$$

which generalizes the conservation law (17) for $\bar{k} \neq 0$.

A necessary condition for SR is that the photon escape time is much shorter than the interaction time, or *radiation suppression condition*

$$\bar{k} = S/G \gg 1 \ . \tag{26}$$

The previously discussed ASE regime is valid in the opposite limit $\bar{k} \ll 1$. In the limit (26) the field variables can be adiabatically eliminated from the evolution equations. This amounts to dropping the time derivative of the complex field amplitude A in (23), thus obtaining the adiabatic expression

$$A = [(\bar{k} + i\delta)/(\bar{k}^2 + \delta^2)]b \ . \tag{27}$$

By taking the squared modulus on both sides of (27) and going back to the original variables, we derive the following expression of the power radiated out of the electron bunch [13]

$$\begin{aligned}P(t) &= |E_0|^2(t)cS_e/4\pi \\ &= 4\gamma_0^2(B_0^2c/4\pi)(\lambda_0^2/S_e)[1 + (\delta/\bar{k})^2]^{-1}[r_eN|b|(t)]^2 \ , \end{aligned} \tag{28}$$

where S_e is the electron bunch cross-section and $r_e = e/m_0c^2$ is the classical radius of the electron. It is apparent from (28) that when the bunching parameter b reaches values on the order of one, the radiated power exhibits a dependence on N^2. Hence the collective instability and the field radiation reaction, like in ASE, *plus* the radiation suppression condition (26), make the system display the cooperative superradiant behaviour. We stress that no preparation of the system is required, rather the dynamics is such that the electrons self-organize to radiate cooperatively. In this sense the phenomenon should be called more properly superfluorescence [14]. Equation (28) admits a transparent physical interpretation. According to the discussion in Sect. 1, in the electron rest frame the FEL process can be described classically as a Thomson backscattering of the pseudoradiation field into the laser field. In the laboratory frame the PRF has an amplitude $B_{\rm pr}$ and a wavelength

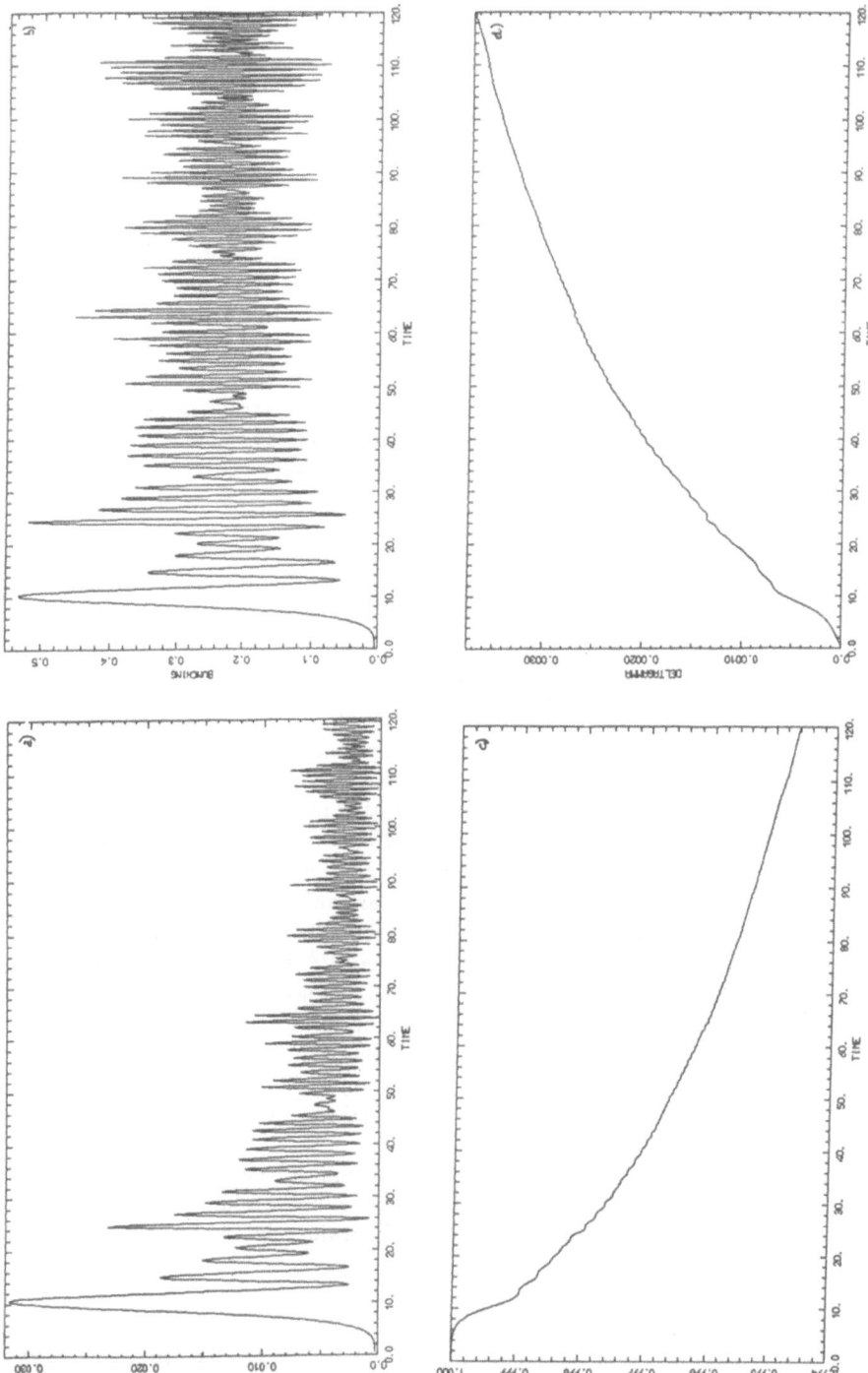

Fig. 2a–d. Time evolution of (a) field intensity $|A|^2$, (b) electron bunching parameter $|b|^2$, (c) mean electron energy $\langle\gamma\rangle/\langle\gamma\rangle_0$, (d) electron energy spread $\Delta\gamma/\langle\gamma\rangle$, from (15a,b) and (23), for $k = 4$ and all other parameters as in Fig. 1

λ_{pr} such that

$$B_{\mathrm{pr}}\lambda_{\mathrm{pr}} = B_0\lambda_0 \ . \tag{29}$$

Hence (28) can be written (on resonance, $\delta = 0$, for the sake of simplicity)

$$P = 4\gamma_0^2(B_{\mathrm{pr}}^2 c/4\pi)(\lambda_{\mathrm{pr}}^2/S_e)(r_e N|b|)^2 \ . \tag{30}$$

Equation (30) is the power radiated by N electrons in a *coherent Thomson scattering* of the PRF: $B_{\mathrm{pr}}^2 c/4\pi$ is the pseudoradiation intensity, $(r_e N|b|)^2$ the coherent Thomson cross-section, $\lambda_{\mathrm{pr}}^2/S_e$ a diffraction solid angle, and $4\gamma_0^2$ a kinematic factor due to the relativistic motion of the radiating charges with respect to the observer [2,15]. Alternative derivations of the result (30), as well as different ways of introducing the field damping constant k, are discussed in [13].

Like in ASE one must impose that (at least) the first superradiant pulse is emitted before interaction is switched off, that is a threshold condition which adds to the radiation suppression condition. Since the SR time-scale can be estimated as [13]

$$\tau_s \simeq \sqrt{k} = \sqrt{S/G} \tag{31}$$

one must impose that $\tau_s \ll \tau_i$, where $\tau_i = 2\omega_0\varrho L_0/c = G$ is the dimensionless interaction time [see (16), (24')]. By putting this condition together with the SR condition (26) we obtain the double inequality

$$S^{1/3} \ll G \ll S \ , \tag{32}$$

which implies $S \gg 1$. Also, note that the SR condition $G \ll S$ can be written as $L_e \ll L_c$ where $L_c = \lambda_L/4\pi\varrho$ is the cooperation length of this problem.

We have suggested values of the parameters suitable for the observation of superradiant effects in a single-passage FEL amplifier, operating in the high-gain regime at sufficiently long wavelength (e.g., in the infrared) and with sufficiently short electron bunches (e.g., $L_e \lesssim 2\,\mathrm{mm}$). Our proposal is supported by the numerical results obtained from the evolution equations (15a,b) and (23), integrated with an initial condition close to the equilibrium condition (19). In particular, the validity of the adiabatic elimination regime has been checked for $\bar{k} \gtrsim 3$, in agreement with the condition $\bar{k}^{3/2} \gg 1$, (32) [13]. In Figs. 2a–d we present the time behaviour of the same quantities of Figs. 1a–d, for the same values of all parameters (except the damping rate $\bar{k} \neq 0$), and for a value, $\bar{k} = 4$, such that the radiation suppression condition is fulfilled. Unlike in the ASE regime, we see that most of radiation is emitted in the first pulse (Fig. 2a); accordingly, the mean electron momentum (spread) decreases (increases) in an almost monothonical way, see Fig. 2c (Fig. 2d), instead of oscillating as in Fig. 1c (Fig. 1d).

In conclusion, SR in high-gain FEL amplifiers appears to be a process capable of generating very short radiation pulses with high peak power. From a fundamental viewpoint, this process is a novel, nice example of synergetic behaviour in the transient radiation-matter interaction.

References

1. See e.g. *Free Electron Lasers*, ed. by S. Martellucci and N.A. Chester (Plenum, New York, 1983)
2. J.D. Jackson: *Classical Electrodynamics*, 2nd ed. (Wiley, New York 1975)
3. A. Bambini, A. Renieri: Lett. Nuovo Cimento **21**, 399 (1978)
4. R. Bonifacio, F. Casagrande: Nucl. Instr. Methods (in press)

5. R. Bonifacio, F. Casagrande, G. Casati: Opt. Commun. **40**, 219 (1982);
   R. Bonifacio, F. Casagrande, G. Casati: in *Evolution of Order and Chaos, ed. by* H. Haken Springer, Berlin, Heidelberg, New York 1982) p. 248;
   R. Bonifacio, F. Casagrande, G. Casati, S. Celi: in *Coherence and Quantum Optics*, vol. V, ed. by L. Mandel and E. Wolf (Plenum, New York 1984) p. 801
6. R. Bonifacio, C. Pellegrini, L. Narducci: Opt. Commun. **50**, 373 (1984)
7. J.B. Murphy, C. Pellegrini, R. Bonifacio: Opt. Commun. **53**, 197 (1985)
8. See e.g. C. Pellegrini [Ref. 1, p. 91];
   W.B. Colson [Ref. 1, p. 189]
9. See e.g. H. Haken: *Laser Light Dynamics* (North-Holland, Amsterdam 1985)
10. T.J. Orzeckowski, B. Anderson, W.M. Fawley, D. Prosnitz, E.T. Scharlemann, S. Yarema, D. Hopkins, A.C. Paul, A.M. Sessler, J. Wurtele: Phys. Rev. Lett. **54**, 889 (1985)
11. E.T. Scharlemann, A.M. Sessler, J.S. Wurtele: Phys. Rev. Lett. **54**, 1925 (1985)
12. R. Bonifacio, F. Casagrande: Opt. Commun. **50**, 251 (1984);
   R. Bonifacio, F. Casagrande: JOSA B2, 250 (1985);   R. Bonifacio, F. Casagrande: in *Instabilities and Chaos in Quantum Optics*, ed. by F.T. Arecchi and R.G. Harrison (Springer, Berlin, Heidelberg, New York) in press
13. R. Bonifacio, F. Casagrande: in *Coherence and Collective Properties of Electrons and Electron Radiation*, ed. by R. Bonifacio, F. Casagrande and C. Pellegrini (North-Holland, Amsterdam) in press
14. See e.g. *Dissipative Systems in Quantum Optics*, ed. by R. Bonifacio (Springer, Berlin, Heidelberg, New York 1982)
15. L.D. Landau, E.M. Lifshitz: *The Classical Theory of Fields*, 4th ed. (Pergamon, Oxford 1975)

Index of Contributors